T0181396

# MAKING SENSE OF DATA III

# MAKING SENSE OF DATA III

## A Practical Guide to Designing Interactive Data Visualizations

**Glenn J. Myatt**
**Wayne P. Johnson**

A JOHN WILEY & SONS, INC., PUBLICATION

Copyright © 2011 by John Wiley & Sons, Inc. All rights reserved.

Published by John Wiley & Sons, Inc., Hoboken, New Jersey
Published simultaneously in Canada

No part of this publication may be reproduced, stored in a retrieval system, or transmitted in any form or by any means, electronic, mechanical, photocopying, recording, scanning, or otherwise, except as permitted under Section 107 or 108 of the 1976 United States Copyright Act, without either the prior written permission of the Publisher, or authorization through payment of the appropriate per-copy fee to the Copyright Clearance Center, Inc., 222 Rosewood Drive, Danvers, MA 01923, (978) 750-8400, fax (978) 750-4470, or on the web at www.copyright.com. Requests to the Publisher for permission should be addressed to the Permissions Department, John Wiley & Sons, Inc., 111 River Street, Hoboken, NJ 07030, (201) 748-6011, fax (201) 748-6008, or online at http://www.wiley.com/go/permission.

Limit of Liability/Disclaimer of Warranty: While the publisher and authors have used their best efforts in preparing this book, they make no representations or warranties with respect to the accuracy or completeness of the contents of this book and specifically disclaim any implied warranties of merchantability or fitness for a particular purpose. No warranty may be created or extended by sales representatives or written sales materials. The advice and strategies contained herein may not be suitable for your situation. You should consult with a professional where appropriate. Neither the publisher nor authors shall be liable for any loss of profit or any other commercial damages, including but not limited to special, incidental, consequential, or other damages.

For general information on our other products and services or for technical support, please contact our Customer Care Department within the United States at (800) 762-2974, outside the United States at (317) 572-3993 or fax (317) 572-4002.

Wiley also publishes its books in a variety of electronic formats. Some content that appears in print may not be available in electronic formats. For more information about Wiley products, visit our web site at www.wiley.com.

***Library of Congress Cataloging-in-Publication Data:***

Myatt, Glenn J., 1969-

Making sense of data III : a practical guide to designing interactive data visualizations / Glenn J. Myatt, Wayne P. Johnson.

p. cm

Includes bibliographical references and index.

ISBN 978-0-470-53649-0 (pbk.)

1. Data mining. 2. Information visualization. I. Johnson, Wayne P. II. Title. III. Title: Making sense of data 3. IV. Title: Making sense of data three.

QA76.9.D343M93 2012

006.3′12–dc23

2011016267

Printed in the United States of America

oBook ISBN: 978-1-118-12161-0
ePDF ISBN: 978-1-118-12158-0
ePub ISBN: 978-1-118-12160-3
eMobi ISBN: 978-1-118-12159-7

10 9 8 7 6 5 4 3 2 1

# CONTENTS

# PREFACE

Across virtually every field in science and commerce, new technologies are enabling the generation and collection of increasingly large volumes of complex and interrelated data that must be interpreted and understood. The changes are pushing visualization to the forefront and have given rise to fields such as visual analytics, which seeks to integrate visualization with analytical methods to help analysts and researchers reason about complex and dynamic data and situations. Visual systems are being designed as part of larger socio-technical environments based on advanced technologies in which work is done collaboratively by various experts. The boundaries between the design of visual interfaces, information visualization, statistical graphics, and human-computer interaction (HCI) are becoming increasingly blurred. In addition, design of these systems requires knowledge spread across academic disciplines and interdisciplinary fields such as cognitive psychology and science, informatics, statistics, vision science, computer science, and HCI.

The purpose of this book is to consolidate research and information from various disciplines that is relevant to designing visual interactions for complex data-intensive systems. It summarizes the role human visual perception and cognition play in understanding visual representations, outlines a variety of approaches that have been used to design visual interactions, and highlights some of the emerging tools and toolkits that can be used in the design of visual systems for data exploration. The book is accompanied by software source code, which can be downloaded and used with examples from the book or included in your own projects.

The book is aimed toward professionals in any discipline who are interested in designing data visualizations. Undergraduate and graduate students taking courses in data mining, informatics, statistics, or computer science through a bachelors, masters, or MBA program could use the book as a resource. It is intended to help those without a professional background in graphic or interaction design gain insights that will improve what they design because many smaller projects do not include professional designers. The approaches have been outlined to an extent that software professionals could use the book to gain insight into the principles of data visualization and visual perception to help in the development of new software products.

The book is organized into five chapters and an appendix:

- **Chapter 1 Introduction**: The first chapter summarizes how visual perception affects what we see, provides a brief history of the use of visualization in data exploration, and outlines the design process.
- **Chapter 2 The Cognitive and Visual Systems**: The second chapter describes how various drawings, maps, and diagrams (known as external representations) are understood and used to extend the mind's capabilities. It introduces the computational theory of the mind in the context of how the mind perceives and processes information from the external world. Based on research from vision science, this chapter describes how the human visual system works, how visual perception processes what we see, and how visual representations can be designed to influence visual perception.
- **Chapter 3 Graphic Representations**: The third chapter discusses the seminal work of Jacques Bertin, a cartographer who applied semiotic theory to statistical graphics. After introducing semiotic theory, the chapter discusses Bertin's ideas of the structure and properties of graphics and his observations on ways to construct graphics that communicate efficiently. The chapter also outlines the grammar of graphics developed by Leland Wilkinson and two grammar-based software libraries: ggplot2 for the System R statistical environment by Hadley Wickham and Protovis for Web browser environments by Michael Bostock and Jeffrey Heer.
- **Chapter 4 Designing Visual Interactions**: The fourth chapter assumes that the designs of visual interactions are for complex data-intensive systems. Beginning with a discussion of how the perception of complexity differs from operational complexity, the chapter then outlines in detail the four stages of the process of design: analysis, design, prototyping, and evaluation. It covers some of the important principles and strategies for designing visual interfaces, information visualizations, and data graphics as well as the time thresholds for various cognitive and perceptual processes that impose real-time constraints on design.
- **Chapter 5 Hands-On: Creating Interactive Visualizations with Protovis**: The fifth chapter provides an in-depth explanation of the capabilities of the Protovis toolkit. The chapter leads you through the creation of a series of visualizations and graphics defined by the Protovis specification language, beginning with simple examples and proceeding to more advanced visualizations and graphics. It includes a discussion of how to access, run, and use the software. Exercises are provided at the end of each section.
- **Appendix A Exercise Code Examples**: This appendix provides the source code for the exercise examples in Chapter 5.

This book assumes that you have a basic understanding of statistics. An overview of these topics has been given in Chapters 1, 3, and 5 of a previous book in this series: *Making Sense of Data: A Practical Guide to Exploratory Data Analysis and Data Mining*.

The book discusses visual interaction design starting with an explanation of how the mind perceives visual representations. Knowing the perceptual and cognitive framework allows design decisions to be made from "first principles" rather than just from a list of principles and guidelines. The "Further Reading" section at the end of each chapter suggests where you can find more detailed and other related information on each topic.

Accompanying this book is a Web site (**www.makingsenseofdata.com/**) that includes the software source code for the examples and solutions to the exercises in Chapter 5.

# ACKNOWLEDGMENTS

In putting this book together, we thank the National Institutes of Health for funding our research on chemogenomics (Grant 1 R41 CA139639-01A1). Some of the ideas in this book came out of that work. We thank Dr. Paul Blower for his considerable help in understanding chemical genomics and the ways in which visualizations could be used in that field, and for allowing us to try new ways of prototyping design concepts. We thank the staff at John Wiley & Sons, particularly Susanne Steitz-Filler, for their help and support throughout the project. Finally, Wayne thanks his wife, Mary, for her support throughout this project.

# CHAPTER 1

# INTRODUCTION

## 1.1 OVERVIEW

Across the spectrum of human enterprise in government, business, and science, data-intensive systems are changing the scale, scope, and nature of the data to be analyzed. In data-intensive science (Hey et al., 2009), various instruments such as the Australian Square Kilometre Array (SKA) of radio telescopes (www.ska.gov.au), the CERN Hadron particle accelerator (http://public.web.cern.ch/public/en/lhc/Computing-en.html), and the Pan-STARRS array of celestial telescopes (http://pan-starrs.ifa.hawaii.edu/public/design-features/data-handling.html) are complex systems sending petabytes of data each year to a data center. An experiment for drug discovery in the pharmaceutical and biotechnology industries might include high-throughput screening of hundreds of thousands of chemical compounds against a known biological target, or high-content screening of a chemical agent against thousands of molecular cellular components from cancer cells, such as proteins or messenger RNA. Data-intensive science has been called the fourth paradigm that requires a transformed scientific method with better tools for the entire research cycle "from data capture and data curation to data analysis and data visualization." (Hey et al., 2009)

In 2004, the Department of Homeland Security chartered the National Visualization and Analytics Center (NVAC) to direct and coordinate research

*Making Sense of Data III: A Practical Guide to Designing Interactive Data Visualizations*, First Edition. Glenn J. Myatt and Wayne P. Johnson.
© 2011 John Wiley & Sons, Inc. Published 2011 by John Wiley & Sons, Inc.

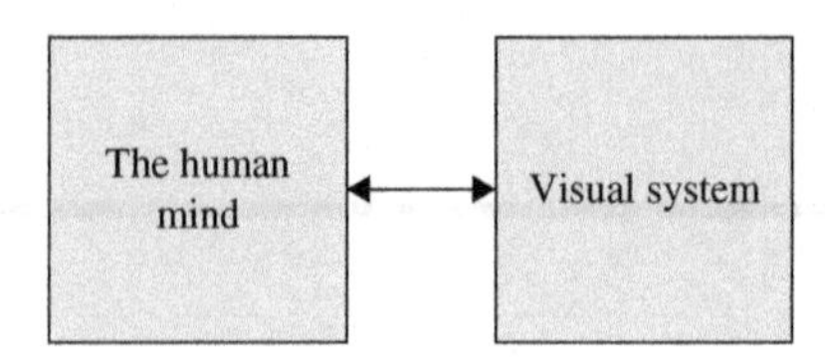


**FIGURE 1.1** Intelligence amplified through visual interaction

and development of visual analytics technology and tools. Its major objectives included defining a long-term research and development agenda for *visual analytics* tools to help intelligence analysts combat terrorism by enabling insights from "overwhelming amounts of disparate, conflicting, and dynamic information." (Thomas & Cook, 2005) This has given rise, more broadly, to the emerging field of visual analytics. The field seeks to integrate information visualization with analytical methods to help analysts and researchers reason about complex and dynamic data and situations. But why the emphasis on visualization as a key element in the solution to helping with the problem of data overload?

In 1994, Frederick Brooks in an acceptance lecture given for the ACM Allen Newell Award at SIGGRAPH said:

> "If indeed our objective is to build computer systems that solve very challenging problems, my thesis is that **IA > AI**; that is, that *intelligence amplifying* systems can, at any given level of available systems technology, beat AI [artificial intelligence] systems. . . . Instead of continuing to dream that computers will replace minds, when we decide to harness the powers of the mind in mind-machine systems, we study how to couple the mind and the machine together with broad-band channels. . . . I would suggest that getting information from the machine into the head is the central task of computer graphics, which exploits our broadest-band channel."

As shown in Fig. 1.1, to effectively design intelligence amplifying (IA) systems requires an understanding of what goes on in the mind as it interacts with a visual system. Clues about how the mind interprets the digital world come from what is known about how the mind interprets the physical world, a subject that has been studied in vision science.

## 1.2 VISUAL PERCEPTION

Imagine yourself driving into an unfamiliar large metropolitan city with a friend on a very crowded multilane expressway. Your friend, who knows the city well, is giving you verbal directions. You come to a particularly complicated system of exits, which includes your exit, and your friend says "follow that red sports car moving onto the exit ramp." You check your rearview mirror, look over your shoulder, engage your turn signal, make the appropriate adjustments to speed, and begin to move into the space between the vehicles beside you and onto the exit ramp. Had this scenario taken place, you would have been using visual perception to inform and guide you in finding your way into an unfamiliar city.

The human visual system, which comprises nearly half of the brain, has powerful mechanisms for searching and detecting patterns from any surface that reflects or emits light. In the imagined scenario, the optic flow of information moment by moment from various surfaces—the paint on the road dividing the lanes, the vehicles around you, the traffic signs, the flashing lights of turn signals or brake lights—creates *scenes* taken in and projected onto the retinas at the back of the left and right eyes as upside-down, two-dimensional (2-D) images. The visual system, through various processes executed by billions of highly connected biological computational elements called neurons operating in parallel, extracts information from a succession of these pairs of images in a fraction of a second and constructs a mental representation of the relevant objects to be aware of while navigating toward the exit ramp and their location in the external world.

The perception of a scene from a single moment in time is complex. A 3-D world has been flattened into a pair of 2-D images from both eyes that must be reconciled and integrated with information from past experience, previous scenes, and other sources within the brain to reconstruct the third dimension and generate knowledge relevant to the decisions you are making. There are several theories about how the various perceptual processes work, the representations of their inputs and outputs, and how they are organized. But a generally accepted characterization of visual perception is as stages of information processing that begin with the retinal images of the scene as input and end with some kind of conceptual representation of the objects that are used by thought processes for learning, recall, judgment, planning, and reasoning. This information-theoretic approach divides the general processing that takes place in vision into four stages as shown in Fig. 1.2.

*Image-based processing* includes extracting from the image simple 2-D features, such as edge and line segments or small repeating patterns, and their properties, such as color, size, orientation, shape, and location.

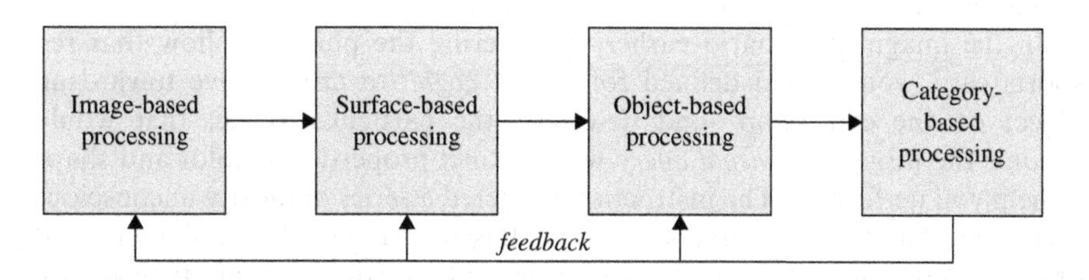


**FIGURE 1.2** Information-theoretic view of visual perception

*Surface-based processing* uses the simple 2-D features and other information to identify the shapes and properties of the surfaces of the objects in the external world we see, and attempts to determine their spatial layout in the external world, including their distance from us. However, many surfaces of objects are hidden because they are behind the surfaces of objects closer to us and cannot be seen.

*Object-based processing* attempts to combine and group the simpler features and surfaces into the fundamental units of our visual experience: 3-D representations of the objects and their spatial layout in the external world of the scene. The representation of an object is of a geometric shape that includes hidden surfaces, the visible properties of the object that do not require information from experience or general knowledge, and 3-D locations.

*Category-based processing* identifies these objects as they relate to us by linking them with concepts from things we have seen before or are part of our general understanding of the world, or that are being generated by other systems in the brain such as those processing speech and language. Classification processing uses visible properties of the object against a large number of conceptual patterns stored in our memory to find similar categories of objects. Decision processing selects a category from among the matching categories based either on novelty or uniqueness.

The visual processing just described is a simplification of the process and assumes a static scene, but the world is dynamic. We or the objects in our visual field may be moving. Moment by moment we must act, think, or reflect, and the world around us is full of detail irrelevant to the task at hand. The optical flow, a continuous succession of scenes, is assessed several times a second by small rapid movements of our eyes called *saccades* that sample the images for what is relevant or interesting. In between, our gaze is fixed for only fractions of a second absorbing some of the detail, for there is far too much information for all of it to be processed. The overload is managed by being selective about where to look, what to take in, and what to ignore. Vision is active, not passive. What we perceive is driven not only by the light that enters our eyes but also by how our attention is focused. *Attentional focus* can be elicited automatically by distinct visual properties of objects in the scene such as the color of a surface or the thickness of a line, or by directing it deliberately and consciously. We can intentionally focus on specific objects or areas of the scene relevant to the task *overtly*, through movement of the eyes or head, or *covertly* within a pair of retinal images, through a mental shift of attention.

In the imagined scenario earlier, by uttering the phrase "follow that red sports car," your friend defined for you a *cognitive task*—move toward an object on the exit ramp—and described the particular object that would become the target of a *visual query* with distinct properties of color and shape to help you perform it. The instruction triggered a series of mostly unconscious events that happened in rapid succession. Based on the goal of looking for red objects along exit ramps and prior knowledge that exit ramps are typically on the outer edge of the highway, the *attentional system* was cued to focus along the outer edge of the expressway. Eye movements, closely linked with attention, scanned the objects being visually interpreted in this region. The early part of the visual processing pathway was tuned to select objects with red color properties. Red objects within the focal area, assuming there were only a few in sight, were identified almost immediately by the visual system and indexed in a visual memory buffer. These were categorized and considered by later-stage cognitive processes, one at a time, until the red sports car was found.

The goal shifted to tracking and following the sports car. Your eyes fixed on the sports car for a moment and extracted information about its relative distance from you by processing visual cues in the images about depth. These cues included occlusion (the vehicles whose shapes obscure other vehicles are in front), relative size (the longer painted stripes of a lane divider are closer than the shorter ones), location on the image (the closer painted stripes are below the farther painted stripes of a lane divider), and stereopsis (differences in location of the same object in the image from the left and right eye that allowed calculation of the object's distance from you). The eyes then began a series of saccades targeting the vehicles in front of and next to you to build up the scene around your path as you maneuvered toward the exit.

Every day as you reach for the handle of a cup, scan the spines of books on the shelves of a library or surf the Web, your eyes and brain are engaged in this kind of interaction and activity to parse and interpret the visual field so that you can make decisions and act. Yet you are mostly unaware of the many complex transformations and computations of incoming patterns of light made by the neural cells and networks of your brain required to produce a visual experience of a spatially stable and constant world filled with continuous movement. Replace the scene of the external world with visual forms that can be displayed on computer screens, and the same neural machinery can be used to perceive an environment of digital representations of data to make different kinds of complex decisions. If the visual forms are carefully designed to take advantage of human visual and cognitive systems, then we will more easily find or structure individual marks such as points, lines, symbols, or shapes in different colors and sizes that have been drawn to support various cognitive tasks.

## 1.3 VISUALIZATION

The scenario in the previous section used visualization—imagining what was not in sight—to introduce the human visual and cognitive systems. The technical fields of scientific, data, and information visualization and visual analytics use the term *visualization* differently to mean techniques or technologies that can be thought of as visualization tools (Spence, 2001) for making data visible in ways that support analytical reasoning. The essence of this definition includes the person doing the analysis, the user interfaces and graphics that we will call visual forms, and the data. We cannot design effective visualization tools or systems without thinking about the following:

- The analytical tasks, the work environment in which these tasks are done, and the strategies used to perform them
- The content and structure of the visual forms and interaction design of the overall application and systems that will incorporate them
- The data size, structure, and provenance

In the earliest days of computer-supported data visualization, the tasks focused on preparing data graphics for communication. Data graphics were the points, lines, bars, or other shapes and symbols—marks on paper—composed as diagrams, cartographic maps, or networks to display various kinds of quantitative and relational information. The questions included how graphics should be drawn and what should be printed to minimize the loss of information (Bertin, 1983).

With advances in computation and the introduction of computer displays, the focus began to shift to exploratory data analysis and how larger datasets with many variables could be visualized (Hoaglin et al., 2000). Three examples follow. John Tukey and his colleagues introduced PRIM-9 (1974), the first program with interactive graphics for multivariate data that allowed exploration of various projections of data in space up to nine dimensions to find interesting patterns (Card et al., 1999). Parallel coordinates (1990), a visualization tool for multidimensional space, showed how a different coordinate system could allow points in a multidimensional space to be visualized and explored just as 2-D points are in scatterplots using the familiar Cartesian coordinate system. SeeNet (1995), a tool for analyzing large network data consisting of a suite of three graphical displays that included dynamic control over the context and content of what was displayed, was developed to gain insight about the sizes of network flows, link and node capacity and utilization, and how these varied over time (Becker et al., 1999).

The last example in the previous paragraph shows the influence of advances from the human-computer interaction (HCI) community that were made in the 1980s and 1990s in the user interfaces of the three displays of the SeeNet tool and how it changed the way work was done. Instead of using traditional methods of data reduction that aggregated large numbers of links or nodes, averaged many time periods, or used thresholds and exceptions to detect changes, the dynamic controls allowed changes to display parameters that altered the visualizations so all of the data could be viewed in different ways (Becker et al., 1999). The user interface techniques of direct manipulation and interaction, important to the tasks of exploration, had taken priority over the quality of static graphics. For data-intensive analysis, data visualization and user interfaces were converging, and exploration was being done not just by the statistician, but also by domain experts, in this case, engineers in telecommunication responsible for the operations of large networks.

Alongside the new directions in data visualization, the HCI community was taking advantage of advances in computer graphics that included a new understanding of the human-computer interface as an extension of cognition; an expanded definition of data, which included abstract or nonnumeric data; and the emerging World Wide Web. Important new user-interface techniques and visualization tools were introduced that gave rise to the field of information visualization. A sample of these techniques and tools include the following:

- Dynamic queries that could be performed through user-controlled sliders in the user interface instead of through text-based queries, and provided immediate and constant feedback of results through a visual form (Ahlberg & Shneiderman, 1999).
- Techniques to support the need for seeing context and detail together and for seeing different information in overviews than for detailed views.
- A general-purpose framework that used panning and zooming—a form of animation—to see information objects in a 3-D space at different scales (Bederson et al., 1994). Google Maps™ mapping service and Google Earth™ mapping service are examples of this approach.
- Information visualization workspaces that allowed direct manipulation of the content so that the user could focus only on what was relevant, reorganize it into new information, or prepare it for presentation (Roth et al., 1997).

Information visualization had broadened the definition of exploratory data analysis. It now included tasks such as searching, dynamically querying information, grouping and reorganizing data, adjusting levels of detail, discovering relations and patterns, and communication. Interactive visual forms could operate with numeric or abstract data, allowing for the results of statistical calculations or data-mining computations to be linked to the information objects from databases or data tables or to other complex objects such as chemical structures from which they were derived.

The past couple of decades have also seen the emergence of data-intensive science. Projects in the physical and life sciences generate large amounts of data that originate from a variety of sources and flow into data centers from complex collections of sensors, robotics, or simulations from supercomputers or grid computing. Data capture, curation, and analysis are done by teams of individuals who are often geographically dispersed and have expertise in IT, informatics, computational analysis, and a scientific discipline. This has given rise to high-throughput data analysis and exploratory tools.

The data comes in all scales. It might be a large dataset with values from a single experiment or a family of datasets from related experiments. For example, the National Institutes of Health (NIH) has carefully defined sets of rules and procedures—protocols—for conducting experiments that allow chemical compounds screened against a set of cancer cell lines to be compared with the values of data from the screens of other cellular parts—for example, genes, messenger RNA, or micro RNA—across the same set of cancer cell lines using microarray technology. The ability to integrate data across experiments provides insight into various mechanisms involved in cancer.

The data to be analyzed can come from one of several stages of processing. For example, in a microarray biology experiment, the amount of each gene expressed in a cell can be measured by the intensity of light at a point on a microarray where the mixture containing the gene has been spotted. From the microarray, a machine produces a digitized graphic image. Image analysis converts the digital image to a matrix of numbers. The image analyst might explore the analog signals from the laser scanner, the statistician might explore the matrix of numbers, and the biologist might explore the genes of a particular group discovered by clustering or the factors of a principal component analysis generated by the statistician.

A broad collection of computational tools and algorithms are available to support high-throughput analysis tasks, which include many of the same tasks described for data and information visualization. These tools and algorithms are drawn from classical statistics, machine learning, and artificial intelligence (AI).

The data is distributed across a network and stored in a variety of formats under control of different data-management systems. Some of it may contain data called *metadata*, that describes the raw data. For example microarray experiments are recommended to contain the minimum information about a microarray experiment (MIAME) needed to interpret and reproduce the experiment. The information includes the raw data, the normalized data, the experimental factors and design, details about each item in the microarray, and the laboratory and data processing protocols (FGED, 2011). Other associated data will also need to be linked. For example, the IDs of genes that reside in the headings of a microarray matrix of numeric expression values can be used to retrieve information about their function or their sequences.

Once again, the definition of exploratory data analysis has broadened. The data to be analyzed is no longer only within a single file or data table. Even if the primary focus is on numeric data, the objects or observations from which the numbers were derived—abstract data—and information about the provenance of the data or details about the experiment are important aspects of exploration that are integrated into the user interface with the visual forms. The data may require stages of processing where the output of any stage may be of interest. It may require subsets to be retrieved through queries to remote servers. The computational tools may run on the same machine as the visual tool, as a service accessible via the Internet, or as a large-scale computation distributed over many computers. Teams of experts with different skills—for example, a molecular biologist, a bioinformatics specialist, and a computational biologist—may be required to analyze the data.

The fields using computer graphics for visual forms—data graphics, information visualization and user interfaces—have rich traditions of design. Interaction techniques are increasingly being added to data graphics, new visualizations are being invented to handle the much larger scale of data being generated by data-intensive methods, and a greater variety of information is being linked into exploration. As this continues, the boundaries between the three kinds of visual forms within a visualization application will eventually disappear if it is designed without seams. The visualization systems for exploration in data-intensive environments will benefit from the considerable body of research and practical knowledge on design that comes from these traditions, and we will draw on each of them in this book.

## 1.4 DESIGNING FOR HIGH-THROUGHPUT DATA EXPLORATION

Whereas the concern of the analyst is exploring the data, the concern of the designer is to create a visualization tool through which the analyst "sees" the data but not the tool. For the work the tool is designed to support, interaction with the tool should be engaging and not interrupt the flow of thought. This is the ideal for usability. The fields of human-computer interaction (HCI) and interaction design (ID) have spent decades studying how to achieve this in interactive computing systems and various products, technologies, systems, and services. For aspects of the visual forms that are static, rules and principles that have evolved from centuries of print tradition can also inform design.

### 1.4.1 The IA (Intelligence Amplified) System

A conceptualization of the IA system for high-throughput data exploration, which is the focus of our design, is shown in Fig. 1.3.

The analyst, referred to in design as the *user*, is the *primary* focus. For those coming from engineering or computer science without a design background, it is important to recognize the shift in perspective from technology to user. To a designer, technology is a design material—something usually selected later in the design process to solve the needs of the user after they are understood. The visual and cognitive systems of the brain have already been introduced. They are the critical part of the overall IA system and these systems interpret and act on the visual information that is a source for analytical thinking. Viewed as information-processing systems, the visual and cognitive systems that comprise the mind have constraints that must be considered in design.

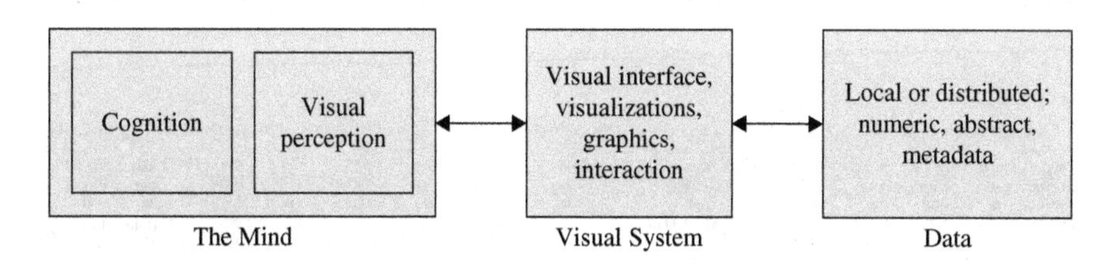


**FIGURE 1.3** The IA (intelligence amplified) system

The *visualization system* is a computational system capable of interacting with the user through high-resolution displays and interaction devices, and of accessing data from one or more sources on the same machine or on the Internet. It might be implemented as a Web browser running on a PC or a handheld tablet, the client application running on the laptop of a distributed system, or a set of components that controls user interaction with a wall-mounted display.

The visual representation displayed by the visualization system is generically called *visual forms* (Card et al., 1999, p.17). Visual forms are compositions of user interface *components* and *controls*, *graphics* of numeric data, or *visualizations* of abstract data. All interact with the user using the same display and interaction devices and are drawn in a visual language of graphical elements—points, lines, planes, volumes, and so on—with visual properties such as position, color, shape, size, orientation, and texture. The graphical elements are designed to relate to one another in a way that communicates information. For example, a visual form might be a scatterplot with sliders to set the limits on the x-axis and y-axis and filter the points displayed in the plotting region, or an interactive visualization showing the groups of a clustering algorithm, each of which can be selected to retrieve more information about the elements in the group.

The data being analyzed and explored by the users of the visualization system can only be accessed through the visual forms. How and where the visual form is stored is irrelevant to the user but of great importance to the designer. Access to all or part of the visual form comes at a cost in time: data stored locally on the same machine that is generating the display is generally accessed much faster than data stored at a remote site. Certain visual forms may require significant computational time to generate. The response time for user actions affects the interaction and must be taken into account during design.

### 1.4.2 Design

Every product, physical or not, has an interface. The *interface* is the point of interaction. Teapots have handles which are part of the physical form of the container. Windows have latches to help you lock them. Figure 1.4 shows the control panel of a dishwasher with buttons to let you choose how to wash and dry dishes. The panel has been designed to access the hidden functions programmed into the electronics of the machine. By looking at and

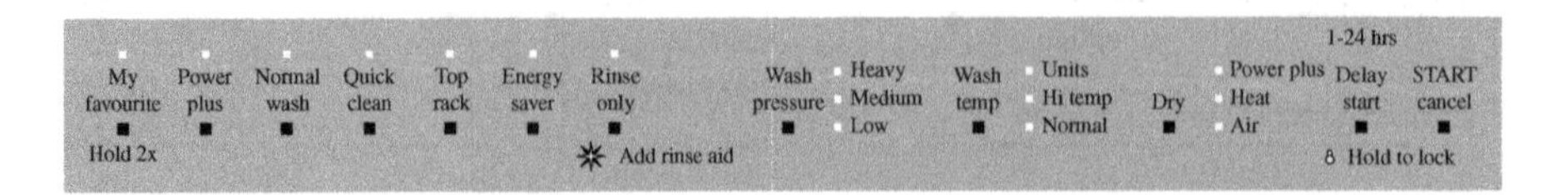


**FIGURE 1.4** Dishwasher control panel

experimenting with the interface, we begin to develop a *mental model* of what functions the dishwasher provides, how it works, and how to interact with it.

Because of the layout of the visual elements, we unconsciously perceive a hierarchical organization: starting from the left there is a cluster of seven groups, followed by a cluster of three groups, and then two groups. Every group has text with a black square below; some groups have one white circle, some have three white circles with text to the right, and some have no white circles. What is the meaning of the various combinations of text, black squares, and white circles?

From prior experience with these kinds of panels, the groups leave the impression of buttons. The text of some buttons use terms familiar to anyone who works in the kitchen such as "wash," "clean," "wash temperature," "dry," or "heat." Every group has text and a black square; these must be buttons. Because the text in the cluster of seven buttons seems to relate to how dishes get washed—"normal wash" or "quick clean," these must be buttons that represent a way to set the method for cleaning dishes. The organization in the cluster of three groups is different. If the text and black square represent the function, then maybe the text and white circles are ways to vary that function by degree. The wash temperature is "sanitize," "hi temp," or "normal." Sanitize is not normally associated with temperature, but because the white circle with the "sanitize" text is above "hi temp," and the white circle with "normal" text is below, "sanitize" must mean very high temperature. Seven buttons relate to a way of cleaning dishes and three to setting different degrees of water temperature, pressure, and drying temperature, but how do we make it work? How are the selections for water temperature and pressure, and drying temperature made? Can "energy saver" and "top rack" be selected simultaneously? Does "energy saver" have any effect on "rinse only"? Which combinations make a difference?

Even in this simple interface, you can see the design questions about how the visual form of the interface relates to function, how the interface reveals and constrains the actions that can be performed, and what the sequence and flow of the interaction should be to support the tasks. No matter how complicated interactive visual forms become, they play the same role for the analyst in their domain as the dishwasher panel does for the cook in the kitchen. The visual forms that the visualization system presents is the only way to see and initiate actions that transform the data, but, as we have seen in the previous example, the design of these forms is critical to how quickly and easily they can be interpreted and understood. So how do we begin to design the interface and interaction with the visualization system?

Interaction design is a highly dynamic process that consists of four basic steps: identify the user's needs, develop alternative designs, prototype, and evaluate. Although it is presented as a sequence of steps, what is learned in each step provides feedback that may change the result of another step. For example, during prototyping, insight may be gained when converting a low-fidelity prototype—for example, one implemented on paper or a PDF document—to a high-fidelity working prototype that reveals software and hardware bottlenecks requiring changes to the requirements. (Note: Because we are focusing only on the design of the system as it pertains to the visualization of data, we will ignore those aspects of design that would be addressed if a commercial product were being created. These include the context in which it will be used and other aspects of design that include standards, integration with other systems, and portability.)

**1. Identify the user's needs, and establish initial requirements.** Getting a feel for the work and how the system may be used requires a thorough understanding of the following:

- Who will use it
- The driving problem behind the analysis
- The analysis tasks, and the concepts and actions needed to perform them
- The source, content, and processing of the data
- The usability goals as they relate to user performance

For many data-intensive projects, there may be several people who directly use the system: an informaticist who manages the systematic collection and organization of data so that it can be retrieved or searched, a computational specialist who is trained in statistical and data-mining methods, and a domain expert with knowledge about the relevant subject area such as biology, marketing, telecommunications, or finance. If, for example, a system is being designed to explore the integrated data of biological targets or chemical compounds generated by high-throughput screening or microarray technology in order to study cellular mechanisms of cancer, the users might be molecular biologists, medicinal chemists, and cheminformatics or bioinformatics experts and the domains would be molecular biology, medicinal chemistry, computational biology, bioinformatics or cheminformatics, and statistics.

Becoming familiar with the domain implies becoming immersed in the field. It may include reading textbooks, technical material, or important papers that have been published; interviewing and talking with experts in the field; or attending a conference or workshop. The ideas for design come from a depth of understanding. An understanding of the field provides a foundation for seeing through the users' descriptions about the steps they take in performing their tasks or the feedback they provide during evaluation to intent: what the user is really trying to accomplish. With many experts, the intent may be buried from years of practice so that it is no longer noticed and cannot be articulated. Knowing the intent or goal that the user is trying to accomplish allows alternative designs to be considered that may do things in a different but more efficient way.

*Usability*, a term used by designers, is a way to think about the usefulness of the analysis system with which the user will interact. Most usability goals include practical measures for assessing them. The goals include the following (Shneiderman & Plaisant, 2010):

- *Time to learn*. How long does it take to learn to do the tasks? The length of time it takes to learn is more of a problem for systems that are used infrequently than it is for a system that will be used daily.
- *Performance*. How long does it take to do each task once the user is proficient?
- *Accuracy*. How many and what kinds of errors are made? Errors refer to any action that does not accomplish what the user intended.
- *Memorability*. How easy is it to remember how to use the system the next time the user interacts with it, for instance, after a day or a week? A complex system that is used infrequently can be as frustrating as a simple system that is used frequently but lacks keystroke shortcuts or functions that allow expert users to adjust parameters. There are trade-offs to consider.
- *Satisfaction*. Did users respond positively or negatively to specific visual forms or areas of the user interface? How did they experience the system as a whole? Qualitative assessments not only provide insight into specific aspects of the design, but psychology research shows that emotions affect how the mind solves problems. Positive experiences open our minds and make us more creative and better able to learn. Negative experiences narrow our thoughts to the problem at hand (Norman, 2004).

The outcome of studying users and their work leads to preliminary requirements that are stable enough to begin design and a set of benchmark analysis tasks that can be used for prototype evaluations.

**2. Develop alternative designs**. For many software developers or programmers, the first step in design is to begin by sketching ideas about what the interface should look like. However, designers begin by designing a conceptual model focused on the tasks and goals of the user that will, in turn, guide the physical design that includes the details of the interface and interaction.

A *conceptual model* is a simplified, high-level model of the system that designers want to communicate to the users to make it easier for them to understand and use. The interface will be designed to convey and support the conceptual model. Preferably the conceptual model is drawn from the task analysis because the task descriptions contain important words that identify the concepts that users are most familiar with; but the conceptual model could also be based on metaphors or analogies. A widespread example is the office metaphor first introduced in the Xerox Star and now widely used in the Macintosh and Windows PCs. This conceptual model includes a desktop, folders, files, pictures, notes, mail, and other objects. Folders or files can be created or deleted. Files are stored in folders. A file can be edited and its content copied onto a clipboard and pasted into another file. The familiarity of the user with physical objects from the office (desktop, file cabinets, folders, files) allows the actions (move, cut, paste, select, edit) associated with a visible representation of the object on the display to make sense.

The most important component of the conceptual model is the object-action model, which identifies each conceptual object that will be visible in the interface, and each object's actions, attributes, and relationships to the other objects.

Two other areas must be considered in the design that may affect the conceptual model: the style of interaction and the type of interface. These are outside the scope of this book, but we introduce them briefly here and refer you to the end of the chapter for sources of further reading about these areas.

An interaction style characterizes how the user interacts with the system and can be categorized as follows (Sharp et al., 2007):

- *Command.* The user directs the system by entering text in a command language, speaking, selecting menu items from a menu tree, and so on.
- *Conversational.* The system has an ongoing dialog with the user through speech or text.
- *Direct manipulation.* The user initiates actions by selecting an object that is displayed on the screen and choosing an associated action; this style is also known as point-and-click.
- *Exploration.* The users explore a virtual or physical space, for example, virtual reality or smart rooms.

We will assume direct manipulation for the visualization system because it is most widely in use. The Protovis visualization toolkit we will introduce in Chapter 5 for creating visualizations in a Web browser supports direct manipulation.

Many new interface types have been invented recently and are becoming available as design material (Sharp et al., 2007). However, these are not widely in use. We will assume that visualization systems have a presentation layer or client application that runs on a desktop PC or laptop with a high-resolution display, a keyboard for entering text, and a mouse or touchpad as a pointing device.

**3. Prototype**. Prototypes are not a replacement for analysis and design. They are visualizations of some or all of the requirements, benchmark tasks, and conceptual model—a way to see what has largely been verbal descriptions of the tasks to be performed and the steps to perform them that were developed in the previous stages of design. For individuals participating in design, they are helpful for thinking through the details of interface and interaction, for quickly generating designs that can be assessed and incorporated or discarded, and for deciding which alternatives are best to advance. The type of prototype and its role changes as the design process progresses. Prototyping is highly iterative.

At the beginning of physical design, *low-fidelity prototypes* are used to explore designs. Low-fidelity prototypes include sketches of single screens with the user interface components and controls along with visualization or data graphics, paper prototypes, storyboards of task sequences, or high-quality digital sketches exported from illustration drawing or slide presentation software applications. Using combinations of these approaches is also effective. Whatever method is used, it should support the ability for users to physically interact with the prototype to execute limited tasks selected from those developed in task analysis.

As the physical design progresses, *high-fidelity prototypes* are used to explore issues related to critical areas or to develop a working prototype with interaction sequences and flows. The working prototype can evolve into a proof-of-concept system that can be demonstrated outside the design team. High-fidelity prototypes take longer to build, so they should be used for areas of the interface where the design is considered stable and changes are infrequent.

**4. Evaluate**. The goals for evaluation for design differ from the goals of evaluation that are intended to assess usability. Evaluation for design is intended to provoke discussion about better ways to the structure the system, or to uncover tasks that aren't necessary or may be more or less important than initially thought. The evaluation stage is iterative and must be a fundamental part of the design process from the beginning.

Usability testing, on the other hand, is done in the late stages of development. It measures the users' performance on a set of predefined tasks. It is intended to uncover small problems or areas that are found to be difficult to understand. The changes made are to polish the interface and improve the interaction to refine the product. If the prototyping has been done well, there should be no major surprises.

### 1.4.3 Data

Data is captured or collected from sensors and sources that range from arrays of charge-coupled devices (CCDs) in telescopic cameras to electronic cash registers in retail stores. The data that is collected—*raw data*—is stored in many ways: into files in proprietary or standard formats, with or without metadata; and into all sorts of databases with various schemas. Before trying to use the raw data for visualization, it is helpful to transform it into a standard *data table* as shown in Fig. 1.5. The data-processing pipelines that convert raw data to data tables depend on the kind of data collected. For example, preparation for the mining and analysis of numeric data often includes cleaning, transformations, reduction, and segmentation, which results in tabular data (Myatt & Johnson, 2009).

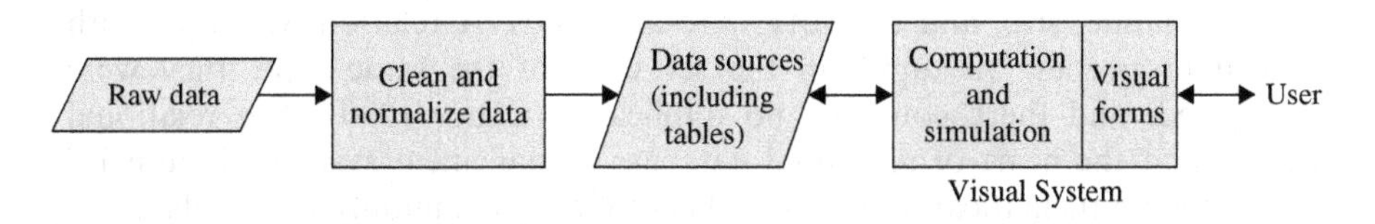


**FIGURE 1.5** Data-processing pipeline

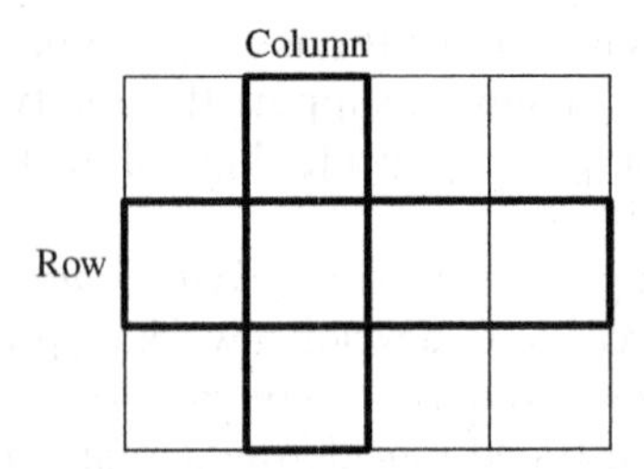


**FIGURE 1.6** A simple table and its nomenclature

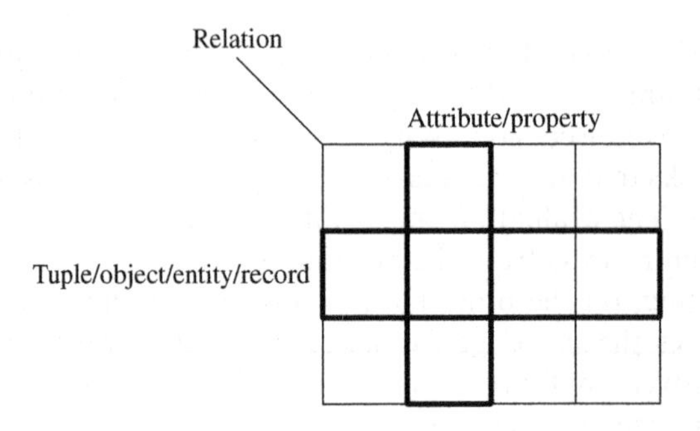


**FIGURE 1.7** A relational table and its informal nomenclature

Different methodologies and techniques for data preparation are used in various scientific disciplines, computer science, and statistics (Myatt & Johnson, 2009). When referring to tables, different terms are used for similar concepts or the same term for different concepts. Spreadsheet, row, column, entity, object, relation, table, tuple, record, attribute, property, dataset, case, observation, variable, matrix, and metadata are terms used to describe different perspectives on what is commonly understood as a table of rows and columns as shown in Fig. 1.6. As a guide through the literature and to clarify our use of the term *data table*, we provide a summary of the different ways terms are used and a definition.

In relational databases, data is logically stored in tables of rows and columns. The terms *relation*, *tuple*, and *attribute* are from the mathematical theory of relations that underlie the relational model developed by E. F. Codd. A tuple represented a thing or object in the world and its associated attributes. For example, a machine part in an inventory system was a tuple, and its serial number, name, size, and quantity were attributes. A relation was a set with special properties: all tuples in the set were of the same type, they were unordered, and there could be no duplicates (Codd, 1990). The relational model is at the heart of relational database management systems. Figure 1.7 shows the mathematical terms and others that are commonly used today.

In statistics, *dataset* refers to a collection of information often in a tabular format. For example, the dataset might be a collection of data on patients or cars. The patients or cars are objects. In the dataset, each item in the collection is an observation, and there may be many *observations* for a particular object. The observations may be described in various ways. For example, a car has a vehicle identification number, a manufacturer's name, and a weight. Each of these features describing the car is a statistical *variable*. The variables play a role similar to the one attributes play in database relations. Smaller datasets are often stored in files. In these files, the first row is often made up of the names of column headings, and the first cell of each row is an ID of the object whose values are represented by the remaining cells in the row. The terminology used in statistics is shown in Fig. 1.8.

In visualization, the dataset of statistics is called a data table, but the variables are rows, and the cases are columns as shown in Fig.1.9 (Card et al., 1999). Jacques Bertin, whose work we will study in Chapter 3, uses this orientation of the data table, but he refers to variables as *characteristics* and cases as *objects*. The Protovis toolkit that we will explore in Chapter 5 also uses this orientation of the table and expects input to the visualization pipeline as a "variables by cases" array.

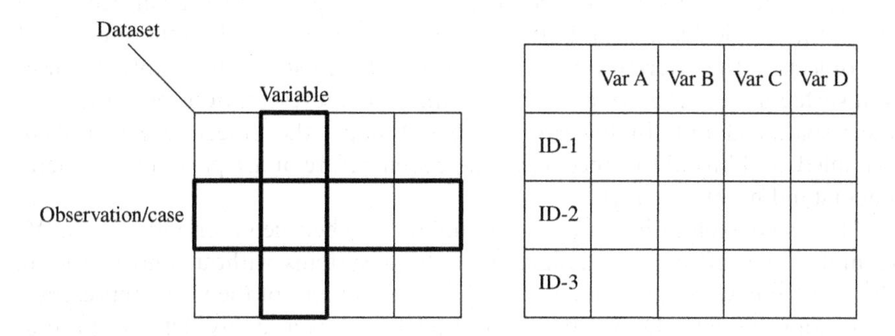


**FIGURE 1.8** A dataset and its nomenclature; a dataset with row and column headings

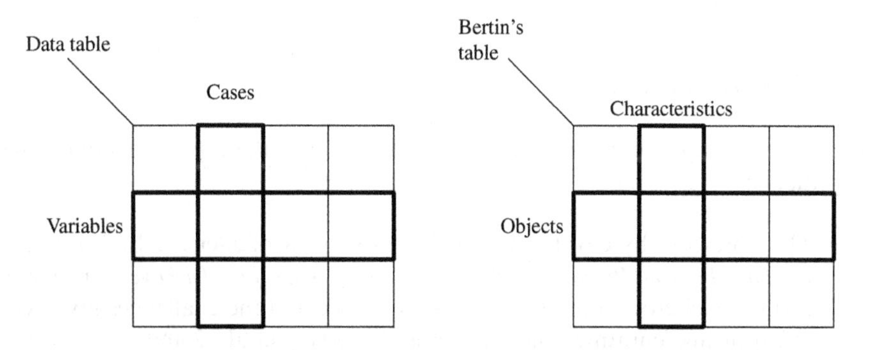


**FIGURE 1.9** A data table and Bertin's table with terminology

The definition of data table used in this book adheres to the one commonly used in statistics and data mining: the observations are rows, and the variables are data columns. If datasets include row and column headings, they are treated as metadata. If the first column contains the IDs of the observations and these IDs are not considered values of the dataset the ID values are treated as metadata.

## 1.5 SUMMARY

Data-intensive systems are changing the scale, scope, and nature of the data to be analyzed. The human visual system, which comprises nearly half of the brain, has powerful mechanisms for searching and detecting patterns. This capability has given rise to the field of visual analytics, which seeks to exploit these capabilities.

To effectively design IA (intelligence amplified) systems requires an understanding of what goes on in the mind as it interacts with a visual system. Information-based theories divide visual perception into four stages that transform the image in the retina of the eye to concepts that we understand. In the image-based stage, simple features are extracted. In the surface-based stage, the simple features and other information are used to find the surfaces of the objects in the external world. In the object-based stage, the simpler features and surfaces are combined and grouped into 3-D representations of objects and their spatial layout. In the category-based stage, the objects are identified, classified, and linked to concepts we have seen before or are part of our general understanding of the world.

The visual exploration of data is an interaction between user and system. We cannot design effective visualization tools or systems without understanding the work, the work environment, the form and content of the visual representations, and the data. Designing visual interactions is a highly iterative process that consists of identifying the user's needs, developing alternative designs, prototyping, and evaluation. This results in visual systems that, even if complex, are understandable and allow the tasks they were designed to support to be efficiently performed.

## 1.6 FURTHER READING

There are several references that go more deeply into the topics introduced in this chapter.

- The Foreword by Gordon Bell and the overview of eScience by Jim Gray in *The Fourth Paradigm: Data-Intensive Scientific Discovery* provide further background on data-intensive science and the challenges involved in capturing, curating, and analyzing data (Hey et al., 2009).

- The idea of AI was presented in a lecture by Frederick Brooks given in acceptance of the ACM Allen Newell Award at SIGGRAPH in 1994 (Brooks, 1996). The lecture emphasized the role of computer scientists as makers of tools and the importance of using an interdisciplinary collaboration on a real and complex problem as a driving problem for research. It also provides insights into issues that arise when doing user-centered design.
- The first chapter in *Readings in Information Visualization* (Card et al., 1999) provides a detailed overview of information visualization. This is essential reading.
- Chapters 1, 3, and 5 of *Making Sense of Data: A Practical Guide to Exploratory Data Analysis and Data Mining* (Myatt, 2006) provide an overview of data preparation and the use of statistics in exploratory data analysis and data-mining applications.
- The first chapter in *Vision Science: Photons to Phenomology* (Palmer, 1999) gives an introduction to vision science that includes a description of the problems in scene recognition, the human visual system, and visual perception.
- Two chapters of *Illuminating the Path: the Research and Development Agenda* for *Visual Analytics* provide helpful background. In the context of providing support for analyzing security threats to a nation, Chapter 2 discusses the science of analytical reasoning, and Chapter 3 discusses various visual representations and interaction technologies.

# CHAPTER 2

# THE COGNITIVE AND VISUAL SYSTEMS

Physical tools have long been crafted from natural material to overcome our inherent physical limitations and do things—such as cut down trees, move heavy objects, and shape and fasten pieces of wood into shelters—to improve our well-being. Cognitive tools, although they have a much shorter history, were invented to extend our mental capacities. Even before written language emerged, depictions were being used in abstract ways that included recording transactions and ownership of property or keeping track of the balance of accounts (Tversky, 2003).

In maps, space and elements referred to things that could be seen in the world such as geographical areas and major cities or trade routes. The elements were symbols of the things they stood for, and their position in space represented the relationships between them. Just as a hammer is a physical tool that increases and directs the force of the arm through the hammer's head into the nail, a map is a cognitive tool that extends memory, organizes and archives knowledge, and reduces the mental effort in drawing certain conclusions about the world. For the hammer, the material it is made of—wood or metal—conveys the force; for the map, the force is conveyed through a representation of the world.

*Making Sense of Data III: A Practical Guide to Designing Interactive Data Visualizations*, First Edition. Glenn J. Myatt and Wayne P. Johnson.
© 2011 John Wiley & Sons, Inc. Published 2011 by John Wiley & Sons, Inc.

## 2.1 EXTERNAL REPRESENTATIONS

Cognition, the process of thought, derives its power from *representation* and *abstraction*. A representation consists of symbols, which refer to things in the world, and a specification—rules, constraints, or relationships—for how the symbols are structured or relate to each other. The *representing world* consists of the symbols encoded in some physical system—for example, marks on a piece of paper, the magnetic orientation of bits on a computer disk, or the beads on an abacus—that refer to something in the represented world. The *represented world*—physical or abstract—is what is being represented in the external world or *environment*. For the representation to be effectively used, it must make only the relevant details explicit; in other words, it must be at the right level of abstraction (Norman, 1994).

Imagine a soccer coach with a playbook of plays similar to the one shown in Fig. 2.1. Each play is a diagram where *X* represents a member of the team, *O* represents a member of the other team, and the rectangle represents part of the playing field. The *X*s and *O*s are marked on a diagram of a soccer field drawn to scale. The coach's intent is to teach the team some plays. The *X*s, *O*s, and lines are the symbols. The condition of the field, such as the kind of grass used for the field or whether it slopes in one direction or another, is irrelevant and not shown in the diagrams. For how it was intended to be used—to convey positions and the movement of players on the field—the play diagram is at the right level of abstraction.

*External representations* are cognitive tools that exist outside our minds, and we use them all the time. The map and the playbook just described, shopping

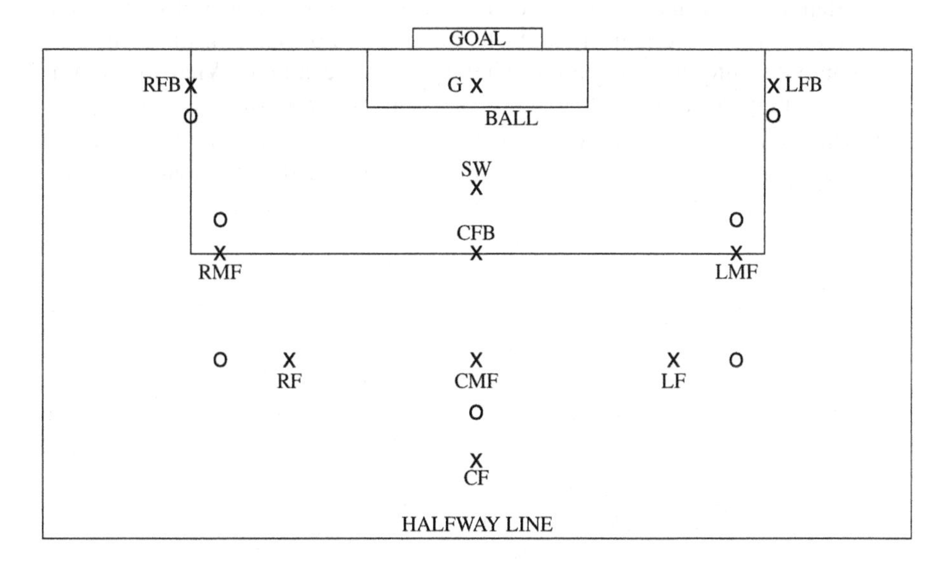


**FIGURE 2.1** Example of an external representation from a soccer playbook

lists, multiplication tables, bank statements, and various diagrams, graphics, and pictures are all examples of external representations that make explicit specific kinds of information structured in various ways. *Internal representations* are the representations inside the mind that are created by one of the senses—seeing, hearing, touching, smelling, tasting—and encoded in the biological machinery of the mind which is made up of networks of neural cells. The information these contain is all we can know of the world.

For the external representations that we design, any particular representation makes explicit certain details at the expense of others. This imposes a trade-off: one representation may make some operations easier to do than another and make other operations more difficult. Different representations of the represented world can result in dramatic differences in how a cognitive task is done and how quickly it can be performed. For example, an unordered grocery list sitting next to the refrigerator may make it easy to write down items as they run out, but these items then have to be mentally reordered at the grocery store to efficiently navigate the aisles. A list categorized by the aisle or area the product is in has the opposite effect. Which you use depends on how frequently you shop and how many items are on the list.

External representations are not just inputs to the mind as are the scenes we see of the world; instead, they are deliberately designed to convey meaning. They can be referred to, studied, or shared; the symbols can be manipulated or operated on independently of what they represent. For many cognitive tasks, they guide and constrain the work itself, and they play many important roles such as the following:

- *Aiding memory*. When asked how to spell a word that we know but don't often use, we will often write down the word in different ways to see which "looks" right. Icons in the buttons on toolbars or items in the menus of computer interfaces remind us what actions we can take. What we cannot recall may be retrieved through perception by seeing the right form.
- *Making abstract concepts visible.* The concept of a right triangle is more easily understood by the diagram shown in Fig. 2.2 than by saying "a right

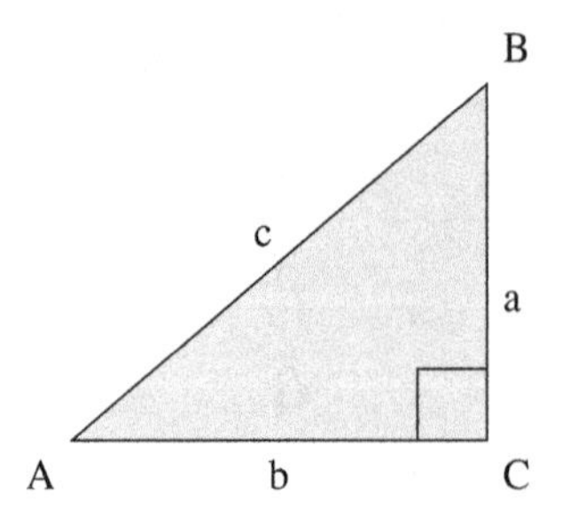


**FIGURE 2.2** A right triangle

triangle is a triangle where two sides are at angles perpendicular to each other." The diagram takes advantage of the natural 2-D space of the medium to show the relationship. Similarly, spatial distance could be used to show the distance of other abstract relations as in, for example, a dendrogram that shows the similarity of items and groups that have been clustered by some algorithm. The dendrograms at the top and right of the heatmap in Fig. 2.3 are used to determine which rows or columns belong in a group and how closely they are to each other or other groups.

- *Solving problems.* Before the advent of computers, the mental effort of solving a linear equation was made easier by using pencil and paper and a representation of numbers and variables. The use of pencil, paper, and

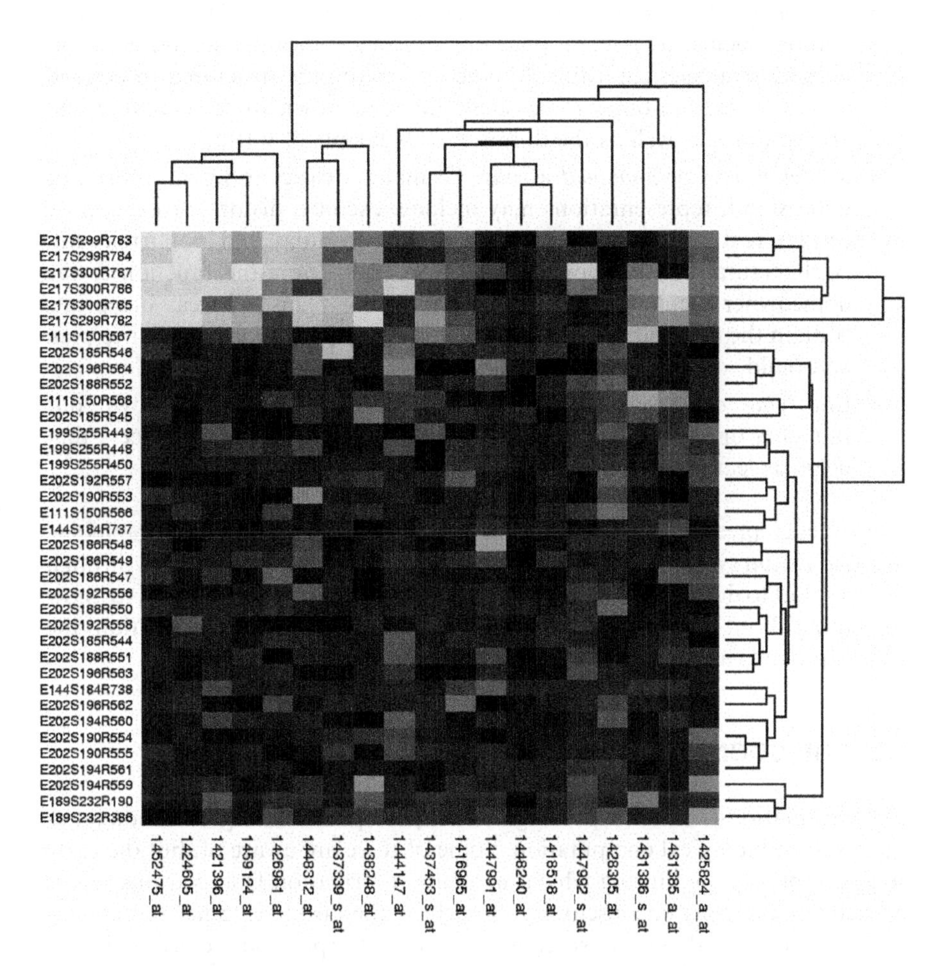


**FIGURE 2.3** Heatmap (Image courtesy of Wikipedia: http://en.wikipedia.org/wiki/File:Heatmap.png)

**TABLE 2.1 A Structured Presentation of a Flight Schedule**

| Flight | Departure | Arrival | Duration |
|---|---|---|---|
| AA 631 | 7:30 AM EST | 10:10 AM EST | 2:40 |
| UA 732 | 12:30 PM EST | 3:45 PM EST | 3:15 |
| US 1250 | 7:05 PM EST | 8:25 PM EST | 1:20 |

representation as a cognitive tool made the process straightforward and largely error free. The steps could be done in any order; and the person solving the equation could be interrupted without having to recalculate intermediate results.

- *Supporting decision making*. Different representations of the same information, such as a table, graphic, or list, can significantly change decision-making strategies. In Table 2.1, the information is structured to be read across to see the times for a single flight or down to see depature and arrival times as well as the duration of segments of a trip.
- *Modeling real or imagined worlds*. To make it easier to learn, inform, or understand, representations may include, exclude, distort, or exaggerate certain features so that what remains is essential and not irrelevant. Architectural design drawings used to explain form and function to the homeowner focus on outside views and floor plans, whereas blueprints contain the detail about construction that is required for a builder to plan and build.
- *Clarifying or sharpening our thinking*. The act of writing down our thoughts or sketching out ideas can be part of a reflective process that helps us learn and discover.

External representations—the user interface, visualizations, or data graphics—in visualization systems are the starting points for what will be perceived through the eyes. But to design a visualization system as a cognitive tool, we need to know something about the human mind that wields it and how the representations the system presents are processed by the human visual system.

## 2.2 THE COGNITIVE SYSTEM

The information-processing paradigm views perception, feeling, and thought as the result of biological computation. Some of it we are aware of, but the early stages of processing, during which incoming information from various senses is encoded, happens unconsciously. Somehow the visible external representations, such as a diagram, spreadsheet, or scatterplot, are converted into representations inside our minds. These *internal representations* are eventually transformed into a language of thought, which is the structure for the knowledge that we remember, reason about, and reflect on. The word "computation" elicits analogies to the computer: the input and output channels; the various forms of memory: disks, computer memory, and computer caches; and the graphics, central processing units, and other components that transform input to output. Although there are similarities, there are many more differences that make the computer a weak metaphor for what happens in the mind (Pinker, 1997).

### 2.2.1 The Matter of Thought

Until the 1950s, the brain as a biological organ was just gray matter, and its functions were not understood. It had taken many years to determine that neurons, the fundamental cells of the nervous system that includes the brain, were only loosely connected to each other and that they communicated through electrical and chemical signals.

A *neuron*, as shown in Fig. 2.4, is a complex cell in the nervous system that is still not fully understood. The features of interest are dendrites, axons, the cell body, and the connections between neurons called synapses. Incoming signals travel along *dendrites*; an outgoing signal from the cell body travels along an *axon* to other connected neurons. (Although only one axon extends from the cell body, the axon may branch many times, and each branch may form a connection.) There is a gap where the ends of the outgoing axon meet the ends of the dendrites of the neurons it is connected to. The gap is part of a specialized junction called a *synapse*. The way the signal passes through the synapse—how much and whether it is positive or negative—is controlled chemically. The synapse can be altered to strengthen or weaken the connection for short or long periods of time based on how much the connection has been used and other

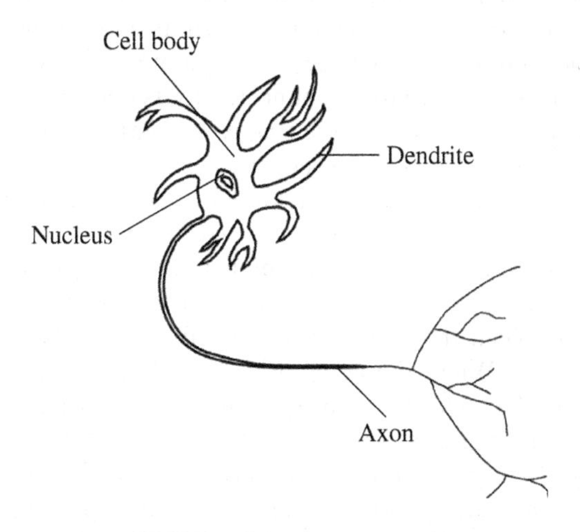


**FIGURE 2.4** A neuron cell

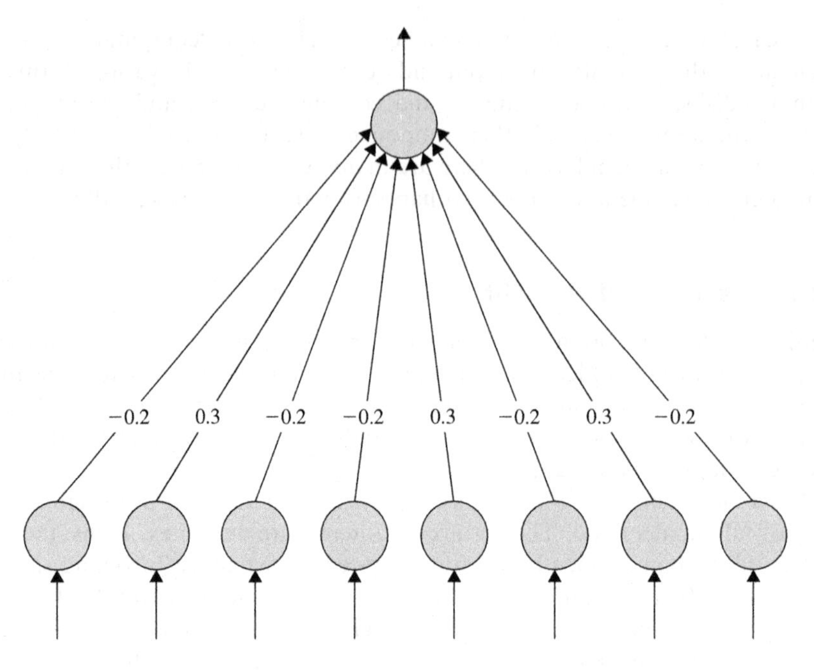


**FIGURE 2.5** A model of a neuron

factors. This alteration is considered to be one of the foundational mechanisms for learning and memory.

The most significant thing a neuron does, conceptually, is add a set of quantities and indicate whether the sum is greater than some threshold. A model of a neuron is shown in Fig. 2.5. Only when the threshold is exceeded will the neuron send out an electrical signal through its axon. When this happens, the neuron is said to have *fired.* The activity level of a neuron is affected by the incoming axons from other neurons connected to its dendrites and the synapse at each connection. The strength of the synapse ranges from negative to zero to positive. The synapse *inhibits* if it's negative, has no effect if it's zero, and *excites* if it's positive. The activation level of each incoming dendrite is a product of the synaptic strength and the activation level of the axon connected to it. If the sum of the incoming levels is higher than the threshold, the neuron fires, which propagates a signal along its axon to any connected neurons. During the period when the neuron is not firing, it is in its resting state and considered to be off. When the activation levels become elevated above the threshold, it is on. The neuron is always firing at some rate, either slower or faster. The firing rate represents the intensity of the stimulus.

A set of connected neurons is called a *neural network.* The drawing shown in Fig. 2.6 is of a small neural network in a pigeon's brain. There are estimated to be up to 100 billion neurons in the human brain, and each may have hundreds

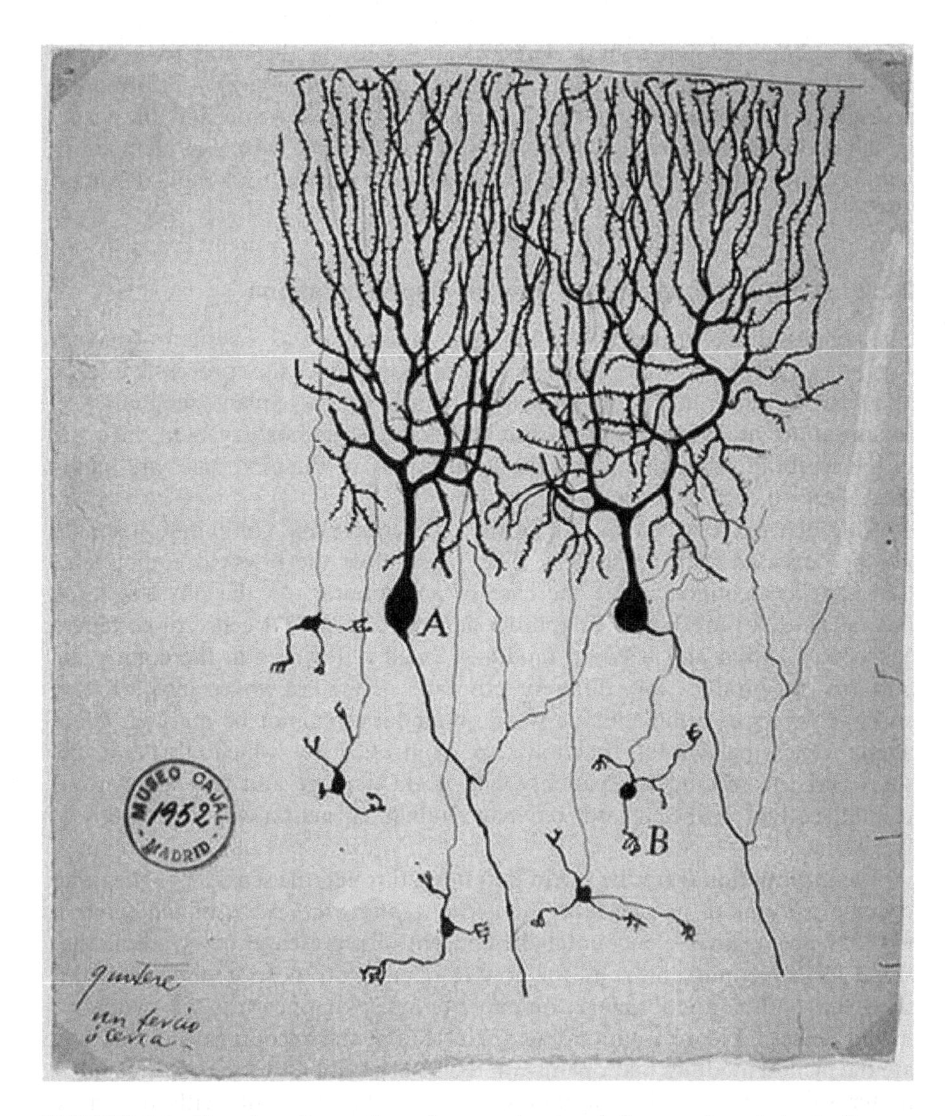


**FIGURE 2.6** Drawing of a portion of a neural network from a pigeon's brain (Image courtesy of Wikipedia, http://en.wikipedia.org/wiki/File:PurkinjeCell.jpg)

of connections. There are trillions of synapses, each too small to be seen as something other than a point under a light microscope, although they can be seen clearly with an electron microscope.

It was not until the advent of the computer and the development of physiological techniques that allowed the recording of electrical impulses of a single neuron's response to stimuli that the brain began to be seen as an information-processing organ. In 1951, the mathematician John von Neumann directly compared the neural electrical spikes along the axon to a digital code. These events, among others, attracted interest in biological information processing that eventually led to an understanding of vision and the visual system. Although few neuroscientists today would agree with Von Neumann's analogy, the idea that the brain processes information took hold (Palmer, 1999).

### 2.2.2 Mental Processes and Internal Representations

The computational theory of the mind views the brain as a system of mental organs—or *modules*—that processes information. Those that process language and what we see or hear and that help us remember, make plans, set goals, and focus our attention are mostly spread across the uppermost layers of the brain in the cerebral cortex just inside the skull. Many of these modules are highly specialized to carry out specific functions.

Although the term "module" is used to describe them, computation should not be confused with computers. They are unlike the physical components that can be identified inside the case of a computer. At the physical level, mental modules are highly adaptable networks of neural cells connected to each other within and between modules. Small differences in the connection patterns can produce very different programs. They are woven into a folded sheet of layers of cellular tissue whose boundaries cannot be marked. Clues about which behavioral functions are controlled by which areas of the sheet come from studying patients with brain damage and through various techniques that probe, record, or image individual neural cells or regions of the brain.

The information from the world that flows through the senses into the mind become patterns of neural data — *internal representations* — of the external world. Once again, we encounter the concept of representation: symbols that stand for something and rules for how to operate on those symbols. But this time, the symbols are not marks on paper or a display screen, as they would be in some external representation, but are instead neural encodings; and the rules are the mental modules that operate on them. The symbols have been created as patterns of activity by physical things in the world that interact with the senses, for example, photons of light reflected off surfaces or sound waves from a violin. The mental modules process information by mapping one representation to another. Through a series of transformations, the initial input of the senses becomes the language of thought.

How the mind works has been studied for more than half a century, and we are just at the beginning. Although there are other theories, the computational theory of the mind is still widely held, and the approach has been successfully used to develop computational models and various algorithms that have helped to understand the human visual system. Figure 2.7 shows a high-level diagram of the cognitive architecture that emphasizes the stages of processing involved

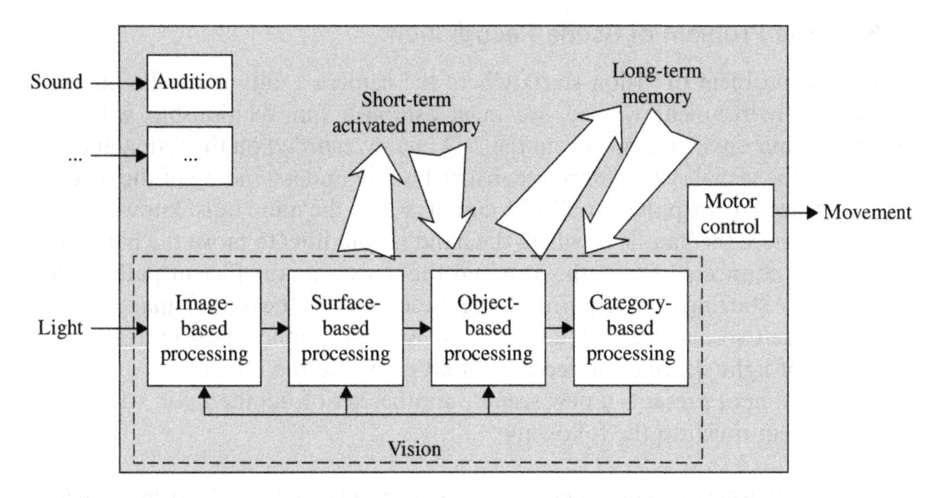


**FIGURE 2.7** The cognitive architecture emphasizing vision

in visual perception. The designer needs to be aware of the processing pathways and key modules that affect the interpretation of the visual forms.

## 2.3 VISUAL PERCEPTION

"Vision is a process that produces from images of the external world a description that is useful to the viewer and not cluttered with irrelevant information" (Marr, 2010). This definition by David Marr, the neuroscientist and psychologist who pioneered much of the early work on computer vision, emphasizes the active role that the visual system plays in shaping the information of the world we see to our advantage. So prevalent is the camera as a device for recording images of scenes we have taken that we assume our own visual system works in much the same way. The lens (eye) faithfully records an image (sees) by allowing light to strike an array of receptors (retina) at the back of the camera. This much of the processing is largely governed by optics, and there are similarities between the two systems. But here the similarities end, for there is nothing in the camera that takes the image and produces from it knowledge that can be used.

The emphasis on usefulness rather than fidelity should not be surprising because we do not see the world in the same way as do other species with visual systems. The compound eye of a common housefly, for example, is designed to be hypersensitive to motion to help it avoid in-flight obstacles and allow it to land, sometimes inverted, toward the center of surfaces that loom nearby (Marr, 2010). A key role of the visual system is to create an internal model of the environment—a description—that helps us predict what will happen and know how we should respond.

### 2.3.1 The Problem of Scene Recognition

The hard problem of vision starts where the camera's duty ends. To interact with the environment we see, we must establish our relationship with the objects in our environment. From the 2-D image recorded on the retina at each moment, the visual system must reconstruct a 3-D understanding of the objects in the scene. To grasp the handle of a cup of coffee, the mind must know at each moment where the handle is, where the hand is, and how to move the hand and everything connected to it—the arm and the body—in an efficient path to the handle. The starting point for processing these scenes is the retinal image, which is in essence two matrices—one from each eye—of numbers that represent the intensity of light striking the receptors that comprise the retina.

Each moment presents a new scene—another set of numbers—in which the visual system must do the following:

- *Identify where an object ends and the background begins.* The scene is made up of many colored patches like a quilt or a stained glass window. Having access only to the variations of intensity, such as where big numbers appear next to small numbers, the visual system must determine the effects of shadow, the overlay of one object on another, and whether the object is opaque or transparent.
- *Determine what objects are made of.* The amount of light reflected from a lump of coal in broad daylight as measured by a light meter is greater than the amount of light reflected from a snowball inside a dimly lit room, yet we see the coal as black and the snowball as white.
- *Reconstruct the third dimension.* The image is two dimensional. The same-size shape projected on the image can come from a large object far away, from a smaller one that is nearer; or from an object in one of many orientations. From the pair of images that come from the left and right eye, the depth must be reconstructed so that we can determine how far away the object is from us in order to know how to interact with it.
- *Recognize the object.* A kitchen table seen from different perspectives and at different times of the day projects onto the retinal image shapes of different sizes and shades of color. Yet we recognize all those differently shaded shapes on the image as the same kitchen table. However, recognition is deeper than just knowing what the object is. The mind must assign meaning to the shapes so that we know, for example, that the trapezoid on the image is the surface of the table that we will set with dishes and where food will later be served.

### 2.3.2 Levels of Explanation

Understanding any complex information-processing system requires explanations at different levels. Any theory of information processing can be broken into three levels (Marr, 2010):

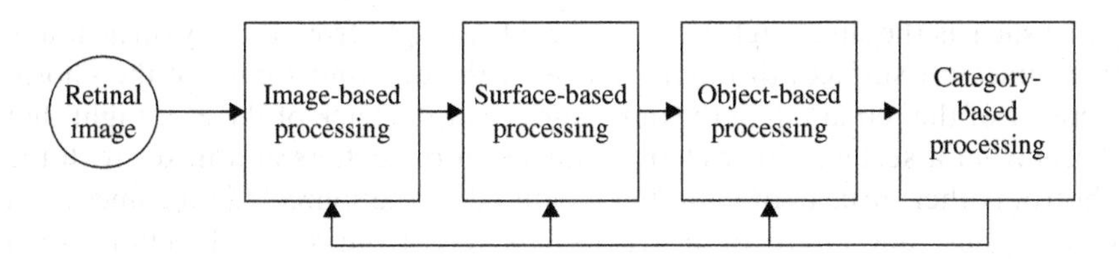


**FIGURE 2.8** The four stages of visual perception

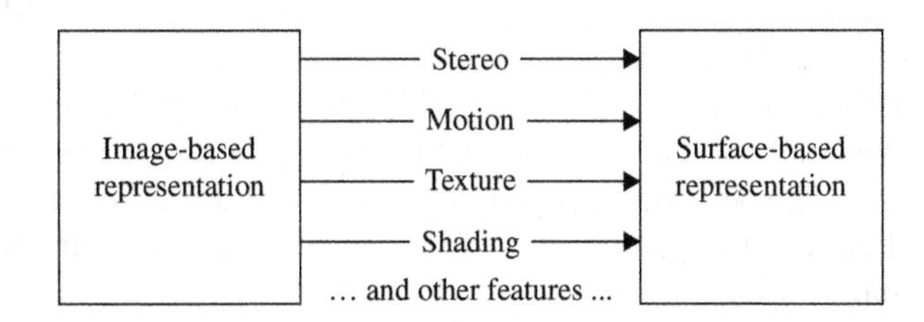


**FIGURE 2.9** Flowchart of major operations performed during visual perception

- At the computational level—the highest level—the system is characterized only by a mapping from its input to the output. It describes what needs to be computed and the information on which the computation will be based, without saying how it will be done. The visual system can be characterized as having four stages of processing as shown in Fig. 2.8.
- At the algorithmic level, the specific algorithm that performs the operation is described as well as details of the representation of the input and output. The algorithm is defined in terms of specific operations that will be performed on the input. Each of the preceding four stages includes various algorithms. An example of algorithms is shown in Fig. 2.9. These algorithms are implemented by the modules that transform the retinal image into a surface-based representation of the scene. Algorithms will be alluded to but not discussed in detail.
- At the implementation level, the specific physical system that will implement the algorithm is described. For example, the same algorithm could run as computer-based models or in the biological system. We will describe the anatomy of the eye and visual pathways through regions of the brain, but only to the extent that it helps to understand the representations and processing relevant to designing visual forms.

### 2.3.3 Illuminating the Environment

Seeing requires the interaction of light with surfaces that reflect it and a visual system that can perceive what is reflected or emitted. The visual processing begins when photons—small wavelike packets of light energy—from some source such as the sun, a lightbulb, or the diffuse light from the sky on a cloudy day bounce off surfaces through the lens of the eye, and stimulate the photoreceptors of the retina lining the inner surface at the back of the eye. Light that reflects from a surface either changes direction or scatters and mixes with the light from other surfaces or light that is emitted by the source. Light comes from all directions, and more of it reaches the eye from some directions than from others. We see only what reflects from the surfaces of the objects facing us.

The relationship of the geometry of the surface of objects in the environment is preserved in the retinal image because the light focused by the lens onto the retinal photoreceptors obeys the perspective projection laws. The 3-D object's surfaces are projected into the 2-D image upside-down.

However, information is lost about the hidden part of the object and objects that lie behind the surfaces we see. The light converging on the eye is a complex optical structure. The perception psychologist J. J. Gibson called it the ambient optic array (Palmer, 1999), and the eye is responsible for converting these arrays to retinal images.

### 2.3.4 The Eye and Visual Pathways

The eye is a complex organ. It is the portal for the optic array from a scene or external representation that the eye converts into neural activity. The wave of activity—the electrochemical signals that travel through the optic nerve into the visual centers of the brain—that the eye generates is the retinal image. The retinal image is the initial internal representation that is the input to the first stage of visual perceptual processing. The image contains values of light intensity for each point in the image.

A simplified cross section of the human eye is shown in Fig. 2.10. Light enters through the circular opening—or *pupil*—in the *lens*. The lens can be adjusted by *ciliary muscles* to bring close or distant objects into sharp focus on the *retina* as shown in Fig. 2.10. The *iris* controls the amount of light that enters the eye by increasing and decreasing the size of the opening of the pupil. It adjusts the pupil to compensate for lower or higher levels of illumination. The retina is a layer of tissue containing *photoreceptors*. It lines the curved surface at the back of the eye. *Rods* and *cones*—the names describe their shapes—are two kinds of photoreceptors. Rods, which are extremely light-sensitive, are spread throughout the retina except for a small area at its center called the *fovea*. There are more rods (about 120 million) than cones (about 8 million). The rods contribute little unless there are low levels of light such as at twilight or in dimly lit areas. The cones are responsible for the light and colors we experience in normal lighting conditions. We will discuss how in a later section on color.

Unlike the receptors in a digital camera, the receptors in the retina are not uniformly spread across the surface. The fovea is critical for our ability to see fine detail sharply. It is densely packed with cones, with many fewer cones spread in the areas outside. The fovea, which covers an angle of about 2 degrees

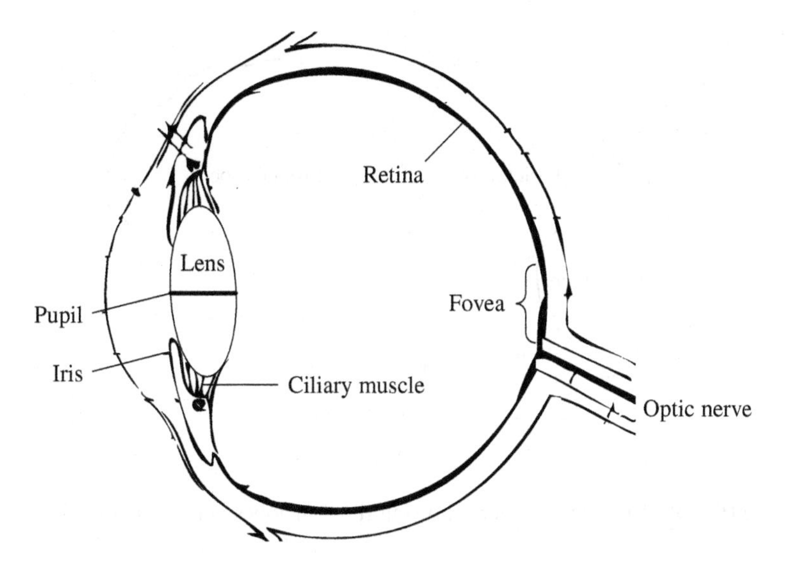


**FIGURE 2.10** A simplified cross section of the human eye

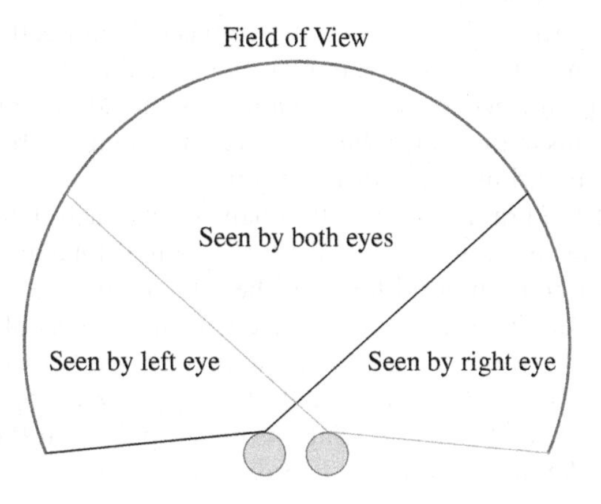


**FIGURE 2.11** The field of view is what we see with both eyes

of the field of view or about a thumbnail at arm's length, provides almost all of the detail needed for tasks that require visual acuity such as reading, driving, or texting. In the areas outside the fovea, the image is blurred because the density of cones is much smaller. Most of the retinal image comes from about 5% of the field of view, and the mind creates an abstract description of the entire scene through rapid eye movements that scan the environment.

The *field of view* is the extent of the world we can see from both eyes as shown in Fig. 2.11. The objects in this field of view, by the way light is directed into the eye, project an upside-down image. The patterns in the image of each

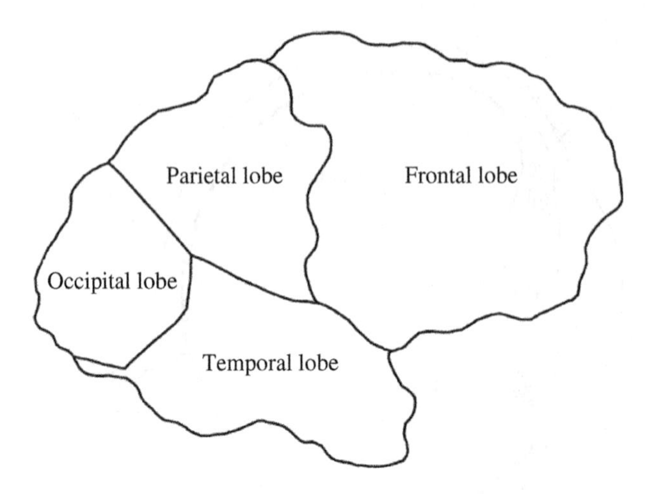


**FIGURE 2.12** Hemisphere showing the four lobes without pathways

eye are mostly similar, although the image in each eye has patterns from its inner edges that are not in the other image because they come from the periphery of the field of view.

The photoreceptors in the retina are connected indirectly to a layer of ganglion cells in the retina that are, in turn, connected to the optic nerve fibers. The processing performed by the ganglion cells include refining images and identifying locations in the image that have recently changed. Information from the image flows to various areas of the brain.

The cortex of the brain is split in two halves, and each half is divided into four broad regions as shown in Fig. 2.12: the frontal lobe at the front of the head, the parietal and temporal lobes in the middle, and the occipital lobe at the back of the head. The visual cortex refers to those areas of the cortex that process information from the pair of retinal images. One of the distinguishing characteristics of a visual area is that it is organized topographically to preserve, qualitatively, the spatial relationships of points on the retinal image. The effect is similar to sketching a line drawing on a flat balloon and then inflating it: the lengths of the lines change, but they still keep the same spatial relationship with each other. The visual cortex has many small topographical maps that it uses to encode aspects of the image such as brightness, color, shape, texture, motion, and depth. Most of the early-stage visual processing is done in the primary visual cortex occipital lobe, but later-stage processing is done in the temporal and parietal lobes.

The optic nerve that enters the brain from the back of the eye has more than 100 million connections. The optic nerves from both eyes meet at the chiasma as shown in Fig. 2.13, where the circuits are combined and then rerouted. The circuits that correspond to the left half of the visual field are directed to the right side of the brain, and those that correspond to the right half go to the left side of the brain. Two pathways lead out from the chiasma. The smaller

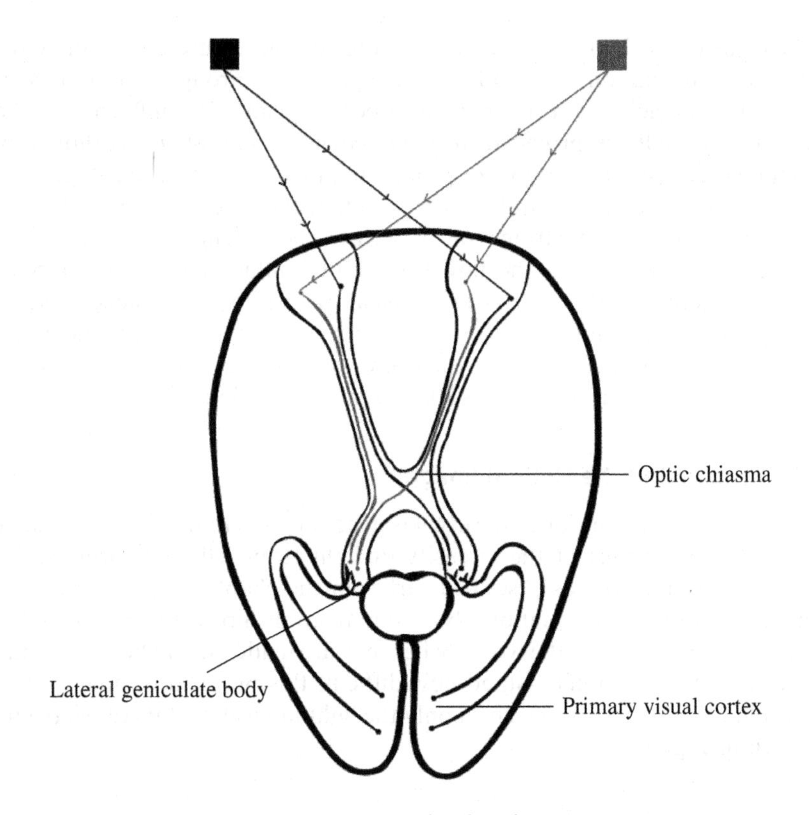


**FIGURE 2.13** Visual pathways

pathway goes to a nucleus in the brain stem—the *superior colliculus*—which is thought to extract information about where objects are in the environment and contribute information needed for eye movement and other responses. The larger pathway goes through the LGN (lateral geniculate nucleus), which relays information to the primary visual cortex known as area V1 where the earliest stage of scene recognition takes place. V1 is the most studied of the visual areas.

Ignoring the enhancements to the image that may already have been made in the retina, the process of reconstructing mental descriptions of concepts from patches of light begins in V1. Information flows through processes in different areas, and the patterns become increasingly complex. A widely accepted theory known as the "two streams hypothesis" is that from V1, the processing separates into two pathways. One pathway—the *dorsal* pathway or the "where" or "how" pathway—culminates in descriptions about the location of the object relative to the viewer so that actions such as eye movements or reaching can be planned and guided. This pathway ends in the parietal lobe. The other pathway—the *ventral* or "what" pathway—culminates in descriptions that allow the recognition of objects or concepts that we have in our memories. This pathway ends in the temporal lobe.

The connections between visual centers are complex. Throughout the visual processing, information flows to later-stage processes through *forward connections*, but feedback from later-stage processes can also influence earlier processing by feedback provided through *backward connections*. Although we will summarize the forward flow of processing, the more advanced stages have the ability to tune earlier neural networks so that they are activated only when there are certain combinations of patterns. The extent to which various processes can be influenced and how they can be influenced is debated. Some argue, for example, that early-stage visual perception is impenetrable by cognitive processes. But whatever the mechanism, there is agreement that later-stage thought processes, such as where we focus our attention, can affect what we see. We will return to this topic in a later section.

### 2.3.5 Processing the Retinal Image

From some scene in the world, the eye has generated the retinal image: an array of numbers that represent the intensity of light from reflected surfaces. The hard problem of vision, as described earlier in the chapter, is how to recover from these numbers information about our relationship to these objects that includes where they are in the world, what they are made of, and how we should interact with them. To understand how difficult this problem is, imagine how you would reconstruct a scene given only the light intensity values taken from it as shown in Fig. 2.14.

**FIGURE 2.14** A numerical array of values from the intensity of light

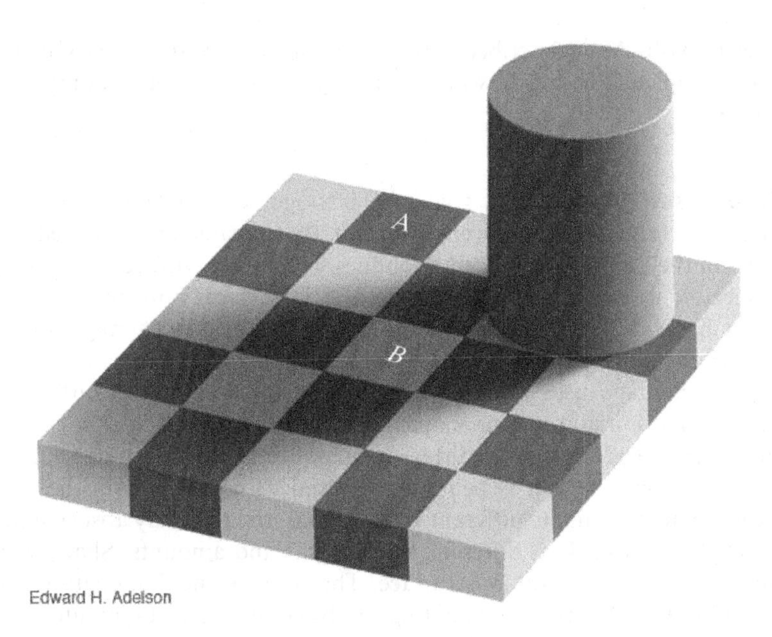


**FIGURE 2.15** The checkershadow illusion (Image courtesy of Edward H. Adelson: http://persci.mit.edu/gallery/checkershadow)

In the following sections, we describe the reconstruction of a scene through classical information-processing theory using the framework developed by Marr and extended by Palmer. (Note that although our experience of motion comes from the visual processing of changes in the images that occur over time, understanding how design of real-time interaction affects this experience is outside the scope of this book. The analysis of motion, although it makes important contributions throughout the stages, will not be discussed in the following sections.) At the algorithmic level of explanation, vision is a process of four stages: image-based, surface-based, object-based, and category-based. The names reflect the kind of descriptions that each stage generates. The stages presume a sequence of representations that build progressively from simple graphical primitives into a complex pattern that describes an object independent of the direction in which that object is being viewed. We will describe the processing using the simple scene created by Adelson called the checkershadow illusion as shown in Fig. 2.15.

***Image-Based Stage.*** The goal of this stage is to find and combine small patterns in the retinal image. The retinal image is an array of light intensity values for each point in the image. This stage looks for spatial structure in the image by analyzing variations of intensity—low numbers next to high numbers—across the image. The small, local patterns it finds are called *features*. The features that are spatial primitives include edge segments or small repeating or random patterns called textures. The primitive features are further combined into longer lines, line contours or blobs. Marr described the output of this stage as the *primal sketch*. It is like an incomplete 2-D line drawing defined by the contrasts in light intensity in the image.

***Surface-Based Stage.*** The goal of this stage is to generate descriptions about the surfaces in the 3-D environment that created the lines, edges, or contours in the sketch. We perceive volumes, but we see surfaces. We see them in perspective as orthogonal projections of 3-D surfaces onto the 2-D retinal image, in many shades of light and color illuminated by direct sunlight or under the clouds of an overcast day, at many different angles, and as one in front of another. This stage uses specialized processes to identify the surfaces and extract information about their shape and depth.

An object can project many different shapes onto the image depending on the location from where it is being viewed and its distance from the viewer. Surfaces are made of many different kinds of material that may absorb, scatter, or reflect the light back at different wavelengths and amounts. Shadows may fall across parts of a continuous surface. The color of the illuminating source may be different. Materials and lighting can be combined in an infinite number of ways to produce the reflected value.

The checkershadow scene illustrates some of the problems that must be solved. The image has 36 shaded shapes—29 trapezoids and 7 irregular shapes, 5 of which are partially hidden trapezoids under the cylinder. (This may be difficult to see because our vision automatically processes the shapes so that they appear as objects. Try to look at each differently shaded area separately from the others.) A shadow is cast by the cylinder, which darkens some of the shapes. The squares labeled A and B are printed in the same color ink. The side of the cylinder is partly in light and partly in shadow. From only the edges, lines, and contours, the regions within them must be separated into 36 shapes and other blobs as if they were mosaic tiles with similar color and texture. The processes in this stage must label each shape or blob with information about which surface it likely belongs to and determine its distance from the viewer and orientation in a 3-D space.

Depth, our sense of distance to the surface of the object, is recovered through specialized modules. Some of these analyze stereoscopic information that come from differences in the location of the overlapping areas of the visual field in the images from the left and right eye. Other depth information comes from tension in the muscles that focus the eyes on an object: closer objects have a higher angle of convergence than do objects farther away. The thickness of the lenses also provides feedback because closer objects require a thicker lens. Other processes embody various kinds of heuristics or assumptions:

- Nearly parallel lines converge if they extend toward the horizon line (railroad tracks).
- Objects closer to the horizon line, below or above, are farther away, for example, the trees of a forest below the horizon and the clouds above.

- Objects of similar height are smaller on the image if they are farther away. The members of a marching band in a parade are about the same height. Their relative sizes on the image conveys their relative distance.

Other modules help recover depth by analyzing edges or edge intersections. Illumination edges separate areas of a surface that fall into shadow, reflectance edges separate areas where the amount of light changes as a result of reflection from different material, depth edges result from a closer surface hiding part of a surface behind it, and orientation edges show where surfaces meet. Edge intersections appear in the image as Ls, Ts, Ys, crow's feet, or as gradients, and these help analyze whether the surfaces that intersect at the junction are convex or concave.

Surface orientation affects the way light is reflected from the illuminating surface back toward you as shown in Fig. 2.16. A flat surface may *slant* toward one side or another or *tilt* away from or toward you. A continuous rounded surface has an infinite number of orientations. The analysis treats surfaces as composed of many small flat surfaces that, in the case of curved surfaces, can be reduced to the size of a point. Some processes use the patterns of variation of the amounts of reflected light from these points to recover information about depth and orientation (Pinker, 1997).

**FIGURE 2.16** Checkboard and cylinder showing how surface analysis might analyze local regions. (Redrawn. Original image courtesy of Edward H. Adelson: http://persci.mit.edu/gallery/checkershadow)

The result of this stage of processing is a 2.5-D representation that includes information about only those surfaces that can be seen. The representation of a surface is as many small flat elements whose properties include color, slant, tilt, texture, and distance from the viewer. It is called 2.5-D because the representation lies between the 2-D image representation and a full 3-D representation of objects.

***Object-Based Stage.*** The goal of this stage is to organize the edge and surface elements into a meaningful hierarchy of 3-D objects, parts, and groups. In the preceding example, we see a checkerboard with four edges—two that our mind fills in—and a solid gray cylinder casting a shadow over the board. We see equally sized squares even though the shapes on the image are trapezoids and the lengths of the sides of the trapezoids in the back are shorter than the ones in the front. We see squares that alternate consistently between dark and light gray even when some of the light and dark squares have the same intensity of light. We see 25 squares on the board even though only 19 are fully visible, 5 are partially visible, and 1 is hidden. We don't see—unless we attend closely to the details—many edges, shaded trapezoids and ellipses, or other irregular patches of light. How this happens raise questions of interest to the designer about the following:

- *Perceptual organization.* What gives rise to the particular structure of the scene that we perceive when there are many alternatives that could be reconstructed from the retinal image?
- *Visual interpolation.* How do we see shapes that are partially obscured? Or what patterns of shapes give rise to objects that don't exist?
- *Visual constancy.* Why do we see properties, such as color, as if they were intrinsic to the object, even though lighting conditions and other objects in the environment, such as the cylinder casting a shadow on the squares, create conflicting properties on the corresponding shapes in the image?
- *Shape recognition.* How do we recognize groups of shapes as an object?

Max Wertheimer, one of the founders of the Gestalt school of psychology, described the problem of perceptual organization in a seminal paper (Wertheimer, 1923). He argued that we do not experience the world as many individual elements, but as complex wholes consisting of elements separate from but related to each other. The elements have been structured by perception in ways we cannot control. Further, he argued, this could only be explained by looking at the entire context and not just at the elements. A simple example is shown in Fig. 2.17. On the left, we see four individual circles; on the right we see two groups of two circles. In the image on the right, it is almost impossible to see the image as a group of one circle next to a group of three. We cannot explain this by saying that the whole—a group of two circles—is the sum of its parts—a circle plus a circle. It can only be explained by knowing the larger context: it is

**FIGURE 2.17** Example of the Gestalt proximity principle

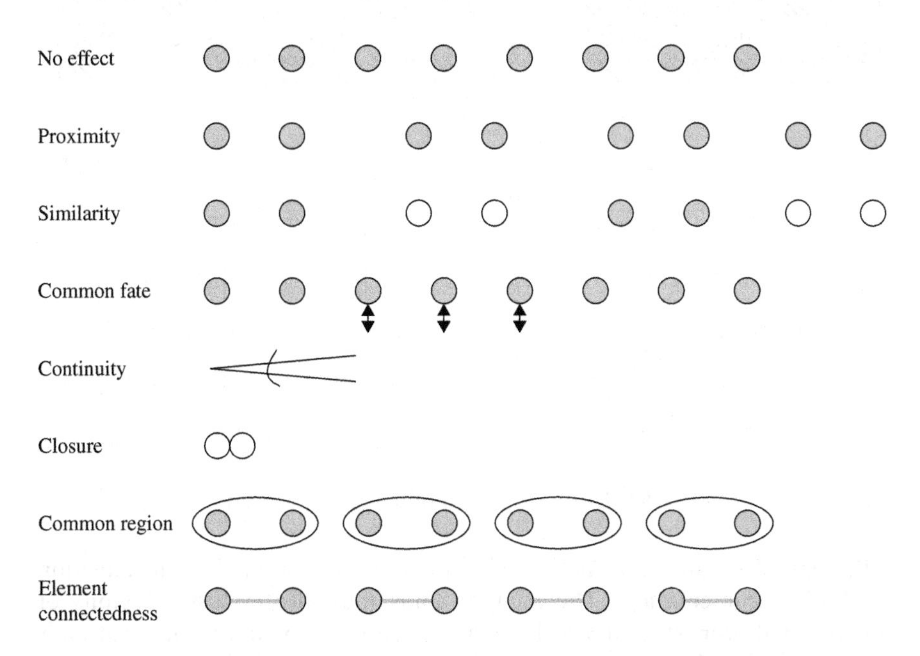


**FIGURE 2.18** Gestalt principles with extensions by Palmer

the proximity of a circle to the other circles that makes the difference. Proximity was the first of Wertheimer's laws of *perceptual organization.*

The classical principles of perceptual grouping that grew out of the work of the Gestalt psychologists and the extensions proposed by Palmer are shown in Fig. 2.18. Groups are perceived when elements are close to each other (proximity); are similar in color, size, or orientation (similarity); have symmetric or parallel patterns; or move together (common fate). Elements that have a smooth continuous boundary are grouped (continuity). In the figure, the arc is seen as a single line and not as three separate segments. Elements that form a closed figure are grouped (closure). In the figure, the two circles with edges touching are perceived as two circles instead of a "figure eight." Palmer proposed additional principles. Elements within the same region are grouped (common region). Elements connected by other elements are grouped (element connectedness). These principles apply if only one is operating at a time. There is no set of rules that predictably describes what will be perceived if combinations of the factors are present. How the visual system structures the scene when more than one factor is involved is not yet fully understood.

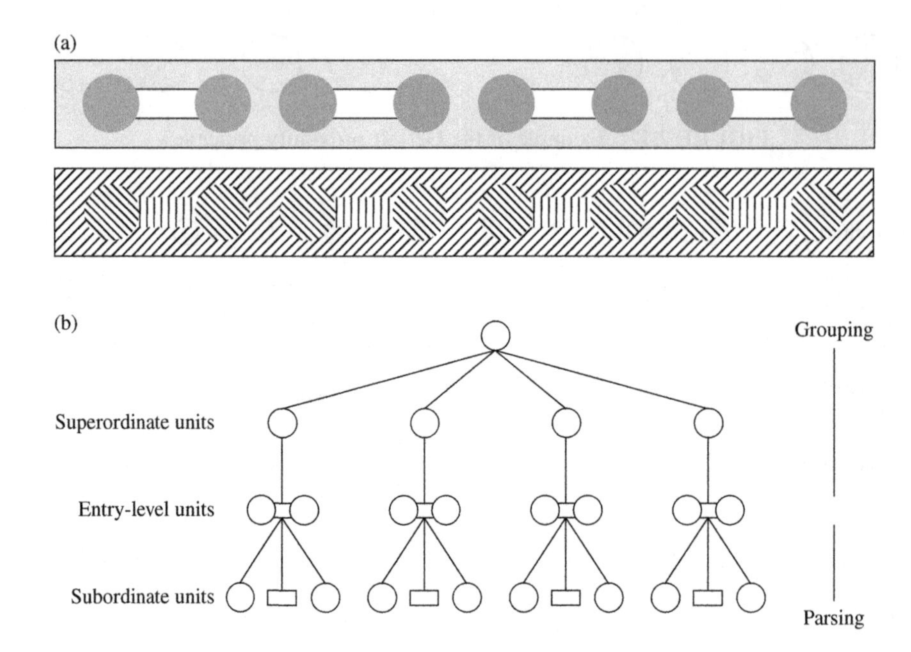


**FIGURE 2.19** Separation of figure and ground

Perceptual organization begins by performing region analysis to partition the 2.5-D representation. The analysis finds mutually exclusive regions of similar visual properties such as luminance, color, or texture. Each partition is further organized by separating the object that appears to be in the foreground (*figure*) from the background (*ground*), grouping elements into a complex object, or parsing an object into its parts. Figure 2.19a shows examples of two scenes. In each scene, two circles are connected by an edge. Either scene could produce the whole/part hierarchy in Fig. 2.19b. The figure—everything not gray or textured—is separate from the ground. The elements are grouped together. Grouping combines the circles and edge into a complex object of circle-edge-circle. Finally, the complex object is parsed into the parts we perceive as being separate but also as belonging to the object.

Other combinations of shapes may result in different perceptual structures. Sometimes perception fills in missing parts of an object or creates the outlines of objects that don't exist, an effect called *visual interpolation.*

When an object is in front of another, the closer object's surface may partially hide—or *occlude*—the surface of the one in the back. In the checkershadow example, the cylinder hides part of five of the squares, but perception fills in these squares.

Certain patterns of shapes can also give rise to contours that don't exist such as the triangle shown in Fig. 2.20.

Translucent objects, such as a tinted window, reflect only some of the light of the objects behind it. Transparency will be perceived only when the shapes combine in certain conditions as in Fig. 2.21a but not the others.

**FIGURE 2.20** An illusory triangle

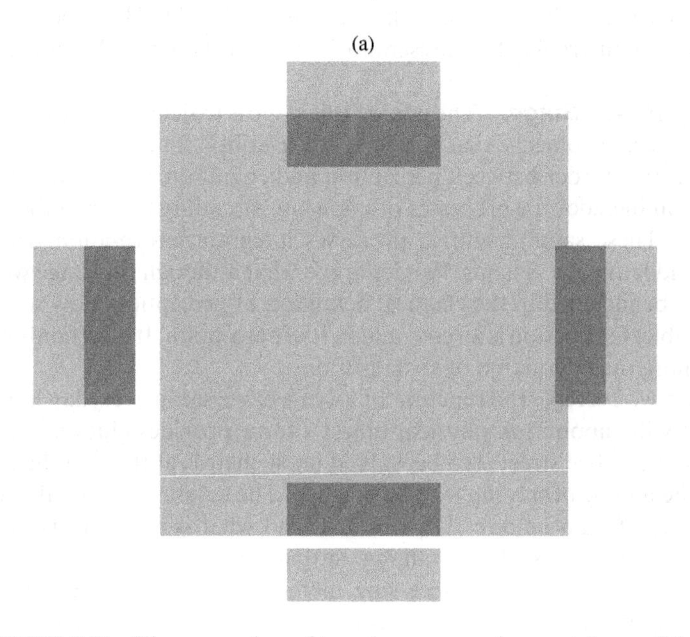


**FIGURE 2.21** The perception of translucency requires certain conditions

So far, we have been describing the role perception plays in organizing the features, surfaces, regions, and shapes it has recognized. Perception structures the environment into the 3-D internal objects that correspond with objects in the environment. But another important role it plays is to make the properties of these objects appear stable and constant even though much of the underlying retinal image changes as we or the objects move about in space. In the checkershadow example, if we were to move around the board, changes in internal representation are taking place at all stages of perception—light intensity, length and orientation of edges, and the shapes. Yet we perceive the board and cylinder at all times as having intrinsic properties that have not changed: the board has 25 light gray and dark gray squares, and the cylinder is a solid gray. This is known as *visual constancy*. The perceptual processing makes adjustments to internal representations to achieve this. We have already seen an example of color constancy: how the square labeled B appears a lighter gray color than the light intensity would predict that it should when compared to the square labeled A. There is also constancy of size, shape, orientation, and position of objects. The processes involved in making these adjustments are spread across all stages and rely on assumptions, which if violated, result in illusions, which are perceptions that don't match reality (Adelson, 2011).

The result of the object-based stage is a 3-D internal representation corresponding to objects in the environment that include descriptions of shape primitives. These representations are not internal 3-D replicates of the objects we see. We experience the cylinder as a round barrel and the checkerboard as having a particular thickness as well as width and height. The experience is the content of the internal 3-D representation but not its form (Pylyshyn, 2003).

***Category-Based Stage.*** The goal of this stage is to determine how to interact with the perceived objects; that is, what is their utility, function or meaning? We are now at the border between perception and cognition because the definition of cognition includes the processes of knowing: attending to, remembering, and reasoning. These same cognitive processes integrate information from other sensory and language systems. Palmer argues that although this stage would not ordinarily be included in the classical definition of perception, how we come to know an object's function is an essential part of perception. It is certainly essential for designing the interaction of visualizations.

How do we perceive the function of an object we see in everyday life and the actions it will support? A physical object's form provides clues about how it should be used. The street is to be walked on, a chair is at the right height to sit on, and the handle of the cup is to be grasped. The actions we can take for these kinds of objects can be directly perceived, and what we do with them largely involves motor skills such as walking, sitting, or reaching. These actions are done unconsciously without accessing memory, and visual perception takes place largely in the "where/how" pathway, which leads to an action-oriented system that coordinates motor activity. The visual processing is very fast and automatic.

For the control panels of modern appliances, mobile devices, or application interfaces, the functions are not so directly and easily apprehended. Their functions have been learned by experience and are stored in our memories with the objects that we see. Using primarily the shape of the object, but also its other visible properties such as color and size, the object is perceived indirectly in a two-step process: classify, then retrieve. The object is classified as a member of one of the many categories stored in memory. The category provides access to the large amounts of information associated with these objects, including their functions. Categorization requires that a perceived object and the categories stored and organized in memory have internal representations that can be compared and that there are processes for comparing and deciding on which category best describes the object. The visual processing takes place in the "what" pathway and is slower than for direct perception.

Perceptual classification has been measured and includes the following results of interest for design:

- Categories have a canonical perspective that allow us to more quickly and accurately classify an object. For example, a horse seen from the side will be classified more quickly than one seen from the front.
- An object being seen again, even if the first time it was seen was a few hours prior, will be classified more quickly the second time. The effect is known as *priming*.
- Objects in normal positions will be classified more quickly than if they are rotated. The difference in reaction time increases with the angle of rotation, as if the viewer was mentally rotating the object back to its normal position to classify the object.
- In a complex object with parts, the parts are critical to object classification.
- Context plays an important role in classification speed and accuracy when what is being classified is ambiguous or is a part of an object. For example, we do not always recognize the face of someone we know from our home city if we were to see them in a place far away because we are not expecting to find them in the remote location. Appropriate context facilitates categorization, whereas inappropriate context hinders it.

### 2.3.6 Color

We see blue skies, a forest of green trees, or a blazing orange sun at sunset, and we attribute the color to the skies or trees or sun. But color is not an intrinsic property of the object; rather, it is our experience of light. Isaac Newton discovered this in the seventeenth century when he used the prism to split sunlight into a rainbow of hues and saw that it was composed of many colors. In his treatise on optics, he wrote that the rays were not colored, but they had a power to stir up the sensation of color (Palmer, 1999).

Color is a byproduct of the way light is processed by the visual system. The energy of light is converted into channels of color information that comprise the retinal image. Later-stage visual processing associates color properties with surface and object representations. These augmented representations give rise to the experience that color is an intrinsic property of the objects we see.

***The Physics of Light.*** Photons, the elements of light, are small packets of undulating electromagnetic radiation that are both particle and wave. When we described them earlier in their role in forming the optic array, it

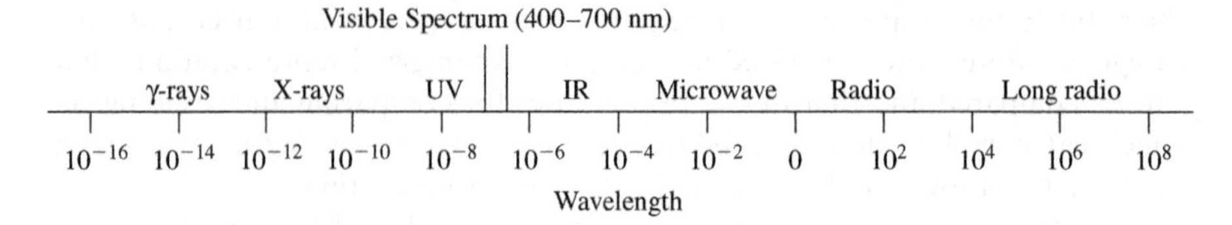


**FIGURE 2.22** The electromagnetic spectrum

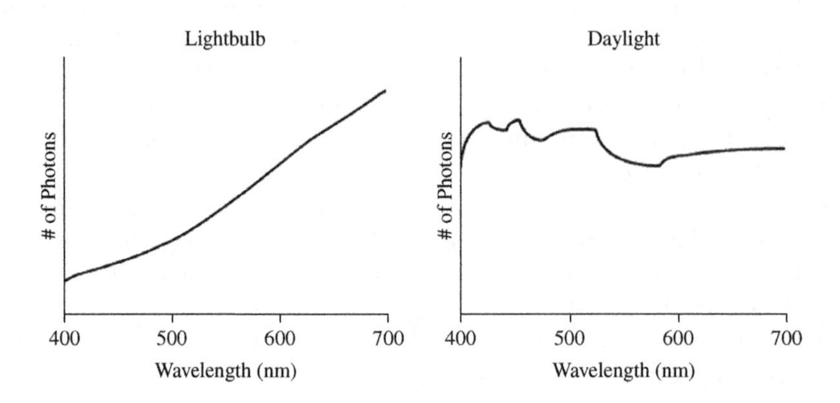


**FIGURE 2.23** Spectra for a tungsten lightbulb and daylight

was as particles bouncing off surfaces. Here we are interested in their wavelike nature because it is the wavelength—the distance between each crest of the wave—that determines which colors we can see. Wavelength is used to classify the electromagnetic spectrum. It starts with small gamma rays ($10^{-16}$ m—meters in scientific notation) and ends with long radio waves ($10^{8}$ m). The part of this spectrum that is visible, called the *visible spectrum*, falls in the range from 400 to 700 nanometers, where a nanometer is 0.000000001 meters ($10^{-9}$ m). The electromagnetic spectrum is shown in Fig. 2.22.

A beam of light consists of a number of photons at each wavelength of the visible spectrum, and the count of photons can be plotted against the wavelength to produce a spectral diagram for that beam of light (assuming it is not mixed with light from any other source). The spectra for an incandescent lightbulb and sunlight are shown in Fig. 2.23. A laser beam has only one wavelength, and it emits *monochromatic light*. Its spectra is a flat line. The bulb and sunlight emit *polychromatic light*. From the perspective of physics, the visible light from a particular source is a mixture of photons of different wavelengths, and its spectrum is its physical representation.

***The Experience of Color.*** The colors a person with normal vision perceives is often represented as a color space in three dimensions: *hue*, *saturation*, and *lightness*. (Saturation is sometimes called "chroma," and lightness is

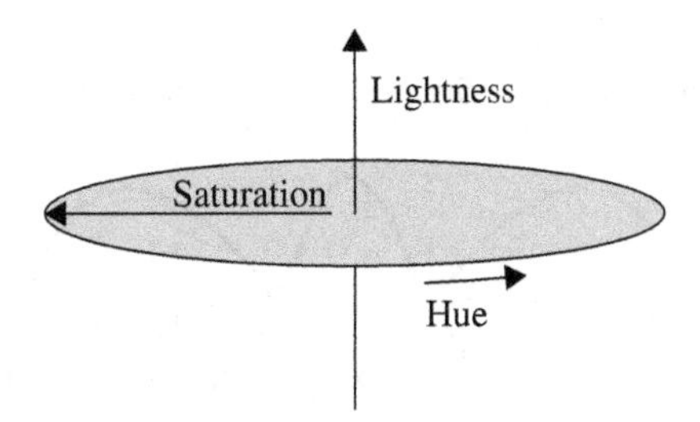


**FIGURE 2.24** The Munsell color model

sometimes called "value.") This color model, shown in Fig. 2.24, was devised by the artist and art teacher Albert Munsell (1905). Hue is the essential color we associate with a surface and references a particular wavelength of light in the visible spectrum. Saturation is the intensity of the hue: how much or how little red or yellow there is in the color, for example. Lightness is the amount of light reflected by the surface: the more the light is reflected, the lighter it is. A black-and-white image has only degrees of lightness. (Lightness is the amount of reflected light. The term *brightness* is used when describing the amount of emitted light.) The actual colors we perceive are a subset of this space because we cannot make as fine a distinction as the model allows.

In the behavioral experiments conducted after Newton's discoveries, various phenomena about color perception were observed that had to be explained by the theories of color:

- *Light as mixture.* Only a small number of the colors we see correspond directly with monochromatic light in the visible spectrum, which are colors of the rainbow. From Newton's discovery forward, efforts were made to understand the relationship between light and color. Newton developed a color wheel that had three primary colors opposite what he thought to be their complements—green–magenta, blue–yellow, and red–cyan. It was a way to visualize how the primary colors could be added together to produce other colors. Sir Thomas Young (1802) hypothesized that there were three types of color receptors in the eye that produced the red, green, and blue primary colors we see. James Clerk Maxwell (mid-1800s) picked up Young's work and conducted experiments to understand the laws of composition, the smallest number of standard colors, and the relationship of the light's spectrum to the standard colors. He found that many different combinations of wavelengths of light—*metamers*—could produce the same color sensation. In other words, light was an infinite-dimensional space that had to be mapped to the 3-D perceptual color space. Hermann von Helmholtz, a contemporary of Maxwell, postulated that each of the receptors responded to different wavelengths as shown in Fig. 2.25. One receptor was most sensitive to

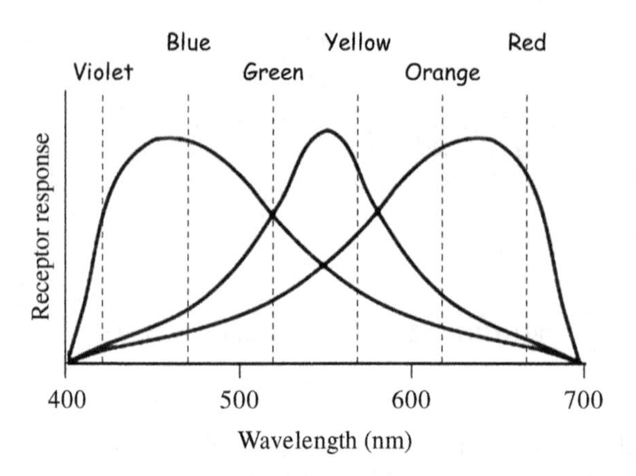


**FIGURE 2.25** Trichromatic theory: color arises from overlapping receptor responses

short wavelengths, the second to medium wavelengths, and the third to long wavelengths. Young's initial theory combined with the later extensions to it eventually became known as the Young–Helmholtz *trichromacy theory.*

- *Color afterimages.* Staring at intense colors for an extended time—for example, up to half a minute—and then looking at a white background, will produce an image of the initial colors' complements. If the intense colors were red and blue, the afterimage should appear as green and yellow.
- *Color blindness.* A small percentage of people—8% male and 1% female—cannot match colors produced by a mixture of three colors that the rest of the population sees. Those with normal vision—*trichromats*—can see all colors; *dichromats* can see a mixture of two; and *monochromats* see only in black and white. The fact that the loss comes only in pairs of colors, such as red and green or yellow and blue, and never as individual colors left the trichromatic theory without an explanation.
- *Simultaneous contrasts.* The color of a foreground shape may be affected by the color of its background. The strongly colored background of a gray circle that passes through a square subdivided into squares of red, green, yellow, and violet will pick up the complement of the color of the square. A similar effect happens with lightness contrasts as shown in Fig. 2.26. The inner rectangle has the same gray print color in both squares, but the perceived color is adjusted by the perceptual processing.

Besides the significant fact that in color blindness only pairs of colors were lost, other phenomenon began to create problems for the trichromacy theory:

- We can sense colors that mix red and green, red and yellow, red and blue, blue and green, and yellow and green. But we cannot sense colors that

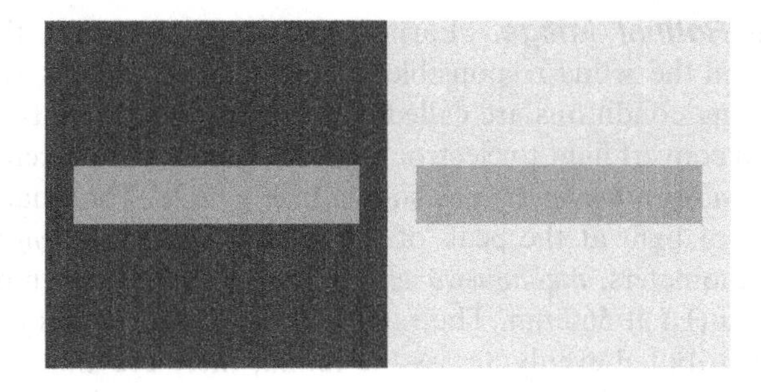

**FIGURE 2.26** The perceived color of an object depends on contrast

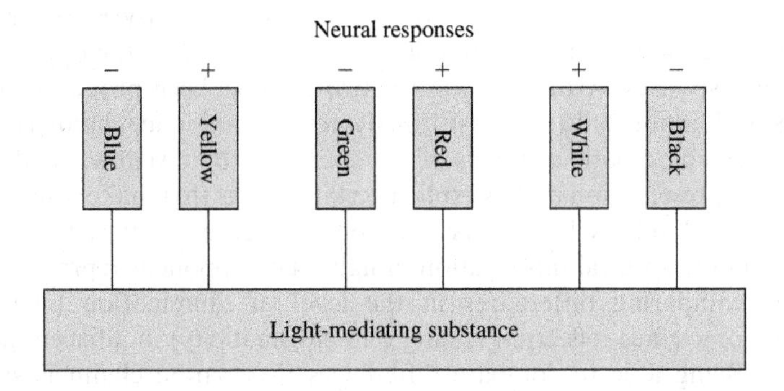


**FIGURE 2.27** Mechanism for opponent process theory

result from mixing red and green, or blue and yellow (ignoring the mixing of inks or paint that work by filtering light—subtracting—rather than combining it—adding).

- Yellow was supposed to result from mixing red with green, but the experience of yellow is not the same as that of purple, which results from mixing red and blue. Subjectively, yellow is experienced as a primary color.

Ewald Hering (1878) developed the *opponent process theory* as shown in Fig. 2.27 that explained these and other discrepancies. The theory held that there are three basic mechanisms, and shifts in the signal in each mechanism between positive and negative poles create the experience of a color or its complementary color, or of light and dark. The theory also stated that blue/yellow and red/green are the chromatic channels, black/white is the achromatic channel, and there are four primary colors.

The opponent process theory explained many anomalies of the trichromacy theory but was subjective; the trichromacy theory, on the other hand, predicted important observed behavior. Eventually, the dual-process theory emerged. Experiments have shown that networks of cells in the retina perform both stages of processing.

***Creating the Retinal Image.*** Earlier, we mentioned that the types of photoreceptors in the retina responsible for the light and colors we experience in normal lighting conditions are called cones. There are three types of cones that absorb and convert light to electrochemical signals. The percentage of light absorbed is the *absorption spectra* as shown in Fig. 2.28. Their names indicate the wavelength of light at the peak of the signal. *Short-wavelength cones* (S) peak at 440 nanometers, *medium-wavelength cones* (M) at 530 nm, and *long-wavelength cones* (L) at 560 nm. The ratio of L to M to S cones is 10:5:1, and they are not distributed evenly across the retina; there are almost no S cones near the very center of the fovea.

The signals from these three types of cones combine as the trichromatic theory predicted. But the discovery of color-sensitive cells in the retina and visual pathways whose responses corresponded to the opponent process theory led to questions about how the three types of cones could produce the opponent-like patterns of responses. Artificial neural networks have been proposed showing how the S/M/L input could be transformed into one luminance channel of black/white, and two chromatic channels of red/green and blue/yellow. As discussed earlier, any representation makes explicit certain things that make some calculations easier and others harder. A trichromatic representation is useful for transforming the spectral information in light. The opponent representation is useful for comparing differences in the level of illumination (changes in luminance) or surface reflection (changes in chromaticity) in adjacent areas of the image. Being able to compare differences in separate channels supports processing that, for example, identifies features in the image or makes adjustments for shade falling across a surface. The opponent representation is the basis for the input to later stages of visual processing.

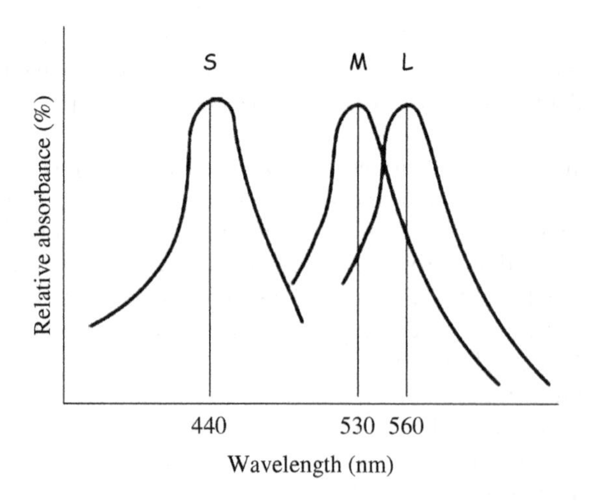


**FIGURE 2.28** Absorption spectra

***Processing Color in the Retinal Image.*** We see the surfaces of objects as having color, but color is a psychological property, not an intrinsic property of the object. Look at a color sample inside a paint store under a normal lightbulb and take it into the sunlight outside, and its color will change. What is intrinsic to the surface is how much (percentage) of the incident light is reflected at each wavelength. The amount of light falling on the surface from the light is the *illumination spectrum*. The percentage of light that will be reflected regardless of the light source is the *reflectance spectrum*; it is a function of the material and composition of the surface. The eye sees the light illuminated by a particular light source reflected from the surface. The resulting spectrum is the *luminance spectrum*, which is a product of the illumination spectrum and the reflectance spectrum:

$$\text{Illumination Spec. } (I_w) \times \text{Reflectance Spec. } (R_w) = \text{Luminance Spec. } (L_w)$$

$I_w$ is the number of photons emitted, $R_w$ is the percentage of photons reflected, and $L_w$ is the number of photons reflected. An example of the spectra and their relationships is shown in Fig. 2.29.

The visual system tries to maintain our perception that the color-related properties of objects are invariant: white snowballs always look white regardless of whether they are in bright sunlight or shadow. This is known as *lightness constancy* and *color constancy*. To determine the properties of the surface—the material it is made of or the color it was painted—the visual system must somehow recover just the reflectance spectrum. The processing is done in the surface-based stage.

The edges between regions play an important role in recovering information about reflectance and illumination. Edges are created by contrasts in the amount of light from one region to another. Even if the overall luminance

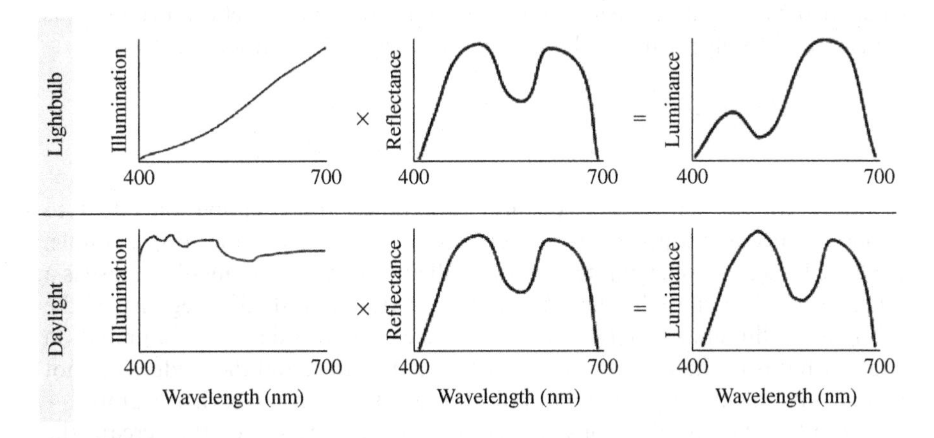


**FIGURE 2.29** (# photons emitted) × (% photons reflected) = (# photons reflected)

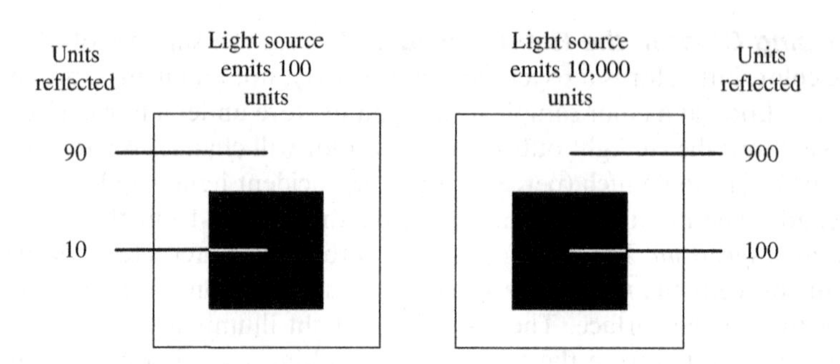


**FIGURE 2.30** Lightness constancy depends on ratios

across all regions increases because of a brighter light source, the relative amounts of light between regions remains mostly the same. In the example shown in Fig. 2.30, the ratios of the luminance in the first and second images remain the same.

The retinex theory (Land and McCann) provides an explanation of how the relative lightness between two regions can be recovered by integrating the luminance ratios of just the edges between them as shown in Fig. 2.31. The emitted light, as measured by the illumination level, is stronger at the top (100) than at the bottom (40). Surface A reflects 20% of the light, B reflects 40%, and C reflects 80%. $L$ (luminance value) at the top and bottom of each region is obtained by multiplying $I$ (illumination level) by $R$ (reflectance percentage) as follows:

$$L_{\text{A-top}} = I_{A\text{-top}} \times R_A = 100 \times 20\% = 20;\; L_{A\text{-bottom}} = I_{A\text{-bottom}} \times R_A = 80 \times 20\% = 16$$
$$L_{B\text{-top}} = I_{B\text{-top}} \times R_B = 80 \times 40\% = 32;\;\; L_{B\text{-bottom}} = I_{\text{A-bottom}} \times R_A = 60 \times 40\% = 24$$
$$L_{C\text{-top}} = I_{C\text{-top}} \times R_B = 60 \times 80\% = 48;\;\; L_{C\text{-bottom}} = I_{\text{A-bottom}} \times R_A = 40 \times 80\% = 32$$

The relative reflectance of A to C is 1/4 (20% / 80%). The global reflectance calculated by the visual system, which matches the known reflectance ratio, is generated by multiplying the luminance edge ratios as follows:

$$\begin{aligned} \text{global reflectance ratio}_{\text{A-C}} &= (\text{A/B}_{\text{L-ratio}}) \times (\text{B/C}_{\text{L-ratio}}) \\ &= 16/32 \times 24/48 = 1/4 \end{aligned}$$

Edges are further categorized into reflectance and illumination edges to make appropriate adjustments to lightness, saturation, or hue. For example, the edge between a blue painted window sill and the white siding of a house is a reflectance edge; the edge between the shaded and unshaded regions of the siding is an illumination edge. The reflectance on both sides of an illumination edge—what is in shadow and what is not—is the same, and these edges are not included in calculating the global reflectance ratio as in the earlier example.

Although the visualizations we will design are not scenes that create the kinds of variations that require continual adjustments to color and lightness, it

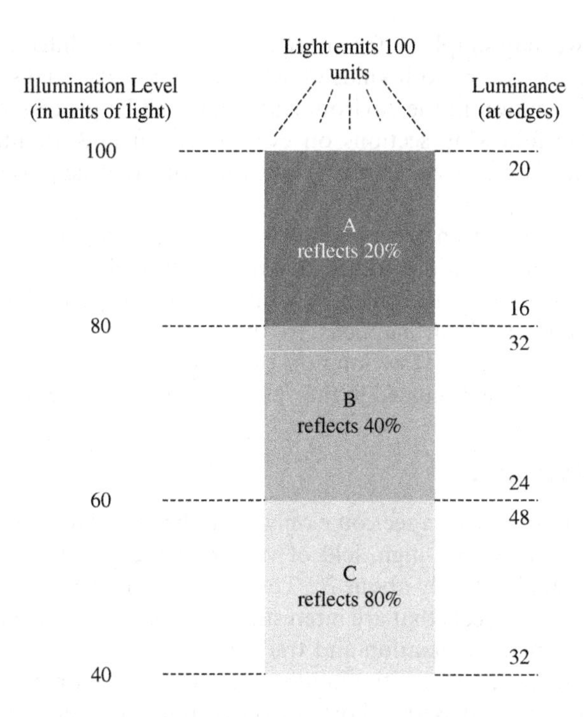


**FIGURE 2.31** Global reflectance can be calculated by ratios of differences at the edges

is important to know that built into perception are adjustments that will be made to try to maintain light and color constancy.

## 2.4 INFLUENCING VISUAL PERCEPTION

So far, we have described the flow of information as a bottom-up process from the light reflected from a scene to the language of thought that is the input to cognition. This is a view of perception as a data-driven process that begins with the retinal image. Each stage of processing takes as input a lower-level representation and creates or modifies a higher-level representation as output. But perceptual processing is much more complex. What we want to accomplish—our tasks, goals, or motivations—strongly influence what we see. The bottom-up process is influenced by a top-down process that takes higher-level representations and creates or modifies lower-level representations (Pylyshyn, 2003; Velmans, 1999; Ware, 2008; Johnson, 2010).

Tasks and activities determine the objects we focus on or give our attention to. Some of the acts we do as we focus our attention are physical such as turning our head or moving our eyes as we scan a page. These acts are called *overt*. Acts that are internal and cannot be noticed by others are called *covert*.

For example, we may simply shift our attention from the architectural structure of a building to its color. Both kinds of acts are a form of visual selection. The amount of information in the environment cannot all be processed; it must be sampled. In the following sections on eye movement and attention, we will discuss how the visual system makes visual selections that help us do our tasks and activities.

After the image has been processed, it becomes an interrelated set of concepts and objects as representations in the language of thought. We have described the bottom-up processing as if there were only a pair of images, but the scene is constantly changing. What happens to these concepts and objects as our attention turns elsewhere? How long do they last and where are they stored? These questions will be discussed in the "Memory" section later in this chapter.

### 2.4.1 Eye Movements

The sharp details of what we see come only from the foveal region of the retina. Of the 180°-wide and 130°-high field of view that the eyes take in, the fovea covers a visual angle of only about 2°. To capture more information, the eyes must be directed to objects that are interesting and relevant. Eye movement has two important functions: fixation and tracking.

*Fixation* is the result of positioning the eyes so that the object of interest is centered in the fovea. It is the short period of time between eye movements when the scene is taken in. For each fixation, significant changes are made to the retinal image, but we are unconscious of the integration of those changes into the resulting representation of the 3-D scene.

*Tracking* moves the eyes to keep an object centered in the fovea when the object, head, or both are moving. Again, perceptual processes integrate these changes so that changing scenes appear smooth and continuous.

***Types of Eye Movements.*** Several types of eye movements play different roles. Small eye tremors happen automatically all the time. Without these movements, experiments have shown that in as little as a few seconds after movement is frozen, the retinal image disappears. The visual system uses the primitive features in changing images—edges or contours, for example—to regenerate the visual experience. Other movements help keep the eyes fixed on moving objects or focusing the eyes on closer or distant objects to perceive distance. But the movements of greatest interest for designing visualizations are those involved in visual selection, which are called *saccadic* and *smooth pursuit* movements.

Saccadic movements are used to sample the scene for new objects of interest. They are quick and abrupt and, once started, cannot be stopped until the eyes are positioned at the destination. The plan and execution of a saccadic movement takes about 150 to 200 milliseconds, and the fixations between them last on average 300 milliseconds. We are not conscious of image motion during the saccade because perception is partly suppressed, and constancy mechanisms adjust the low-level representations to eliminate the movements in the image. Most visual perception takes place in the sequences of fixation. Both the structure in the image and the task determine how the scene will be explored. Fig. 2.32 shows the effects of a task on saccadic eye movements in experiments conducted by Yarbus (1967). Sequences of fixations, called *scan paths*, produce small parts of the scene, like pieces of a jigsaw puzzle, that are integrated by visual processes into a coherent whole.

Smooth pursuit movements track a moving object—such as a ball thrown in the air—to keep it centered in the fovea. Eye movement is smooth and continuous. Constant visual feedback from the image keeps the object stationary on the image so that the visual system can extract the most information possible during the object's movement. Maximum speeds are almost 10 times slower by comparison to saccadic movements. The part of the scene not in the fovea is smeared by motion, and the visual acuity in the foveal region decreases as the speed of the object's movement increases. The ability to track objects with clarity at increasingly faster speeds varies by individual but can be improved with practice.

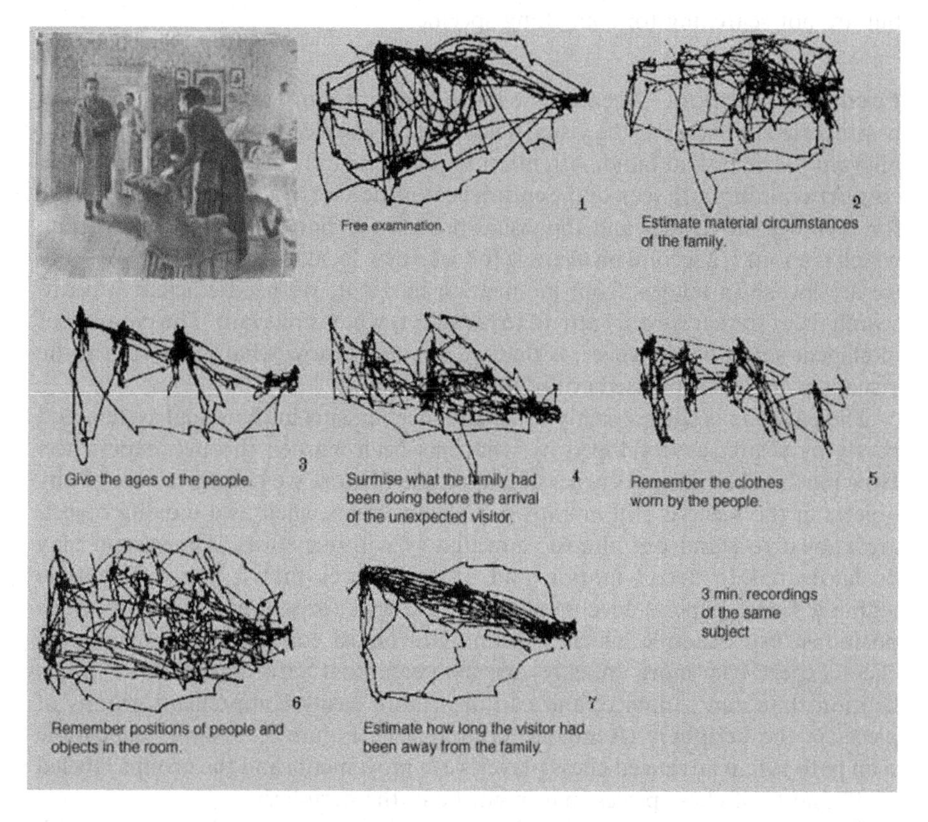


**FIGURE 2.32** The effect of goals on saccadic patterns. (Image courtesy of Wikipedia: http://en.wikipedia.org/wiki/File:Yarbus_The_Visitor.jpg)

### 2.4.2 Attention

The typical scene is far too rich with information to absorb it all at once. The saccadic eye movements are necessary for finding the information that is interesting or relevant. If saccades are the means for carrying out a visual search strategy, attention is the mechanism for directing the strategy. Where the eyes are positioned depends on whether we are looking at the overall structure of a scene, a group of objects, a specific object, or some of its parts and properties. *Visual attention* is defined as the processes that marshal perceptual resources to allow specific objects and properties of the retinal image to be explored while ignoring others.

Attention has two properties: capacity and selectivity. *Capacity* refers to the resources that are available and is affected by our state of mind, such as motivation, wakefulness, and time of day. *Selectivity* is the ability to select different subsets of information—for example, different properties or features of an object—to be processed. Two kinds of attention are important for design. *Focused attention* occurs when an object has been selected. *Distributed attention* occurs when we expect to find objects somewhere in the visual field but are not searching for something specific.

***Focused Attention.*** Because the scene contains more information than can be processed at one time, the visual field must be sampled to find the information relevant to the task at hand. Attention requires mental effort, which comes at a cost. At a minimum, a scan of the entire scene is the sum of a small amount of time for each eye movement and the visual processing that takes place at fixation, which is about 1/2 second on average for each new location in the image. Because we are awash in images from moment to moment, we need efficient ways to search them so that we only attend to (or select) what is relevant. The paradox of intelligent selection, however, is that we can only know what is relevant if the visual system has processed everything in the image.

The paradox is solved partly by innate mechanisms in the visual system and partly by strategies developed by what has been learned through experience. How much of the image is processed in parallel before we become aware of any objects in the scene is still unknown. Certain things, such as a moving object, are known to stand out almost immediately without effort. The system may be hardwired to detect motion and other features that it can differentiate within a scene. Experts develop efficient strategies for scanning familiar board positions. For example, when a meaningful board configuration is shown, a chess expert can more quickly extract perceptual information in a single fixation than can a novice, and can also more easily comprehend groups of pieces in the periphery (Reingold, et al., 2001). Figure 2.33 shows a possible scan path for an advanced chess player's eye movements and the groups labeled A, B, and C of chess pieces that might be in the periphery.

Except for involuntary shifts of attention that are triggered by objects with certain properties, what we consciously perceive is a result of where we have

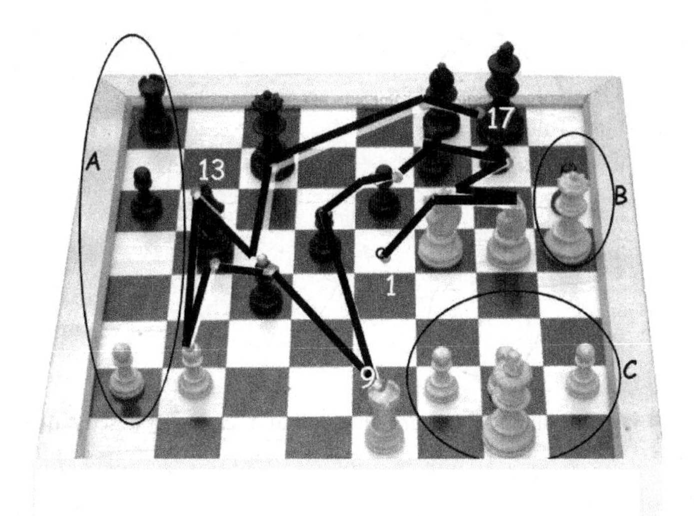


**FIGURE 2.33** A possible scan path of the eye movements of an advanced chess player. (Image courtesy of Wikipedia: http://en.wikipedia.org/wiki/File:Eye_movements_of_a_chess_champion_nc.jpg)

focused our attention. We tend to perceive only what is in our focus of attention. There are several forms of what we miss by inattentiveness to what is not in focus:

- *Inattentional blindness.* We tend to see only what we expect or are specifically looking for.
- *Attentional blink.* When a second object of interest is presented within half a second of an object we are attending to, the second object will be missed because our perceptual resources are already committed. The lack of resources prevents attention to both objects at the same time.
- *Change blindness.* Areas of an image that change are missed unless these areas affect the objects of our attention.

***Nonspatial selection.*** So far, we have discussed how we scan an image and what we perceive in the image as a whole when we focus on an object of interest. Another aspect of attention is the ability to select for the properties of objects: Can we attend to just one or two, or are certain groups of properties perceived all at once? The question is about how properties (or features), such as shape, color, or size of an object are bound together in the object's representation and whether these properties can be selected separately without requiring the perceptual processing of other properties that are not of interest. The answer depends on whether the pair of properties are *integral* or *separable.*

When properties are integral, we tend to see them as a whole, not as separate properties. For example, the width and height of an ellipse are integral with its

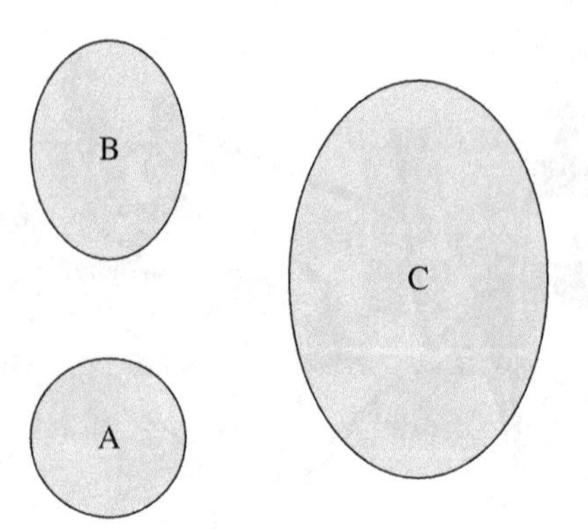


**FIGURE 2.34** Width and height are integral and cannot be separated to make comparisons

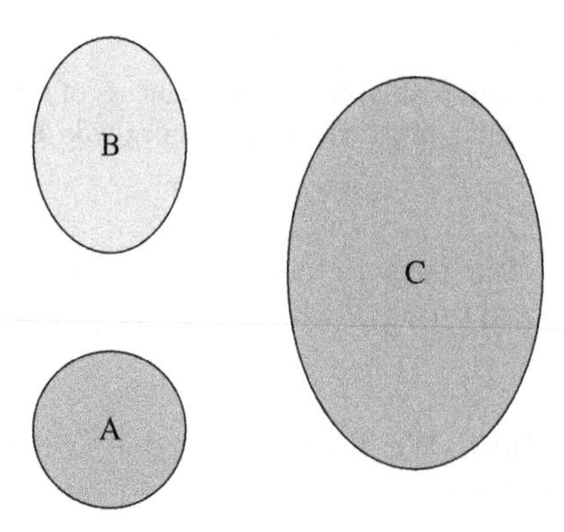


**FIGURE 2.35** Lightness can be separated from size and used to group objects

shape. In Fig. 2.34, the ellipses are more alike even though B has the same width as A. Width and height are integral and cannot be separated to discriminate by just width or height.

When properties are separable, judgments can be made about each property separately. In a field of objects with different colors and shapes, we could select by color, shape, or both. In Fig. 2.35, circle A and ellipse C can be seen as similar if we are looking for the same lightness.

***Distributed Attention.*** When we are expecting to see something in the visual field but not necessarily something specific, our attention is distributed but not focused (Palmer, 1999). The processing that takes place for distributed attention is automatic, performed in parallel on the whole image, and is very fast. By contrast, processing guided by focused attention is serial and is performed as a series of fixations on specific regions or objects.

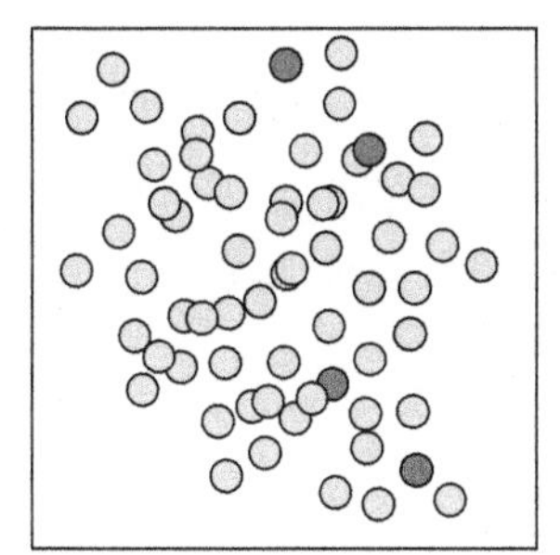
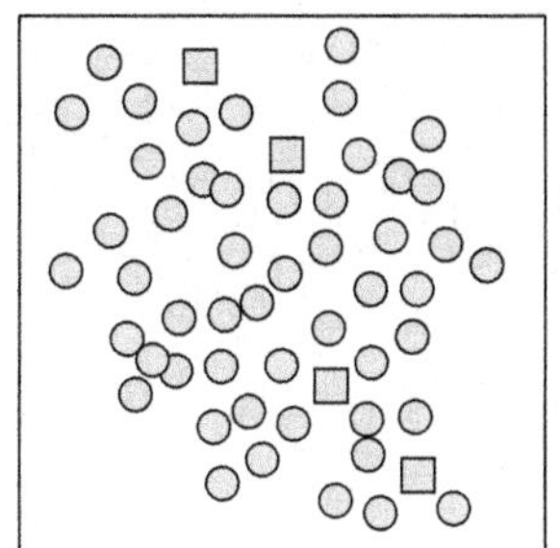
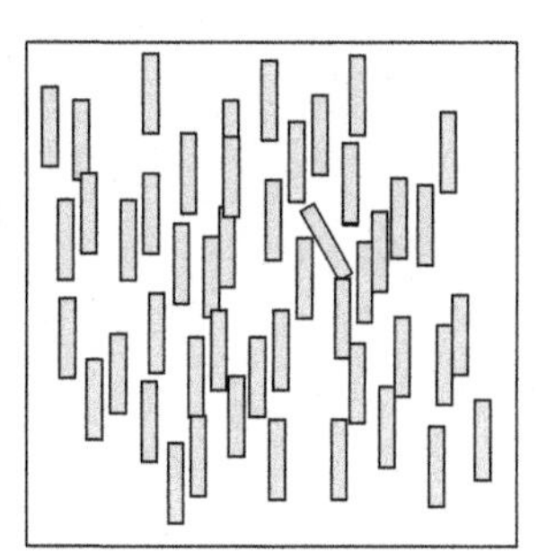

**FIGURE 2.36** Target detection using lightness, shape, and orientation to discriminate

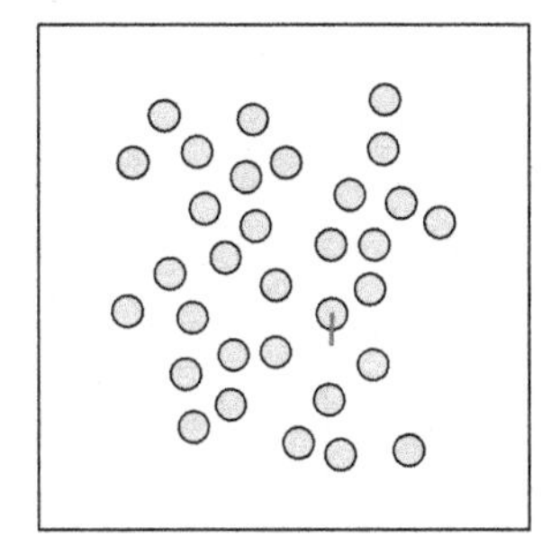
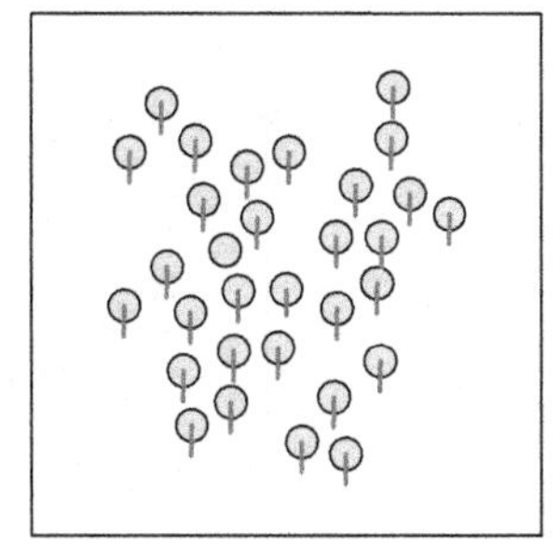

**FIGURE 2.37** Target with line added (left) and line removed (right)

When our attention is distributed, a phenomenon known as *visual pop-out* occurs in which objects with certain features or combinations of features can be easily picked out or separated into groups within a field of similar objects. (These features are also called preattentive features because they are processed before we become aware of them.) Distributed attention allows certain visual tasks to be performed quickly such as:

- *Target detection.* Finding an object within a field of similar objects can be done without focus if the object can be distinguished from the others by mutually exclusive features such as lightness, color, orientation, shape, size, motion, and stereoscopic depth. In Fig. 2.36, the features that differentiate the target are lightness, shape, and orientation. Features can be added to a target to make them pop out as in the left side of Fig. 2.37, but features cannot be removed as in the right side. When using orientation, not all shapes are equally effective, however. Some shapes with convex or concave contours cannot be easily detected as shown in Fig. 2.38.
- *Boundary detection.* When all the objects in one group have a common feature different from the common feature in the other group, they will separate without requiring that attention be focused. In Fig. 2.39, lightness separates the top group from the bottom.

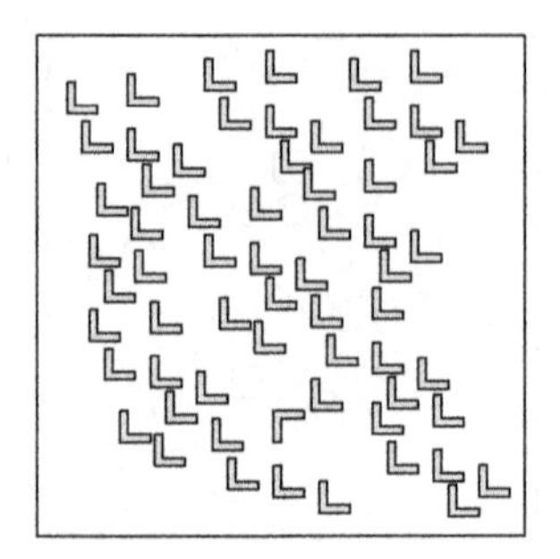

**FIGURE 2.38** The orientation of the target does not distinguish with certain shapes

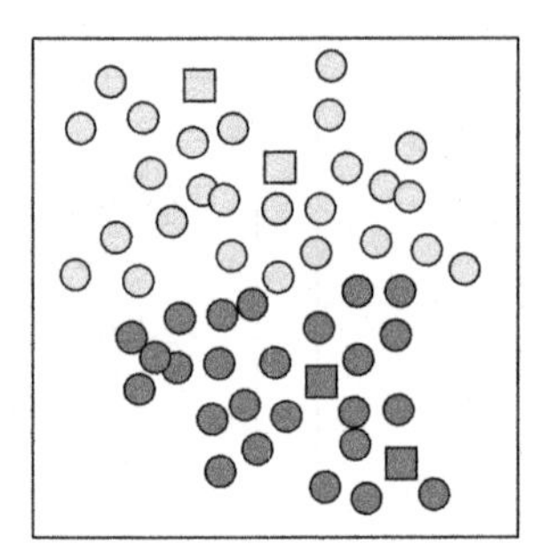

**FIGURE 2.39** Boundary detection using lightness to separate groups

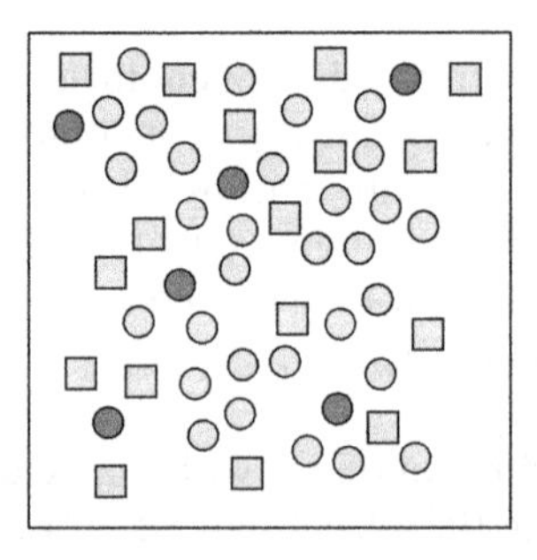

**FIGURE 2.40** Using lightness to distinguish a set of objects

- *Counting and estimation.* Objects with a unique feature can be counted more easily. Lightness distinguishes the objects in Fig. 2.40; it allows the dark circles to be counted.

### 2.4.3 Memory

As shown in Fig. 2.41, many types of memory are spread across the brain and in the major systems that process visual or auditory information, language, and so on (Baddeley, 2010; Luck, 2007; Eichenbaum, 2008). Memory is based

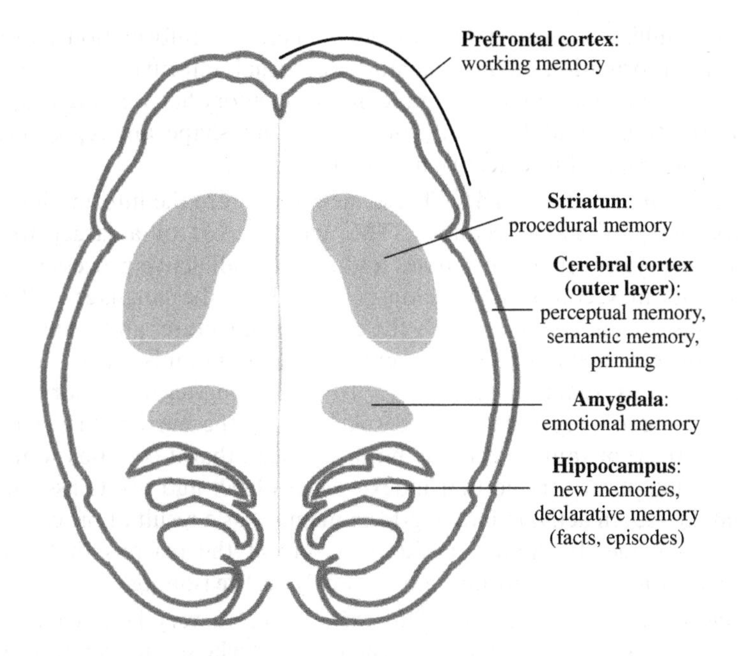


**FIGURE 2.41** A memory map of the brain

on changes to the synaptic connections within the neural networks of a particular memory system. It is not a separate storage area because every neuron processes and stores information. Although the definition of a memory system is hard to pin down, the most important attributes that allow memory systems to be compared include the following:

- *Duration*. Do the contents persist for a short or long period of time?
- *Content*. Does it contain visual, auditory, semantic, or other kinds of information? Is the memory expressed explicitly as with facts and knowledge or implicitly as with skills that are learned?
- *Loss*. Can the information be lost over time or be replaced with other objects while doing a different task?
- *Capacity*. How much information of a given type can be stored? How does capacity vary as objects are forgotten or are replaced?
- *Maintenance*. Does the information need to be refreshed to be remembered?

Information processing theorists organize memory systems into layers by stage of processing:

- *Sensory information stores (SIS)*. These memories can hold a few "chunks" of information for very limited periods of time that last from a few milliseconds to a few seconds. There are information stores for visual information (iconic memory or visual information stores) and auditory information, for example. Iconic memory has very high capacity. Information about location, color, size, and shape are represented in iconic memory, but category information is not.

- *Short-term memory (STM)*. These memories store the information being processed for many seconds. STM may be part of a larger working memory system that coordinates and executes all active processing, or its contents may comprise consciousness. In both the language and visual systems, we understand quickly the concepts of pictures and sentences, but the concepts will not be remembered unless attention is focused on them. Attention is required to leave a trace in long-term memory. Visual memory can only hold what is relevant for the current task which is about three or four objects. By hold, we mean that it contains something like a temporary frame that indexes each object and all its associations that have been activated in long-term memory as a result of processing the retinal image. Its limited capacity is one of the reasons that external visualizations are important aids to analytical reasoning.
- *Long-term memory (LTM)*. This memory has a very large capacity for storing long term the "what," "when," and "where" information of our experiences (*episodic memory*); general and factual knowledge (*semantic memory*) that include categories as diverse as faces, houses, tools, actions, or language; and skills (*procedural memory*). Long-term memory is formed by repeated activation of neurons that strengthen the connections and make them more sensitive to the same patterns of activation the next time they are encountered. For example, we may look through several hundred pictures from recent holidays. Even if on the next day only a handful can be remembered, we will still be able to perform tasks on each picture, such as categorizing, more quickly the second time. The initial exposure is a form of implicit learning, and the effect is known as *long-term priming*.

## 2.5 SUMMARY

Cognition is the process of thought that derives its power from representation and abstraction. External representations are cognitive tools that exist outside our minds. They are designed to aid memory, make abstract concepts visible, solve problems, support decision making, model, and clarify our thoughts. Different representations of the represented world can result in dramatic differences in how a cognitive task is done and how quickly it can be performed.

The computational theory of the mind views the brain as a system of mental organs that process information. The information from the world that flows through the senses into the mind as patterns of data become internal representations. The computational theory of the mind has been successfully used to develop computational models and various algorithms that have helped to understand the human visual system.

The main problem of the human visual system is how to make sense of the image formed on the retina by the light that enters the eye. Within a scene, the visual system must do the following:

- Identify where an object ends and the background begins.
- Determine what objects are made of.
- Reconstruct the third dimension.
- Recognize the object.

Information-based theories on how this is done divide visual perception roughly into four stages. In the image-based stage, simple 2-D features such as edge and line segments or small repeating patterns are extracted from the images. In the surface-based stage, the simple 2-D features and other information are used to identify the shapes and properties of the surfaces of the objects in the external world. In the object-based stage, the simpler features and surfaces are combined and grouped into the fundamental units of our visual experience: 3-D representations of the objects and their spatial layout. In the category-based stage, the objects are classified and linked to concepts we have seen before or that are part of our general understanding of the world.

The perceptual processing of what we "see" is much more complex than just recognizing what is in the image. Our tasks, goals, or motivations strongly influence what we attend to and where we look for the information that is interesting or relevant. Our attention is focused when we are looking for something specific and distributed when we are just attending to the work. When our attention is distributed, there is a phenomenon known as "visual pop-out" in which objects with certain features or combinations of them can be easily picked out or separated into groups within a field of objects. This allows certain visual tasks to be performed quickly and is used to design more effective visual representations.

The characteristics of memory also affect design choices. Short-term memory stores information only for seconds. Visual memory can only hold what is relevant for the current task, which is three or four objects. Attention is required to leave a trace in long-term memory. Long-term memory has a very large capacity for storing our experiences (episodic memory), general and factual knowledge (semantic memory), and skills (procedural memory).

## 2.6 FURTHER READING

In *Spatial Schemas in Depictions* (Tverskey, 2003), Barbara Tversky discusses the history of depictions such as pictures, drawings, maps, and diagrams; the use of space to convey relationships; and the function of these graphical depictions. The cognitive principles and theories about external representations and how they are used in various cognitive tasks have been described elsewhere (Reisberg, 1987; Stenning & Oberlander, 1995;Tversky, 1997; Zhang, 1997; Zhang 2001).

In *How the Mind Works* (Pinker, 1997), Steven Pinker provides an accessible explanation of how the brain works, including chapters on vision and visual perception.

In *Visual Thinking for Design* (Ware, 2008), Colin Ware discusses in more verbal and graphical detail the active role vision plays in using the external world as an aid to cognition and how this affects design choices.

Much of the material on the visual system and perception came from *Vision Science: Photons to Phenomenology* (Palmer, 1999). This work provides considerable detail on the processing performed in each of the stages and includes an extensive bibliography. Palmer notes that the term "preattentive" has been widely used to describe perceptual processing before an object has been selected as the focus of attention. But he points out, citing work by Mack and Rock (2000), that in studies on "preattentive" processing, the observer is told to expect a target (an unusual object) in a field of distractors (objects that are different from the target). He argues that the expectation is a form of attention (he calls it distributed attention) that is not directed at a specific object or localized region but, rather, to the entire visual field. Additional details about the theories of color perception and color models can be found in (Malacara, 2002) and (Ware, 2000).

# CHAPTER 3

# GRAPHIC REPRESENTATIONS

In the previous chapter, we examined how the visual system parses scenes of the physical world into objects perceived as having form and function. The mental representations of objects in the scene derive their meaning from experience, reflection, and reasoning. Graphics as external representations might also be considered scenes, but scenes of abstract worlds rather than the physical world we apprehend directly.

Under the broadest definition, graphics are visual representations of data that rely on graphical elements—points, lines, areas, and volumes—organized in a geometric space. Usually the graphical elements represent objects, while relationships between them are represented by space. Maps are among the oldest graphics. They are schematics that represent a visible world. They rely on spatial distances—sometimes distorted—and the inclusion or exclusion of specific elements to provide different perspectives on the represented information. From the late eighteenth century, graphics began to represent abstract concepts such as quantities, time, space, preference, and qualitative information such as various relationships between objects or sets of objects based on their properties. Interactive graphics and animation in the late twentieth century, made possible with computer graphics and user interfaces, have added the ability to group, classify, filter, superimpose, juxtapose, and permute the displayed elements.

*Making Sense of Data III: A Practical Guide to Designing Interactive Data Visualizations*, First Edition. Glenn J. Myatt and Wayne P. Johnson.
© 2011 John Wiley & Sons, Inc. Published 2011 by John Wiley & Sons, Inc.

## 3.1 JACQUES BERTIN: SEMIOLOGY OF GRAPHICS

*Semiotics* is a field that studies how signs and symbols become associated with meaning and the conventions or code by which signs are organized into systems to communicate and to model information about the external world. The two major contemporary traditions of semiotics originated with the Swiss linguist Ferdinand de Saussure (1857–1913) and the American Charles Peirce (1839–1914). For many semiotic theorists, language was viewed as the central semiotic system, and terms from linguistics have been borrowed and extended in its application to various disciplines outside linguistics. For example, films and television programs are called "texts," which are "read."

The French cartographer Jacques Bertin (1918–2010) extended the reach of semiotics into graphics systems. To help designers and statisticians understand how to create diagrams, networks, and maps with relevant information that could more easily be understood, Bertin developed a semiotic theory for graphic communication (Bertin, 1983). He saw graphic representation as a tool for discovery and as "artificial memory," and he saw both tool and external representation as devices that leveraged the power of visual perception.

### 3.1.1 The Essence of Semiotics

Humans communicate thoughts using systems of *signs*. Meaning is created in our minds when we "read" signs in "text." These carry meaning conveyed by learned conventions or *codes* that become transparent to us. A sign stands for something other than itself and consists of two inseparable parts: a *signifier*, which is the form of the sign; and the *signified*, which is the concept represented by the form.

We cannot make sense of the form unless we can relate it to a code. A code is a set of conventions: an arbitrary set of rules or standards in some domain that have been established to mean certain things. In mathematics, the Cartesian coordinate system is a code. Using this code, we can map latitude to the x-axis, map longitude to the y-axis, and draw maps. The conventions of the Cartesian coordinate system allow us to "read" or interpret a map or a scatterplot. In Fig. 3.1, the form is the connected set of shapes with black outlines and black-and-white shading. So familiar are we with conventions for drawing people that we "read" these shapes (the signifier) as a person with upraised arms (the signified). If this were an airport ramp where ground support is signaling to the pilot, the signifier would be "the upraised arms" and the signified would be "this way."

A sign's vehicle (signifier) may have three possible fundamental modes of relationship with what it represents (the signified object or concept):

- *Symbol/symbolic*. The form of the symbol looks nothing like the thing that it represents, and the relationship is purely arbitrary and must be learned. The word "tree," for example, bears no resemblance to the tree in the world that it represents, nor does the circle in a scatterplot representing a datum of a variable in a statistical dataset.

**FIGURE 3.1** A sign for a person with upraised arms

- *Icon/iconic.* The form resembles the thing it represents. The folder icon (signifier) in the file system explorer applications of the major desktop user interfaces looks like a folder in a file cabinet. The word "cuckoo" (signifier) sounds like the bird call of the bird it represents.
- *Index/indexical.* The form is connected directly or causally in a way that can be observed or inferred: smoke (signifier) means fire (signified), the weather vane points to the direction of the wind, or the mouse arrow on the screen points to a graphical representation of a virtual object.

Signs can be combined into "texts." (To avoid confusion, we will quote the word "text" whenever it is used as defined in semiotics, where it means a set of signs rather than its usual meaning of a string of written words as is meant by a phrase such as "written text.") Anyone standing at the window of an airport terminal watching a plane on the airport ramp being marshaled to the opening of a jetway can read the "text" of hand signals between the ramp agent and the pilot as shown in Fig. 3.2: "this way," "turn to the right," "slow down," and "stop." This "text" has little structure because the signs can appear in almost any order. (It would not make sense, for example, to signal a stopped plane to slow down.) On the other hand, the "text" of language and other semiotic systems has complex structure that separates, generally, into the following levels of organization:

- *Syntax.* Concerned with how signs are recognized relative to each other as, for example, nouns or adjectives in a noun phrase or groups of shapes perceived as a cluster in a graphic.
- *Semantics.* Concerned with how the meanings of signs are understood.
- *Pragmatics.* Concerned with how signs convey the intended meaning within the context in which they are used.

**FIGURE 3.2** The signs of hand signals: “this way,” “turn to left,” “slow down,” and “stop”

Humans have created a variety of semiotic systems to store, communicate, and understand different types of information such as music, language, mathematics, figurative or abstract art, and graphics. The signs in these systems may be associated with one meaning (*monosemic*) as in mathematics, several meanings (*polysemic*) as in language or figurative art, or any meaning (*pansemic*) as in abstract art. We use these systems to communicate by sending *messages*. Messages are *encoded* in some physical medium: words may be spoken, graphics may be drawn as lines and shapes on paper, scenes may be captured as a digital image, or music may be scored. The interpretation of messages is called *decoding*. The word “communication” does not mean, as in common usage, that the content of the message contains the meaning. The content of the message contains the set of signs or “text.” Interpreting the message is a cognitive process that results from the receiver first decoding the message and then, from prior knowledge and context, inferring from the code related to the sign system what the message means.

The ear and the eye are two ways we perceive the message. Because sound varies over time, auditory perception has two sensory variables: sound and time. Auditory perception is linear and temporal. Visual perception can be nearly instantaneous because the entire image is perceptually processed at once. Assuming the “text” is written or displayed on a flat surface such as a piece of paper or a display, visual perception has three sensory variables for each mark: the vertical and horizontal dimensions of the plane that locates the mark’s position and the variation of some visual property of the mark itself, such as shape or color. Visual perception is spatial and atemporal.

For Bertin, graphic representations and mathematical notations were a rational subset of the broader set of images that included verbal, musical, and mathematical notations as well as figurative and abstract images, photographic images, and animated images (film). He saw graphics as a “language for the eye” as opposed to language, music, and mathematics, which he felt were fundamentally systems for the ear, and therefore had to be perceived through a linear process one term at a time. To be rational, a graphics system had to be monosemic: the elements or groups of elements in the system could mean only one thing if a graphic was to communicate information that could be interpreted rigorously and without debate by each reader, assuming the reader knew the code in advance.

Bertin applied the semiotic approach to diagrams, maps, and networks. He outlined three things that must be understood to design graphics that are efficient and effective in communicating the underlying content of the information:

- *The properties and structure of the information.* The information contains what is referred to (signified) by the signs of a graphic representation. An analysis of the information is required before the best representation of it can be considered.
- *The properties of the graphics system.* After the questions to be answered by the graphic are known, the components of the data involved in providing answers to these questions and the elements or groups of elements they contain must be mapped to something visible that includes *marks*—points, lines, and areas—and the visual properties of marks. These are the signs of the graphics system.
- *The rules for constructing an efficient graphic representation of the data.* Data components can be mapped to elements in the graphic in many ways. Some of these are more efficient than others. The rules are guidelines for determining how many components can be included in a single graphical *image* and what mappings between the data components and the marks reduce the effort of perceiving and understanding the graphics. The image—a fundamental unit in a graphic that "the eye can isolate . . . during an instant of perception"—and how to construct it are key concepts of Bertin's theory.

Bertin laid out a theory and process for designing 2-D static graphics primarily to be read from printed material. Although 3-D, animation, and interactive graphics did not yet exist, it is well worth studying for the insights it gives about how the design of graphics should be approached and the kinds of questions that need to be considered. In particular, as shown in Fig. 3.3, he saw the transformation from data to comprehension as taking place in two phases and recognized the central role visual perception had in the design of quality graphics.

To summarize his work (Bertin, 1983), we will use a small, hand-picked subset of 16 vehicles and just a few properties as shown in Fig. 3.4 that have

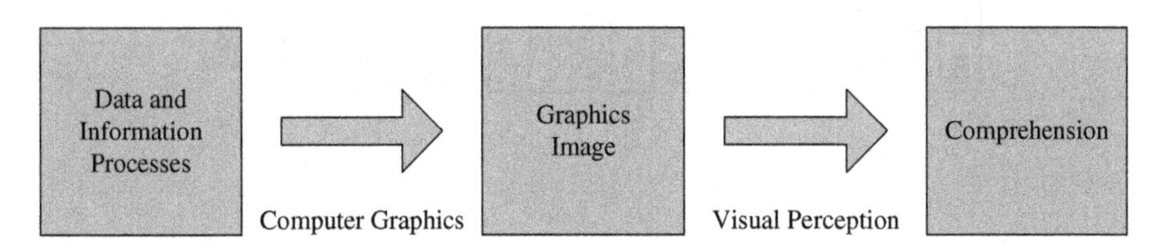


**FIGURE 3.3** Data visualization as a two-phase process

| Mfr Name | Car Line | Car Line Class | Eng Displ | # Cyl | Trans | City FE | Hwy FE | # Gears |
|---|---|---|---|---|---|---|---|---|
| Toyota | Yaris | Subcompact | 1.5 | 4 | Auto(A4) | 29 | 35 | 4 |
| GM | Aveo 5 | Subcompact | 1.6 | 4 | Auto(A4) | 25 | 34 | 4 |
| Ford | Fiesta SFE FWD | Subcompact | 1.6 | 4 | Auto(AM6) | 29 | 40 | 6 |
| Honda | Civic | Subcompact | 1.8 | 4 | Auto(A5) | 25 | 36 | 5 |
| Mazda | Mazda2 | Compact | 1.5 | 4 | Auto(A4) | 27 | 33 | 4 |
| Ford | Focus FWD | Compact | 2.0 | 4 | Auto(A4) | 25 | 34 | 4 |
| Honda | Insight | Compact | 1.3 | 4 | Auto(AV) | 40 | 43 | 1 |
| VW | Jetta | Compact | 2.5 | 5 | Manual(M5) | 23 | 33 | 5 |
| Toyota | Corolla | Compact | 2.4 | 4 | Manual(M5) | 22 | 30 | 5 |
| Toyota | ES 350 | Midsize | 3.5 | 6 | Auto(S6) | 19 | 27 | 6 |
| Chrysler | Avenger | Midsize | 2.4 | 4 | Auto(A4) | 21 | 30 | 4 |
| Volvo | S80 AWD | Midsize | 3.0 | 6 | Auto(S6) | 18 | 26 | 6 |
| GM | Limousine | Large | 4.6 | 8 | Auto(A4) | 12 | 18 | 4 |
| GM | Lucerne | Large | 3.9 | 6 | Auto(A4) | 17 | 27 | 4 |
| Ford | F150 Pickup 4WD | Pick-up 4WD | 6.2 | 8 | Auto(S6) | 12 | 16 | 6 |
| Nissan | Titan 4WD | Pick-up 4WD | 5.6 | 8 | Auto(A5) | 12 | 17 | 5 |

**FIGURE 3.4** A subset of the vehicles from the 2011 EPA Fuel Economy Guide Dataset (EPA, 2011)

been extracted from the 2011 vehicle fuel economy dataset of 1052 vehicles provided by the EPA. Each row of cells contains values for the properties of a vehicle; the heading of the column that contains a cell's value indicates the property of the measured value. Although we will use Bertin's terminology, we will use the standard layout for statistical datasets where rows represent observations, and columns represent statistical variables. We will also restrict the scope of the summary to what Bertin called diagrams, although he discussed networks and maps as well.

### 3.1.2 The Properties and Structure of the Information

The information used for diagrams are datasets or tables of data. An annotation of three columns of data from the example dataset using Bertin's terminology is shown in Fig. 3.5. A diagram depicts representations of *components*, which vary across a fixed set of data called the *invariant*. The invariant is not shown, but it is either assumed or described in the title of the diagram. The basis for understanding the variation can be seen in the representations of those components included in the graphic. For example, if a trend is seen in a scatterplot, it is natural to ask "a trend over what set of data?" For the full dataset from which the example was derived, if the diagram were a scatterplot of miles per gallon versus engine displacement, the invariant might

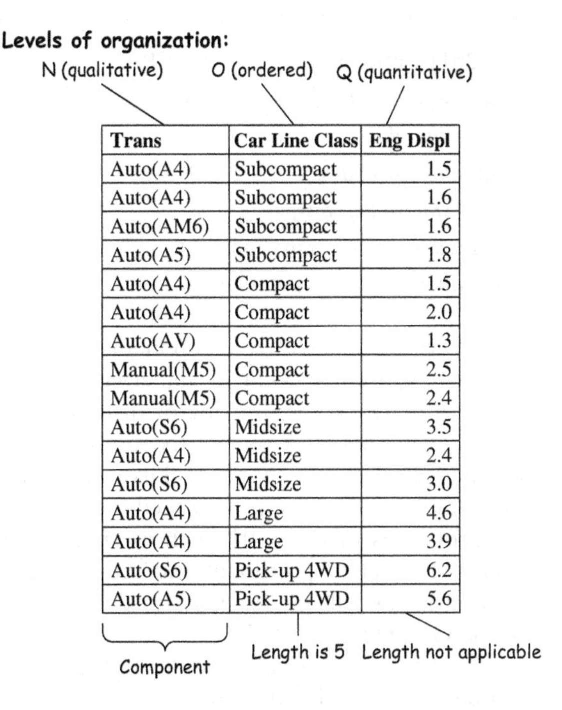

| Trans | Car Line Class | Eng Displ |
|---|---|---|
| Auto(A4) | Subcompact | 1.5 |
| Auto(A4) | Subcompact | 1.6 |
| Auto(AM6) | Subcompact | 1.6 |
| Auto(A5) | Subcompact | 1.8 |
| Auto(A4) | Compact | 1.5 |
| Auto(A4) | Compact | 2.0 |
| Auto(AV) | Compact | 1.3 |
| Manual(M5) | Compact | 2.5 |
| Manual(M5) | Compact | 2.4 |
| Auto(S6) | Midsize | 3.5 |
| Auto(A4) | Midsize | 2.4 |
| Auto(S6) | Midsize | 3.0 |
| Auto(A4) | Large | 4.6 |
| Auto(A4) | Large | 3.9 |
| Auto(S6) | Pick-up 4WD | 6.2 |
| Auto(A5) | Pick-up 4WD | 5.6 |



FIGURE 3.5 Three variables of a dataset annotated with Bertin's terminology

be all vehicles assessed in 2011, all autos, all trucks, or only autos made by a specific manufacturer.

Bertin outlined three things that must be done to analyze the information before deciding how the information should be presented:

- Determine the invariant and the number of components to include in the graphic.
- Determine the level of organization of each included component.
- Determine the length of each included component.

***Number of Components.*** A graphical image or series of images cannot be understood without knowing the invariant and the information components involved in the representations. The analysis results in descriptions of the information of interest. The following are two examples of these descriptions:

**invariant:** all 2011 vehicles assessed by the EPA
**components:** - average *city fuel efficiency* for each
- *car line class* by
- *manufacturer*

or

**invariant:** all 2011 autos assessed by the EPA
**components:** - average *city fuel efficiency* by
- *car line class,*
- *number of gears,*
- *transmission types,*
- *engine displacements,* or
- *number of engine cylinders*

The first of these descriptions has three components; the second has six.

Each data component will correspond to something that can be visually perceived in a graphic: a mark on the drawing or display surface and one or more visual properties of the mark. A mark's location in a diagram has two dimensions, one for its position on the horizontal axis and one for its position on the vertical axis. These two dimensions are called the *planar variables* of the graphic. The visual properties of the mark such as shape, color, or size are called *retinal variables*. The set of positional and retinal variables are *visual variables*. The visual variables will be discussed in detail later in the section "The Properties of the Graphics System."

There must be at least as many visual variables in the graphic as there are data components to be visualized. Further, Bertin's view was that the number of components provided the best classification for graphical constructions. A well-designed diagram with three corresponding data components can be perceived instantly. However, the number of components in a dataset can be quite large. Beyond three components, decisions must be made to construct an efficient graphic. This will be discussed later in the section "Constructing Efficient Graphics."

***Level of Organization.*** In classical statistics, a data variable (one of the columns in Fig. 3.5) is classified into one of four types based on the scale by which the values it contains are measured:

- *Nominal.* The data values are categorical and not numeric. Comparing two observations using the values for the variable, the observations will either be similar or different depending on whether the categorical value matches or not. Using the variable "Trans" to compare yields the result that the "Yaris" is similar to the "Mazda2" but different from the "Insight."
- *Ordinal.* The data values are categorical but ordered. Comparing two observations using the values for that variable, one observation will either be greater than or less than the other observation. If the categorical values of the "Car line Class" variable are an indication of size, then the "Jetta" is greater than the "Yaris" but less than the "Avenger."
- *Interval.* The data values are numeric, but only the differences—distances between—can be compared quantitatively using the basic arithmetic operations (+, −, *, /) and not the values themselves. The values are ordered and may include negative numbers and zero, but zero is arbitrarily selected and is not an absolute reference point. The example dataset does not contain an interval data variable, but if there were a variable in a dataset that recorded the measurements of temperature, it would be classified as an interval variable. If the temperature variable contained the values 40, 60, and 80, we could say that compared with 40°F, 80°F is two times warmer than 60°F (80−40)/(60−40), but not twice as hot because 0°F is an arbitrarily chosen point on the scale.
- *Ratio.* The data values are numeric and include an absolute zero. This allows the values to be compared quantitatively with each other using the basic arithmetic operations. In Fig. 3.5, the engine displacement of the "Volvo S80 AWD" (3.0) is twice that of the "Yaris" (1.5).

Bertin characterized components by level of organization. He called nominal variables "*qualitative* components," ordinal variables "*ordered* components," and interval and ratio variables "*quantitative* components." These are abbreviated N, O, and Q, respectively, as shown earlier in Fig. 3.5. The differences between each type of component influence how they can be used to simplify graphics or make them more efficient perceptually.

Qualitative (N) components can only be used to differentiate: a value is similar to or different from another. In Fig. 3.5, for example, a vehicle with a four-speed automatic transmission—"Auto(A4)"—is similar to all other vehicles that have that type of transmission and different from all vehicles that do not. Qualitative components have the following properties:

- Because they are not ordered, qualitative components can be reordered in different ways to reveal patterns not otherwise apparent as shown in Fig. 3.6. In Fig 3.6a, the ability to permute either the components on the x-axis or the observations on the y-axis in different ways can result in the juxtaposition of elements that reveal various relationships between them. In Fig. 3.6b, the reordering of a qualitative component in relation to an ordered component may show trends.
- The categorical values in qualitative components are equidistant from each other and to display them otherwise distorts the data because the differences in distance will be perceived as being meaningful.

Ordered (O) components can be used to differentiate and to compare whether one value is more or less than another. Its categories cannot be reordered without confusing the reader. Further, as with qualitative components, its categories are equidistant and displaying them otherwise distorts the information.

Quantitative (Q) components are those with countable units. They can be used to differentiate, compare relative values that are more or less than, and quantitatively compare the ratios between values; for example, some value is twice or half another value.

The quantitative, ordered, and qualitative levels of organization are inclusive and overlapping. Quantitative components are also ordered and qualitative.

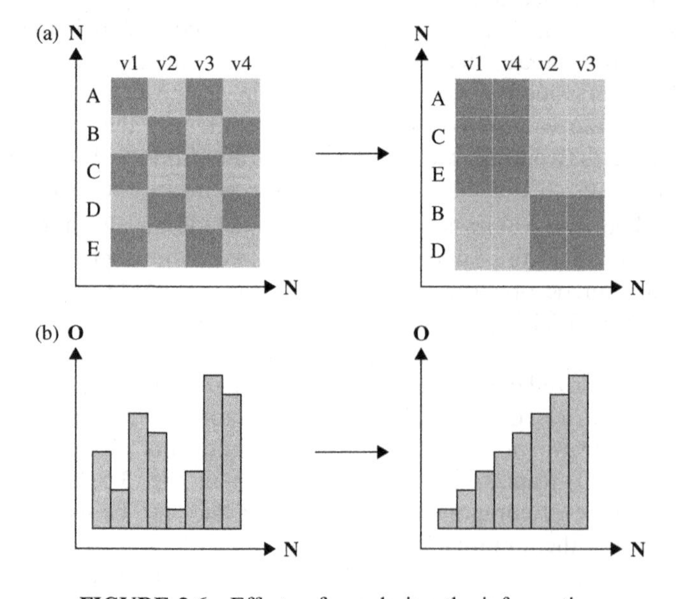


**FIGURE 3.6** Effects of reordering the information

Ordered components are also qualitative. Qualitative components are at the lowest level but may be arbitrarily reordered. The levels may be expressed as $N < O < Q$. As we will see in the upcoming "The Properties of the Graphics System" section, for each data component, the level of the visual variable to which it corresponds must be at least as high as the component's level of organization.

***Length of components.*** Components may be divided. How they are divided depends on the component's level of organization. In Fig. 3.5, the length of the ordered component "Car line Class" is 5 because it can be divided into the categories "subcompact," "compact," "midsize," "large," and "pickup-truck 4WD." Any qualitative or ordered component, or any quantitative component not containing continuous data, can be similarly divided into elements or categories, and the number of these is the component's length. Any quantitative component, such as "Eng Displ," that contains continuous data cannot be subdivided, so length is not applicable. Bertin distinguished between components whose lengths were short (<4) and long (>15). Those with a short length could use special constructions, whereas those with a long length required that a choice of graphics be made from a set of standard constructions.

### 3.1.3 The Properties of the Graphics System

The graphics system is the sign system for encoding the data components in the descriptions that result from an analysis of the information. The physical medium for the "text" is a flat surface—a printed page or a computer display—that reflects or emits the light from *marks* (signifier) that correspond to the content of the data components (signified). The "text" is a graphic. Figure 3.7 shows an example. Graphics are read in three stages: identify what is external to the graphic, identify the mappings between the visual variables and components, and perceive the relevant correspondences between the marks and the subset of data that the graphic represents:

- *External identification (Fig. 3.8).* The first stage is to understand the invariant and the components that are involved in the graphic. Prior knowledge of the domain or conventions is required to make sense of the terms such as "city fuel efficiency" or "car line," as well as the graphical elements, such as colors or shapes, which might be explained in legends. The graphic is showing something about the fuel efficiencies (component 2) of certain car lines (component 1) for a set of subcompact, compact, or midsize autos (invariant).
- *Internal identification (Fig. 3.9).* The next stage is to identify how the components are mapped to the visual variables. In this graphic, the perception is that the horizontal axis corresponds to the "car line" component, and the vertical axis corresponds to the "city fuel efficiency."

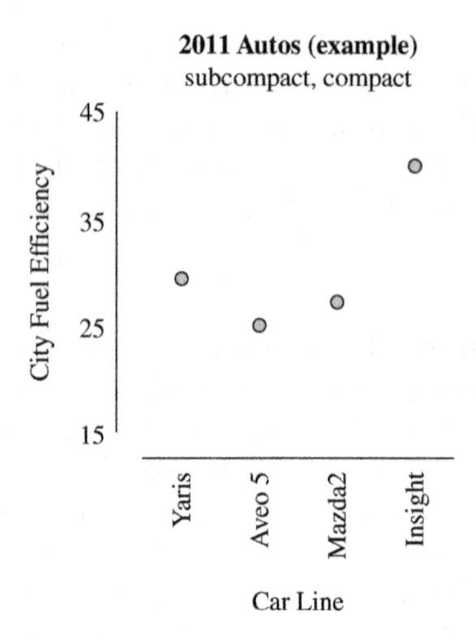


**FIGURE 3.7** A graphical image to be read

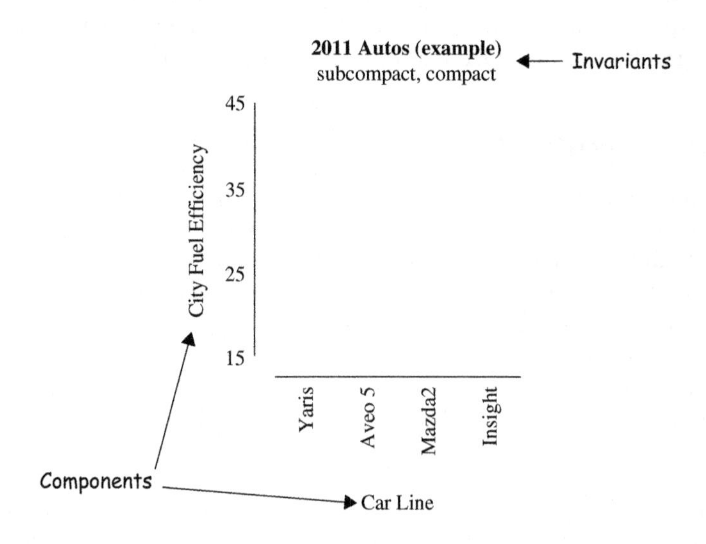


**FIGURE 3.8** Stage 1 of the reading process: external identification

- *Perceiving the marks (Fig. 3.10).* In this stage, the reader perceives the meaning of each mark through its location and visual properties. Each mark corresponds to two planar dimensions—the vertical and the horizontal—which give rise to the *planar variables* X and Y. The mark could be one of three types of signs that Bertin called an *implantation*: a point, line, or area. The point is the implantation used in this example. The mark could also be styled at Z with visual properties that vary based on values in the data component. These properties, called the *retinal variables*, are *size*,

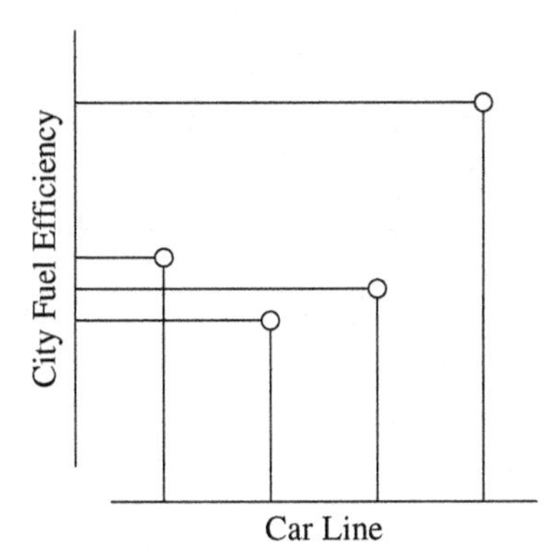


**FIGURE 3.9** Stage 2 of the reading process: internal identification

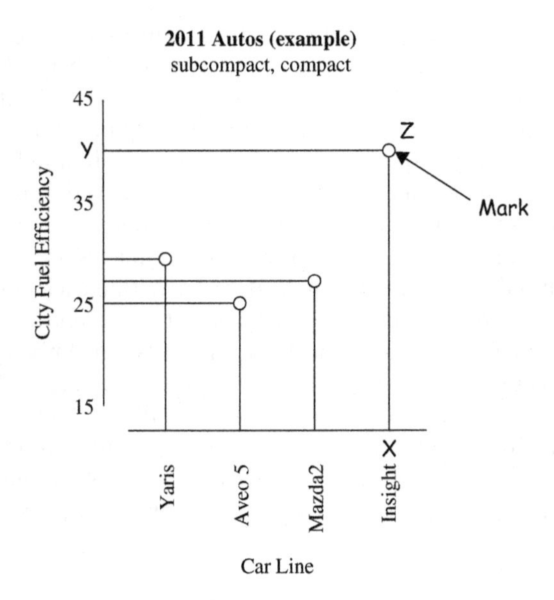


**FIGURE 3.10** Stage 3 of the reading process: perceiving the marks

*value* (brightness), *color* (hue and saturation), *orientation*, *shape*, and *texture*. In all, Bertin identified eight visual variables: two planar and six retinal variables. In this graphic, only the planar variables X and Y have corresponding components; the point is unadorned, but a mark must be visible for there to be a correspondence of components with the planar variables.

The stages of reading are one aspect of what Bertin called the *image theory*. But beyond the explanation for how a graphic is read, Bertin provided a model for how different combinations of mappings of components to planar or retinal

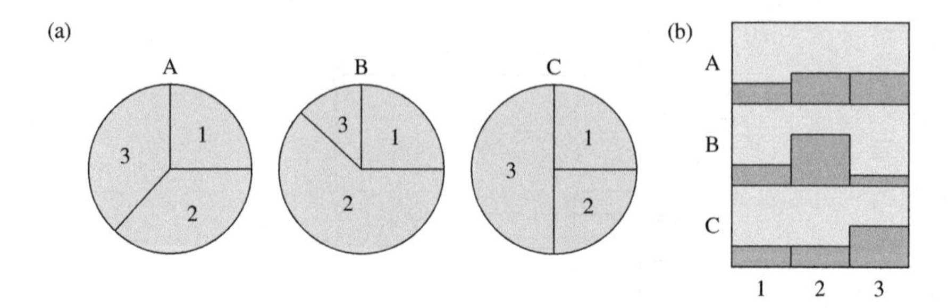


**FIGURE 3.11** The number of images required to read graphics (a) and (b)

variables would be perceived, and he used that model to describe the rules for constructing *efficient* graphics (Green, 1998). An efficient graphic can be interpreted in less time than another.

It is important to understand what Bertin meant by an image. He did not mean the image that results from a graphic as a whole being perceived as a scene. He meant what could be visually perceived in an instant after the visual system has focused attention on some part of the graphic—some specific set of correspondences—because of a question the reader had in mind. Each fixation results in an image. If the question in mind can be answered by perceiving all the relevant correspondences and ignoring the others within that image, the cost is almost nothing. In Fig. 3.11a, at least three images are required to find which sector labeled 2 is the largest in A, B, or C. Within each circle, the sector labeled 2 must be found and compared against the circle containing it to determine its relative proportion, and then it must be remembered as comparisons are made between the sectors labeled 2 in the other circles. In Fig. 3.11b, the question can be answered with only one image: all bars labeled 2 are aligned and can be perceived in one fixation. To know how to map components to variables so that graphics can be read efficiently requires an understanding of the characteristics of the plane on which the graphic is drawn and the visual variables.

***The Plane.*** The plane is two dimensional. Both the space of the plane allocated to content—the signifying space—and the points, lines, or areas drawn on it carry meaning. The space is continuous and can be divided as finely as the resolution of the surface will allow, limited either by the number of dots per inch that can be printed or the width and height of pixels in computer displays. Certain assumptions are made about the signifying space: empty space means there is no data within the graphics frame, differences in visual variables are intentional and have meaning, and a single convention or code applies to all the marks within it.

***Length of Visual Variables.*** The length of a variable is the number of perceptible divisions it supports. Bertin called these divisions *steps*. Figure 3.12

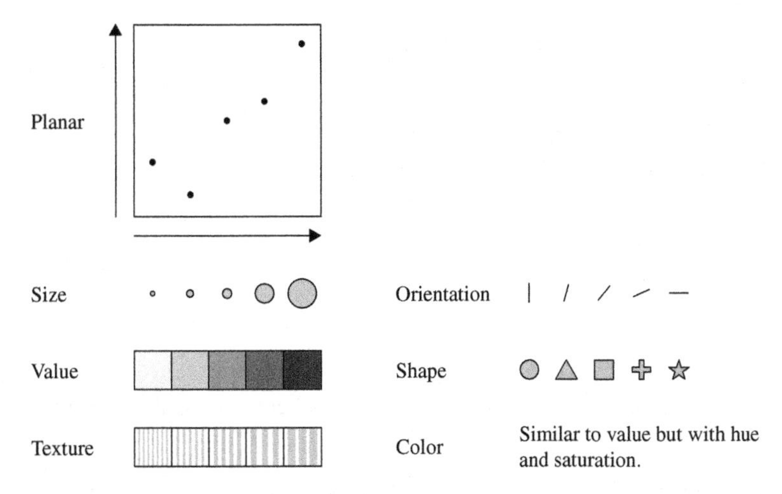


**FIGURE 3.12** Examples of steps of a visual variable

shows examples of steps that can be made with each variable. We can only see a finite number of variations of each of the variables. The variable's length must be at least as large as the length of the component it represents, or some of the steps of the variable will not be uniquely mapped to the values in the component. The plane's variables are the longest and should be considered first for the longest components. Matching the data component's length to the visual variable's length is the first critical factor in determining which visual variable should represent a component.

***Levels of Organization of Visual Variables.*** We have already discussed the three levels of organization of data components: qualitative, ordered, and quantitative. The level of perceptual organization of a visual variable specifies its ability to convey the information of the component it represents. For example, suppose the planar variables were mapped to "car line" and "city FE" (city fuel efficiency), and the design goal was to map some retinal variable of the mark to the "car line class" component as shown in the graphic schema in Fig. 3.13. (In this figure, we use a notation Bertin created for describing the aspects of the design of a graphic. In the schema, the vertical and horizontal arrows represent the two dimensions of the plane. The arrow rising at an angle is interpreted as the type of components mapped to the retinal variables of a mark. The component type is indicated at the ends of the arrows.) Because "car line class" is an ordered component, the reader must be able to compare the relative magnitude of one mark against another. The retinal variables for which order can be perceived are "planar," "size," "value," and "texture." Examining Fig. 3.12 will show why. For planar variables, positions to the right or left of each other on the horizontal axis or above or below each other on the vertical axis allow the comparisons of the values the marks represent; for size variables, larger shapes represent values that are greater than values of smaller shapes;

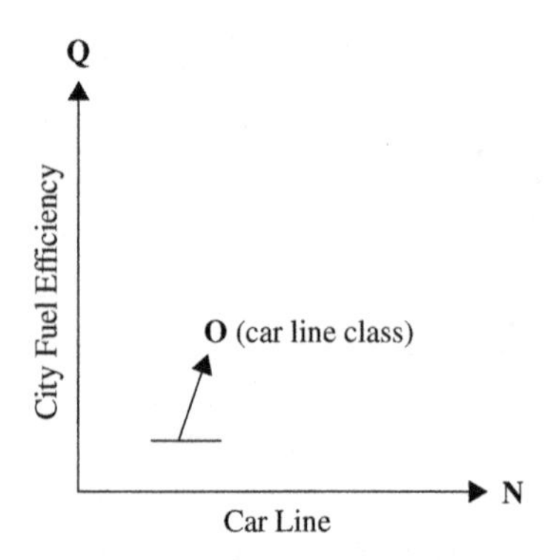


**FIGURE 3.13** Schema for mapping components to visual variables

**TABLE 3.1 Level of Organization of Visual Variables**

| Variable | Associative (≡) | Selective (≠) | Ordered (O) | Quantitative (Q) |
|---|---|---|---|---|
| Planar | yes | yes | yes | yes |
| Size | | yes | yes | yes |
| Value | | yes | yes | |
| Texture | yes | yes | yes | |
| Color | yes | yes | | |
| Orientation | yes | yes | | |
| Shape | yes | | | |

and so on. Comparisons cannot be made in the same way for "orientation," "shape," or "color."

Table 3.1 shows the visual variables and the levels of perceptual organization that each supports. In Table 3.1, the "associative" and "selective" levels have not yet been discussed. Bertin subdivided the qualitative component (N) into two ways in which the values it contains could be perceived, depending on whether the reader wanted to see all the component's elements as a group (associative) or see just those for specific categories (selective). Each of the four levels is really a description for each variable of the kind of visual query that can be perceived automatically: grouping by similarity, differentiating between categories or classes, comparing relative magnitudes, and quantitatively comparing ratios of values.

- *Associative Organization (≡).* This is the lowest level of organization. Variables that are associative allow all the elements of the qualitative component it represents to be instantly perceived as a group. Note that association doesn't mean the ability to group all elements of the component mapped to it; instead, the term describes its effect on the ability to group by other visual properties. Dissociative variables, such as "value," group elements so strongly that it is hard to group the elements by other visual properties. In Fig. 3.14, the marks have been styled by two visual properties: value (light and dark) and shape (circles and squares). Grouping all the squares is much more difficult than grouping all the light or dark shapes because of the strong effect of value.

- *Selective Organization* ($\neq$). This level allows the elements of only a specific category of the qualitative component it represents to be instantly perceived as a group. It is the opposite of association. For the same reasons that "shape" is associative and "value" is dissociative, "value" is selective but "shape" is not, as can be seen in Fig. 3.14.
- *Ordered Organization (O)*. Variables that are ordered allow comparisons of relative magnitude—this is less than or greater than that—to be made about the data values they signify because the steps can be perceived as increasing or decreasing. In Fig. 3.12, the four variables in the left column (planar, size, value, texture) are ordered and those in the right are not (orientation, shape, color). The effects of ordering can be seen in Fig. 3.15,

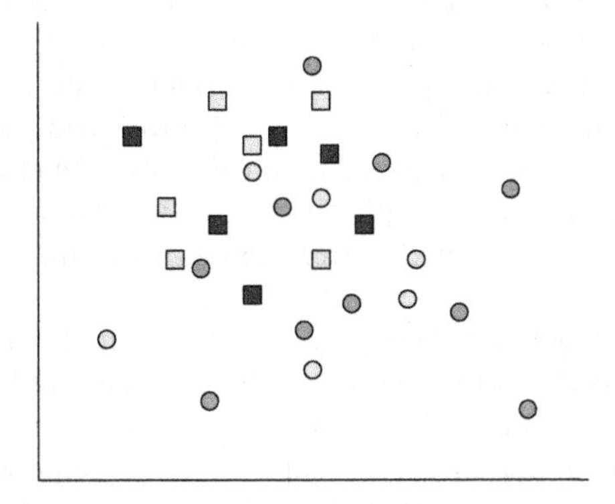

**FIGURE 3.14** Shape is associative, but value is dissociative

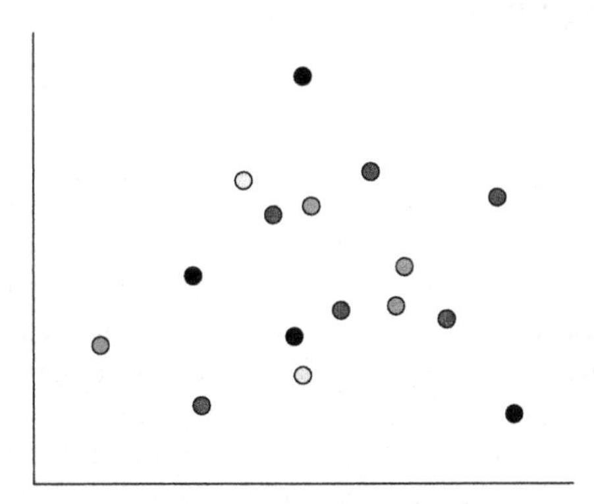

**FIGURE 3.15** A component with four categories mapped to the value variable

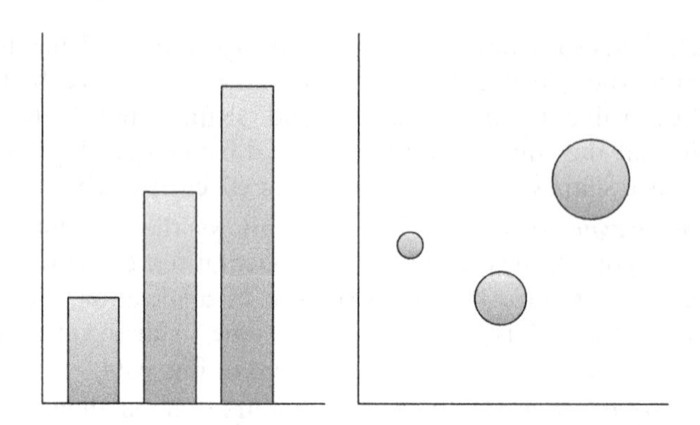

**FIGURE 3.16** Planar and size variables allow direct comparisons of ratios

where four steps of "value" have been used to shade the points to signify four ordered categories of some component.

- *Quantitative Organization (Q)*. This is the highest level. Variables that are quantitative allow the ratios of the values of the component they represent to be compared because of the existence of an absolute zero. These comparisons can be perceived directly without the need for a legend. Only the planar and size variables support quantitative components as shown in Fig. 3.16. The representation of quantities also depends on the type of mark being used. For example, points and lines can vary in size and length, but areas cannot without changing its meaning.

The second critical factor in choosing a variable to represent a component is to make sure that the level of organization of the variable is at least as high as the level of organization of the component. One of the key sources of errors in constructing graphics is the incorrect mapping of a component onto a retinal variable that does not support it. Only the planar variables support all four levels.

### 3.1.4 Constructing Efficient Graphics

A set of data components, two planar variables, six retinal variables, and various marks can be composed in many ways to present the information in a dataset as shown in Fig. 3.17.

Bertin's goal was to create efficient graphics. His definition of efficiency was based on the mental cost of visual perception. He paid close attention to how the eye scanned the graphics to extract the critical information for a set of specific questions that would be asked of the data. The ideal of a good graphic was to present its information so that what was relevant for the task could be perceived in a single image. By definition, an image was what could be perceived immediately at a single glance without mental effort.

An image for a diagram was limited to at most three correspondences between variable and component: two components mapped to planar variables

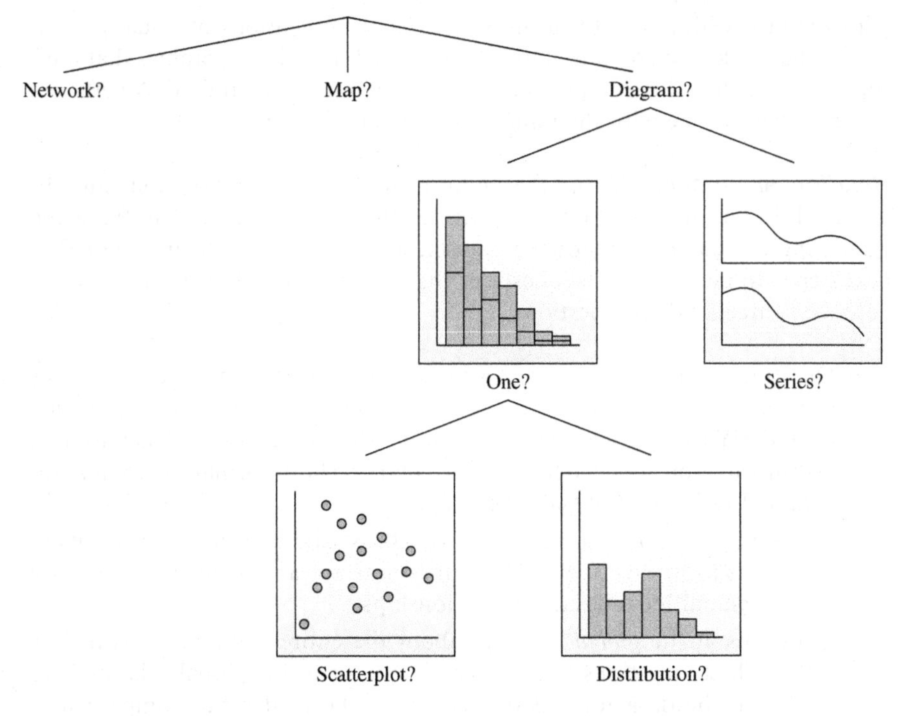


**FIGURE 3.17** Different ways to present information in data

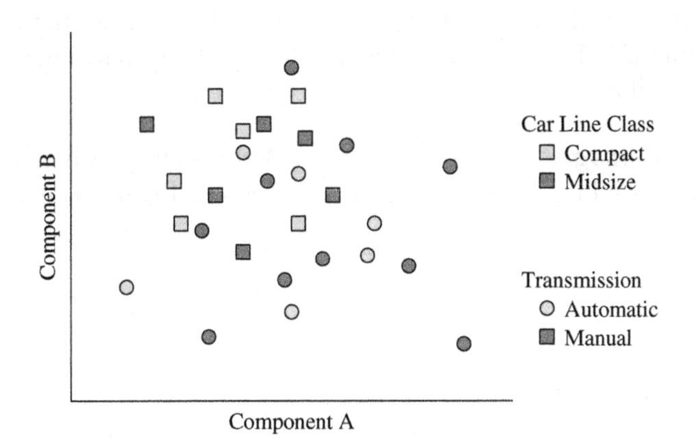


**FIGURE 3.18** Scatterplot with four correspondences

and one to some retinal variable. It is not possible to construct an image with more than three components. For example in Fig. 3.18, with Components A and B mapped to planar variables, "car line class" mapped to the value variable, and "transmission" mapped to the shape variable, it takes effort—scanning sign by sign—for the reader to pick out only dark squares to find midsize autos with manual transmissions. The combinations of visual properties on the marks are not automatically perceived. For those graphics that must support more than three components, a series of images is required. An efficient graphic, then, is one with the minimum number of images.

***Reading an Image.*** Before designing an image, the information must be analyzed. In an earlier section, we described the importance of identifying and classifying the components of the dataset, and determining their length. The next step is to understand the questions that might be asked of the data. Bertin described three kinds of questions:

- Questions at the *elementary level* involved a correspondence of a single element as in Fig. 3.19a. For example, how many city miles per gallon does the Yaris average? These questions use the correspondence to look up information associated with the mark. The information found was expected to be used outside the image.
- Questions at the *intermediate level* were those about groups of elements or categories as in Fig. 3.19b. These questions ask about things in common to understand local trends or relationships.
- Questions at the *global level* are about the entire component as in Fig. 3.19c. These questions try to identify one or a few global relationships across all the data for a component that the reader can compare with other information.

This analysis generates the list of questions, which allows the designer to construct graphics that can answer the questions with a single image or as few images as possible.

***Designing an Image.*** The foundation for Bertin's image was three uniform and ordered variables: two planar and one retinal. Figure 3.20 depicts this

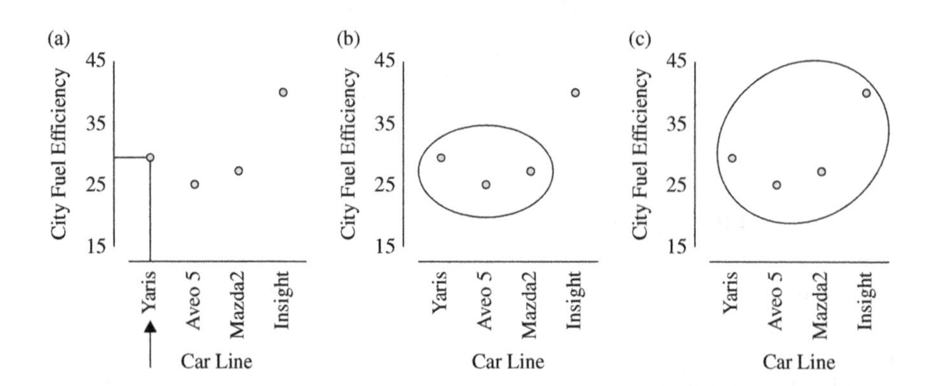


**FIGURE 3.19** Three levels of questions: elementary, intermediate, and global

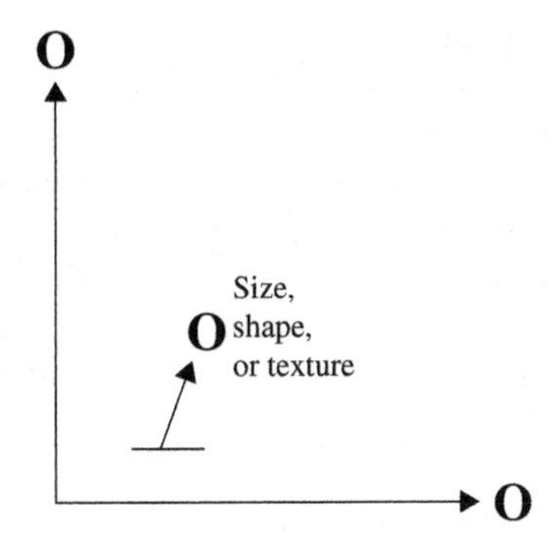


FIGURE 3.20 Schema for creating an image from three components

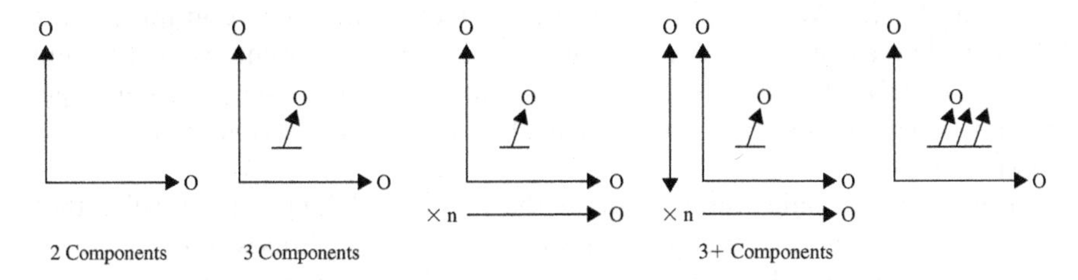


FIGURE 3.21 Standard schemas for diagram graphics

schema. With this schema, Bertin claimed an image could be constructed from three or fewer components that could answer any of the three levels of questions with a single fixation of the eyes.

Bertin separated his discussion of the mapping of components to planar variables from the mapping to retinal variables, suggesting a fundamental difference between these two kinds of variables. He used the organization and nature of the correspondences between elements of components to divide graphics into four classes: diagrams, maps, networks, and symbols. Diagrams were those graphics that could depict the relationships between all the elements or categories of one component with another. For diagrams, the preferred use of space—he called this *imposition*—was by orthogonal axes or the Cartesian coordinate system as shown in the schema in Fig. 3.20, but he also had schemas for other coordinate systems that included polar coordinate systems. His preference for an orthogonal system was based on the notion that the right angle had a psychological advantage in helping to discriminate visually.

Bertin's standard schemas for diagrams are shown in Fig. 3.21. Those with more than three components could be designed, for example, as a series of simpler graphics spread out horizontally or in a matrix or table. Alternatively, retinal variables could be used for additional components. The schemas provide a framework for thinking about the design of graphics in the context of the questions to be answered about the data.

The levels of perceptual organization of the visual variables and the correspondences of components, whether to the planar or retinal variables, play a fundamental role in consolidating the essential information into a single image accurately and efficiently. Bertin's contributions include his insights into the structure of statistical graphics and the critical role perceptual issues play in their design.

Where the previous sections focused on theory, the remaining sections will focus on various software systems that have used the structure of graphics to implement systems to generate graphics from data and a specification.

## 3.2 WILKINSON: GRAMMAR OF GRAPHICS

From the 1980s on, as computer graphics advanced, various research efforts incorporated Bertin's work into visualization and statistical graphics. Among them was Leland Wilkinson, who recognized that quantitative graphics, like language, had a deeper structure. He has spent more than twenty years creating software capable of generating statistical graphics from a descriptive language. He developed a grammar for statistical graphics (Wilkinson et al., 2001; Wilkinson, 2005).

Grammar is usually thought of in the context of language as rules that govern how words can be combined into phrases, and phrases into sentences, and so on. But graphics, as we discussed earlier, also have structure and, therefore, a grammar that constrains their composition. This may not be evident in the plots we see that play a supporting role as evidence in arguments that are being made verbally, nor in the charting tools commonly available in spreadsheets. Charts, Wilkinson points out, are not graphics. They take data as input and generate graphical elements in some specific form, which can then be stylistically altered, but they are not composed from a vocabulary of graphical elements into graphical phrases or sentences that we see. In Fig. 3.22, assuming that the dataset variable X contained values in three equal proportions of categories A, B, and C, a divided interval could be input to either a rectangular

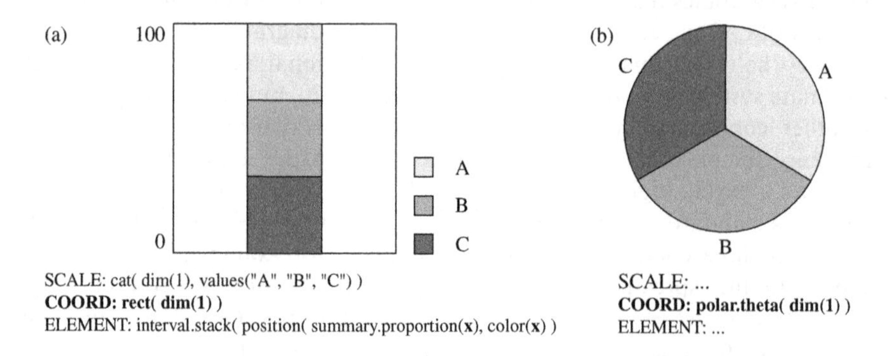
(a)
100
0
A
B
C
SCALE: cat( dim(1), values("A", "B", "C") )
**COORD: rect( dim(1) )**
ELEMENT: interval.stack( position( summary.proportion(**x**), color(**x**) )
(b)
C
A
B
SCALE: ...
**COORD: polar.theta( dim(1) )**
ELEMENT: ...

**FIGURE 3.22** A divided interval in rectangular and polar coordinates

coordinate system as in (a), or a polar coordinate system as in (b). By a simple change in the graphic specification rather than by selecting one of two different types of charts, we get a rectangular divided bar or a pie chart. A graphical grammar allows for much greater variation in what can be expressed visually.

### 3.2.1 The Graphic Pipeline

The generation of a graphic is analogous to the generation of a scene in computer graphics systems. Both transform data into the pixels of a display through use of a pipeline of processes. Wilkinson's pipeline is shown in Fig. 3.23. From a specification and a dataset, a graphic was assembled and displayed.

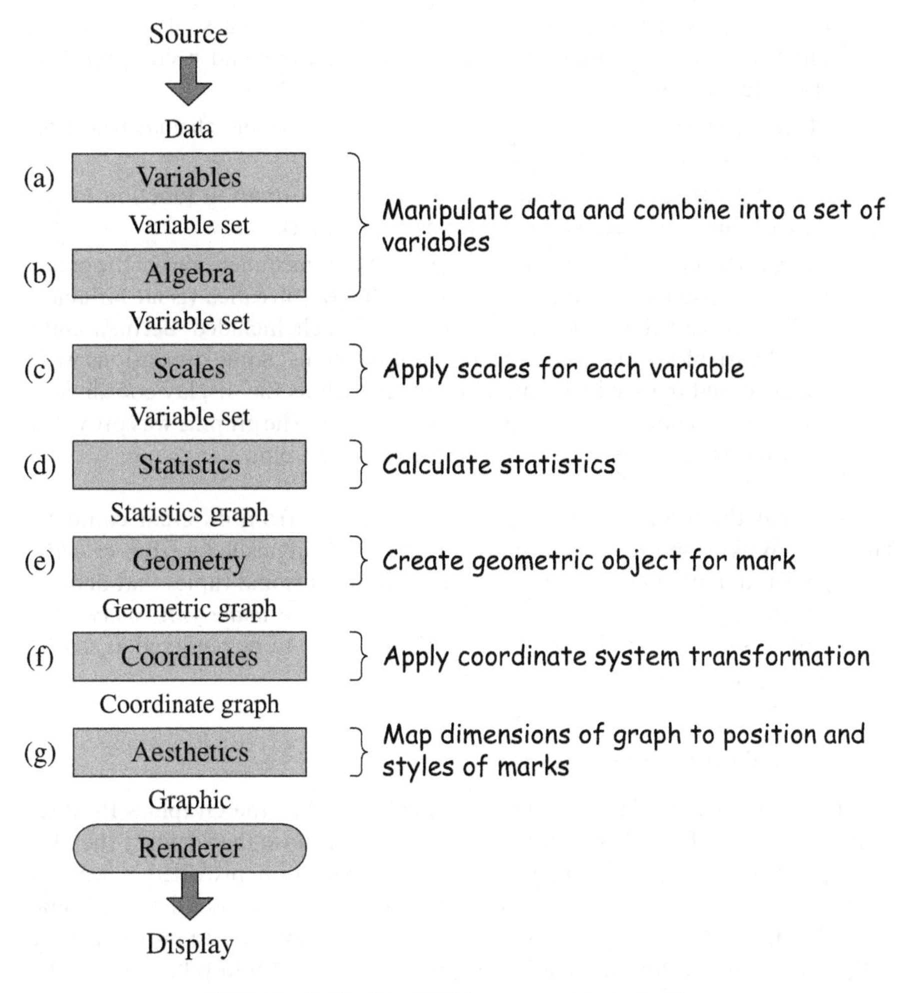


**FIGURE 3.23** The Wilkinson graphical pipeline

Data is assumed to come from any kind of source. The processing is done as follows:

1. In stages (a–b), the data is normalized and combined in various ways through algebraic operators. The result is a single set of variables.
2. In stage (c), scales are applied to the values of variables that would be used in the plot. *Scales* are functions that map a set of domain values into a range of values. For example, a categorical scale might convert string values such as "infrequently" and "frequently" to numeric values such as 1 and 2. Quantitative scales, such as a log scale, would apply the log function to each numeric value.
3. In stage (d), one of many statistical calculations for each variable might be performed. The output of this stage is a graph. The graph produced by (d) is a multidimensional space defined by the ranges of each of its dimensions after scaling and statistical calculations had been applied to the data variables.
4. Stage (e) generated a mark (point, line, area, path, or schema) based on the tuples in the graph.
5. Stage (f) applied the coordinate system's transformation function to the dimensions identified as the positions of the mark.
6. Stage (g) completed the transformation by mapping dimensions of the graph to specific visual properties of the mark. What Bertin called visual variables, Wilkinson called the *aesthetic attributes*, which included Bertin's eight visual variables with modifications and extensions. Some dimensions were assigned and mapped to spatial attributes such as the display coordinates and others to nonspatial attributes such as color. The graphic was provided as input to a renderer, which generates the visible plot.

Note that the term "graph" has several meanings from different contexts. However, Wilkinson used this term precisely and distinguished it from *graphic*. A graph was a mathematical concept: a set of n-dimensional tuples that defined an n-dimensional topological space. A graph was the *underlying space* that could not be seen; a graphic was the 2-D *display space* to be rendered as marks with visual properties on a display or printed page.

### 3.2.2 The Graphic Specification

The language that Wilkinson defined initially was called the Graphics Production Language (GPL). Of the full set of statements, those that defined the plot to be drawn are shown in Table 3.2. The specification provided a way to associate which data variables of the dataset would correspond to aesthetic variables of position and styles of the marks in the plot, and to customize the various transformations and statistical processing that would take place in the pipeline.

**TABLE 3.2 Graphic Specification Statements in Wilkinson's Graphics Production Language**

| Statement | Description |
|---|---|
| SCALE | A scale is associated with each dimension of the graph to be plotted. Scales are functions that operate on dimensions of a mathematical space. |
| COORD | The coordinate system to be used is either rectangular (Cartesian) or polar. |
| GUIDE | Guides are axes, legends, or labels that provide a way to determine the values represented by a mark's position or visual properties. |
| ELEMENT | A description of the mark to be drawn. |

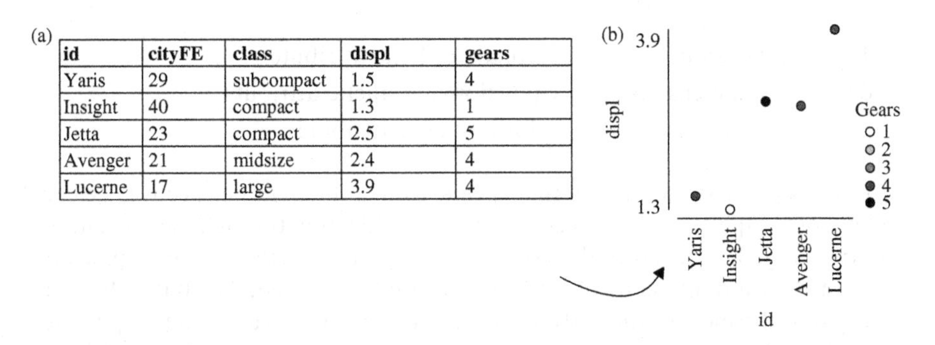

(a)

| id | cityFE | class | displ | gears |
|---|---|---|---|---|
| Yaris | 29 | subcompact | 1.5 | 4 |
| Insight | 40 | compact | 1.3 | 1 |
| Jetta | 23 | compact | 2.5 | 5 |
| Avenger | 21 | midsize | 2.4 | 4 |
| Lucerne | 17 | large | 3.9 | 4 |



**FIGURE 3.24** The final plot resulting from the interpretation of the specification

To illustrate, we look at a simple example. Of the five variables in the dataset shown in Fig. 3.24a, only three will comprise the graph to be generated: `id`, `displ`, and `gears`. The following is a partial GPL specification for the scatterplot using the standard XY(Z) schema for three components. The plot in Fig. 3.24b is generated from the dataset in 3.24a:

```
(1) SCALE:  cat( dim(1) )
(2) SCALE:  linear( dim(2) )
(3) SCALE:  linear( dim(3) )
(4) COORD:  rect( dim(1,2) )
(5) GUIDE:  axis( dim(1) )
(6) GUIDE:  axis( dim(2) )
(7) ELEMENT: point(position(id*displ),color.brightness
(gears))
```

We begin by looking at the statements in the specification. (1–3) The first dimension of the graph is assigned a categorical scale, and the second and third dimensions are assigned linear scales. Remembering that scales are functions that map between data and numeric values, the categorical scale maps terms such as "subcompact" to 1, "compact" to 2, and so on. The linear scale is an identity function that maps a value to itself, so the values of displ and gears remain unchanged in transformations. (4) The rectangular coordinate system is used. The first dimension is assigned to the x-axis and the second to the y-axis. (5, 6) Guides provide a way to see values along a scale, and axes are guides that make positional scales visible. The first dimension of the graph is associated with the x-axis and the second with the y-axis. (7) Elements indicate the marks to be used in the plot. In this case, the mark is a point, and there is one point for each of the five tuples in the graph. The arguments to the "point" function provide the mappings between the data variables and the aesthetic variables. (Implicitly, the variables are assigned to a dimension in the order in which they appear: id to the first dimension, displ to the second, and gears to the third.) The mappings are as follows:

`id` ↔ dimension 1 ↔ x positional aesthetic attribute
`displ` ↔ dimension 2 ↔ y positional aesthetic attribute
`gears` ↔ dimension 3 ↔ color aesthetic attribute

Retracing our steps through the pipeline from a different perspective, we can follow some aspects of the processing of the specification through the pipeline as shown in Fig. 3.25. The variables of the dataset that are used for the graphic are associated with dimensions of the graph as are the aesthetic attributes that are involved in the plot. In the early stages, the `id` variable's values are mapped to natural numbers (countable integers), followed by the creation of the graph from the three components from the dataset (variable set) that were identified in a specification statement (7). In the middle stages, the quantitative values in each dimension are transformed to the values in the space of its associated aesthetic attribute. Because the x-axis and y-axis of display space are each 100 pixels in length, each of the values for `id` and `displ` are scaled appropriately, taking into account the type of scale associated with the dimension. The categorical scale for dim-1 (data space) has evenly spaced the values along the x-axis (display space). The linear scale has mapped the values of dim-2 onto the y-axis. The linear scale for dim-3 has mapped its values into the brightness attribute of color. The values of hue and saturation are set to the default value of 0. The output of the middle stages is a graphic. In the final stage, the renderer draws five points (marks) using the x and y values as display coordinates and the `color` values as HSV (hue, saturation, value) values. The graph and graphic are shown as tables, but each row should be thought of as an ordered tuple `<value1, value2, . . . , valueN>` in an n-dimensional space.

In the ELEMENT statement (7) of the specification, which gives rise to the graph, we see two important aspects of Wilkinson's grammar-based approach: the ability to create, through algebraic operators, an underlying data space, which embeds a graph that is separate from the display space of the graphic; and the ability to specify the mappings between these two spaces.

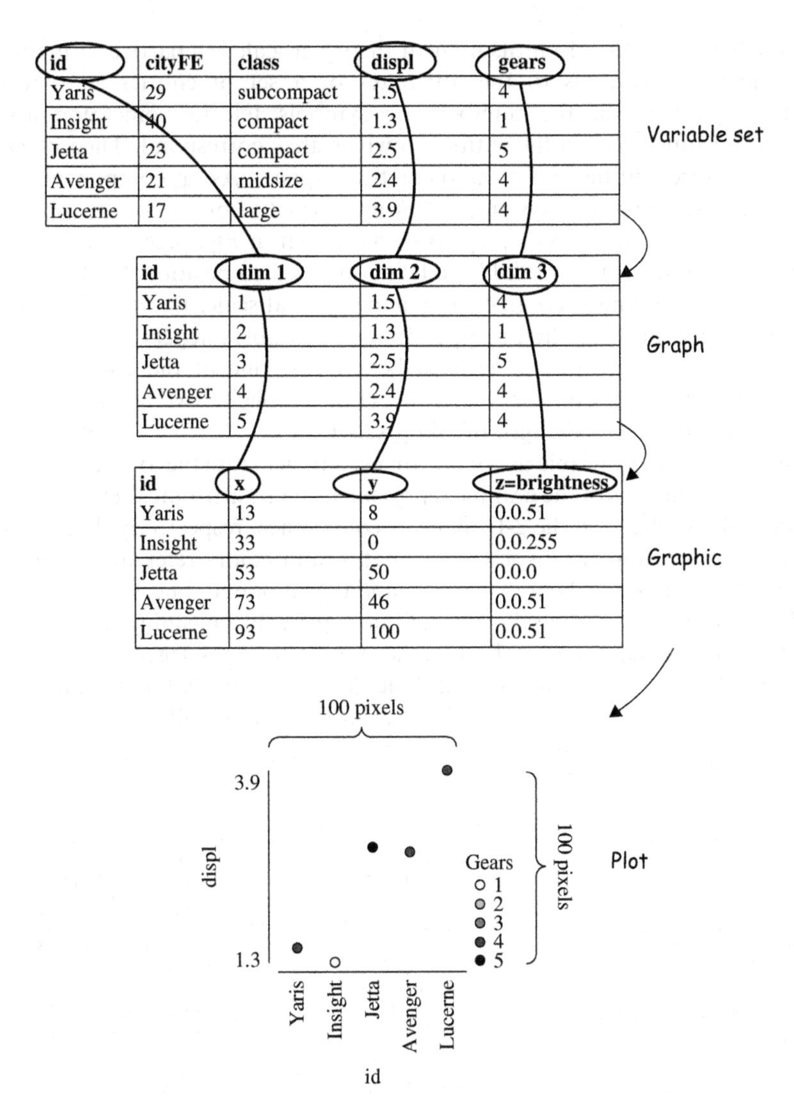

| id | cityFE | class | displ | gears |
|---|---|---|---|---|
| Yaris | 29 | subcompact | 1.5 | 4 |
| Insight | 40 | compact | 1.3 | 1 |
| Jetta | 23 | compact | 2.5 | 5 |
| Avenger | 21 | midsize | 2.4 | 4 |
| Lucerne | 17 | large | 3.9 | 4 |

| id | dim 1 | dim 2 | dim 3 |
|---|---|---|---|
| Yaris | 1 | 1.5 | 4 |
| Insight | 2 | 1.3 | 1 |
| Jetta | 3 | 2.5 | 5 |
| Avenger | 4 | 2.4 | 4 |
| Lucerne | 5 | 3.9 | 4 |

| id | x | y | z=brightness |
|---|---|---|---|
| Yaris | 13 | 8 | 0.0.51 |
| Insight | 33 | 0 | 0.0.255 |
| Jetta | 53 | 50 | 0.0.0 |
| Avenger | 73 | 46 | 0.0.51 |
| Lucerne | 93 | 100 | 0.0.51 |



**FIGURE 3.25** Mapping from data variables to dimensions of the graph to retinal variables used for the graphic

### 3.2.3 Components of the Grammar

This section introduces several important components of Wilkinson's graphical grammar-based system that are also present in other grammar-based systems. We ignore components in the first two stages of the pipeline—data management and manipulation—because they vary widely between systems. We focus on the components in the remaining stages of the pipeline: scales and guides, statistical functions, coordinate systems, marks, and aesthetic attributes.

***Scales and Guides.*** In measurement theory, a scale is a particular mapping of numbers or symbols to the attributes of a set of entities so that the relationships between the numbers or symbols tell us something about the relationships between the entities to which they correspond. Theoretically, scales measure entities in a mathematical space. As a component in a grammar-based graphics system, scales are used to control the mapping of values from data space to display space as shown in Fig. 3.26. In this simple example, the data values of "id" are being mapped to positional values in the display. The mapping is controlled by a categorical scale. Because the scale is categorical, its values are first mapped to the countable integers (a) and then to the x variable, where the values will be used as one of the coordinates in positioning the marks on the x-axis (b). (For simplicity, we assume that the x-axis and y-axis have a length of 100 pixels.)

The x positional variable and scale are used to generate the axis (c). The x, y, and color dimensions of the graphic represent aesthetic attributes. They can only assume values related to the 2-D plane or color space, respectively. Positions in displays are locations in display coordinates, and colors refer to an encoded number that represents the transparency and the amount of hue, saturation, and brightness to use in shading the mark. Guides provide a meaningful way for us to determine the magnitude of the data they represent. Axes, legends, and certain kinds of labels are guides, not scales. Scales are functions that map values from a domain to values in a range. References are kept internally to the values in

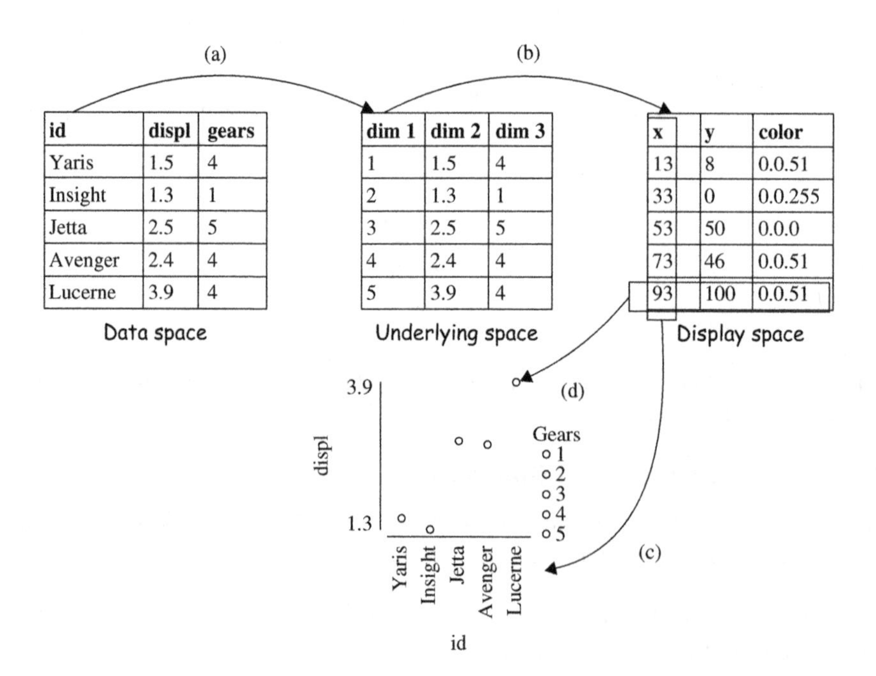


**FIGURE 3.26** The transformations controlled by scaling

data space associated with each tuple (row) in the graphic so that meaningful values can be used in the guides: the x-axis uses the names of autos, the y-axis uses the lowest and highest engine displacement values, and the color legend provides a mapping between the grayscale and the number of gears. For the mark used in this example, each tuple in the graphic supplies the values of the aesthetic attributes needed to render it in the plot (d). Not shown are the defaults that govern other aesthetic attributes such as size and shape.

Beyond the categorical scale mentioned previously, Wilkinson identified several other types of scales. These included linear and time scales, and scales that performed mathematical transformations such as log and power. The scaling transformations were performed on dimensions in the graph with continuous data prior to statistical transformations because certain statistical methods make assumptions about the distribution of data.

***Statistics.*** A statistical component transforms the data into a statistical graph. A graph is a set of tuples. A tuple might represent an individual mark, as in the case of a scatterplot, or a summarization of a subset of the data, as in the case of a mark that represents a histogram bin or a boxplot. Placing statistics under control of the graphics has several advantages.

- Different marks can be used to depict the same graph. Each of the following statements uses the same graph but produces a different graphic:

  ```
  point(position(summary.mean(x*y)))
  line(position(summary.mean(x*y)))
  ```

  In the first case, a point is drawn for the mean of the y values grouped by each category of x, and, in the second case, a line is drawn.
- A graphic can be layered or paneled by executing multiple statistical methods as shown in Fig. 3.27. For example, a smoothing line can be superimposed on a scatterplot.
- The connection between variable sets and elements in the graphic can be maintained to provide support for interaction with the plots.

***Geometry.*** The geometric graph produced by the geometry component is a set of tuples created by graph functions such as `line()`, `point()`, or `polygon()`. Theoretically, the graph is a geometric space because it represents geometric objects that have magnitudes, and, like any graph, it is a mathematical concept that cannot be seen. Wilkinson classified the geometric graphs not by dimensions of the graph, such as 1-D, 2-D, and so on, but by types of data and geometry. He divided graphs into three major categories: functions, partitions, and networks as shown in Table 3.3.

The geometric objects could be positioned relative to other geometric objects by using one of a set of modifiers to the graph functions such as *stack*, *dodge*, or *jitter* to adjust the locations of the objects. For example, bars could be stacked

TABLE 3.3 Classification of Geometric Graphs

| Class | Graphing Functions |
|---|---|
| Function | point, line, area, interval, path, schema |
| Partitions | polygon, contour |
| Networks | edge |

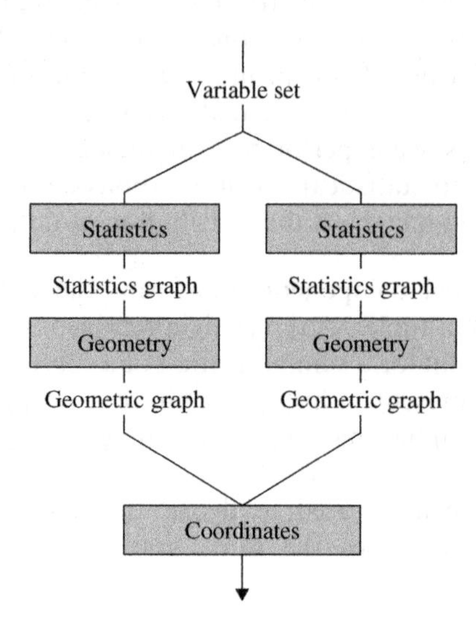


FIGURE 3.27 A pipeline for layers and paneling

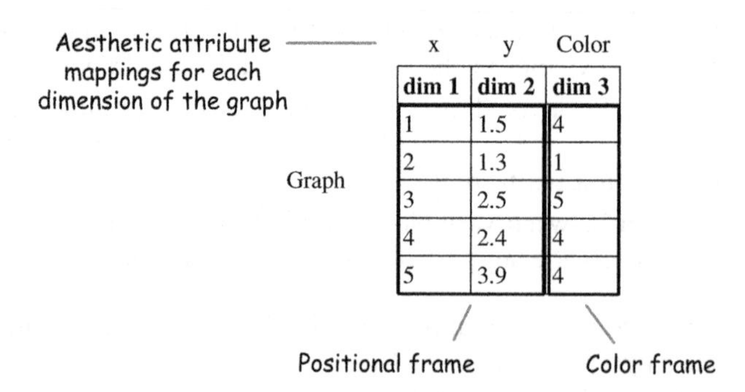


FIGURE 3.28 A graph of three dimensions mapped to aesthetic attributes

vertically or scatterplot points could be constrained to use only open space or to partially overlap.

***Coordinate Systems.*** Coordinate systems organize the mathematical points in a *frame*. A frame is the mathematical space bounded by intervals of each of its dimensions. In Fig. 3.28, the graph has been divided into two sets of tuples called frames, one frame for the coordinates <x,y> and one for color. The bounded regions of the frames in our example are as follows:

positional frame: x:[1, 5] × y:[1.3, 3.9]; color frame: [1, 5]

In other words, any coordinate used to position a mark in this space must have as its x and y values a natural number between 1 and 5 and a real number between 1.3 and 3.9. Similarly, the value that will be mapped into the color space must be a natural number between 1 and 5. Although the coordinate system's transformation could be used for nonpositional attributes such as color, they are most often used for transforming the coordinates of positions into the 2-D coordinates of the plane of the display or page. The Cartesian coordinates comprised of the positional variables <x,y> locate a point on the plane of the surface by measuring distance along the x-axis and y-axis; the polar coordinates <r,θ> locate a point by its radial distance from the center and an angle.

Coordinate systems not only transform the mathematical points in the graphics frame that are drawn as marks in the content area of the plot but also the elements in the guides such as the location of tick marks or the labels of categories. The transformation alters the shape of the geometric object being displayed. For example, a bar is drawn as a rectangle in a rectangular coordinate system but as the wedge of a pie in a polar coordinate system; one axis is straight and the other curved. In addition to transforming the geometric objects to be drawn on the plane, coordinate systems can also apply a variety of transformations to the plane itself that include reflection, rotation, translation, and projection.

Coordinate transformations are useful for several reasons. First, they allow a graphic to be drawn in a way that is most easily perceived. Pie charts may be more difficult than divided bars when comparing a slice across a series but easier for comparing proportions of a whole. Second, adjustments may be needed for the graphic to more closely approximate reality. Maps, for example, require nonlinear projections. Finally, they enable different levels of detail in the graphic to be examined through zooming and distortional transformations.

***Aesthetic Attributes.*** To be seen, geometric objects require location and visible properties that Wilkinson called the *aesthetic attributes*. These attributes, excluding his categories for motion and sound, are shown in Table 3.4. Aesthetic functions are mapped to dimensions of the graph and convert the values in these dimensions to attribute values required by a renderer to draw each mark on the display. For example, the `position()` function converts each vertex in the geometric object to its location based on a pixel-based coordinate system, and color is encoded in a format that indicates the levels of hue, saturation, and brightness.

Although Wilkinson provided a comprehensive explanation of the grammar of graphics and an architectural approach for a grammar-based system, as well as descriptions of a GPL-based and XML-based specification language and a

**TABLE 3.4 Examples of Aesthetic Attributes**

| Category | Aesthetic Functions |
|---|---|
| Form | position |
| | size |
| | rotation |
| | resolution |
| | shape: polygon, glyph, image |
| Surface | color: hue, saturation, brightness |
| | texture: pattern, granularity, orientation |
| | blur |
| | transparency |
| Text | label |

graphics-based data exploration tool, its implementation was not available except as part of the SPSS product line. The grammar-based approach has been adopted in two different environments that are open source and readily accessible: ggplot2 by Hadley Wickham and Protovis by Michael Bostock and Jeffrey Heer. We will take a brief look at ggplot2. We will introduce Protovis in this chapter and describe how to use it in depth in Chapter 5.

## 3.3 WICKHAM: GGPLOT2

Wickham designed ggpplot2 primarily for statisticians (Wickham, 2009). ggpplot2 was based on Wilkinson's graphical grammar and implemented as a loadable package for the System R, a widely used open-source software environment for statistical computing and graphics. R is both a language and an environment. R is a functional language similar to Scheme or Lisp. The R environment has, at its core, an interpreter that understands the R language and provides a command-line interface that allows R expressions to be entered as text and executed without compilation. The "Further Reading" section at the end of this chapter includes references with details about installing the software and getting started.

ggplot2 was designed to support a graduated approach to constructing plots, starting first by plotting the raw data and then adding layers of statistical analysis and annotation. Layers were an extension of Wilkinson's grammar, and Wickham called his grammar the *layered grammar*. In the next subsection, we will describe the changes to the pipeline that resulted from these extensions. Because ggplot2 was implemented in R (language) to run in R (environment), the specification language has an R syntax. Wickham extended the specification language by adding the ability to describe subsets of the data to be plotted as a series of plots, a kind of visualization technique known as small multiples. The extensions to the specification were called a *faceting specification*. The subsection on specification and components will include a discussion of these extensions.

### 3.3.1 The Graphic Pipeline

Figure 3.29 shows the pipeline for ggplot2. Unlike Wilkinson's system, the source for the graphic is assumed to be one or more particular data structures in R called *data frames*, which, for our purposes, can be thought of as datasets. Because the R environment contains many functions for data manipulation,

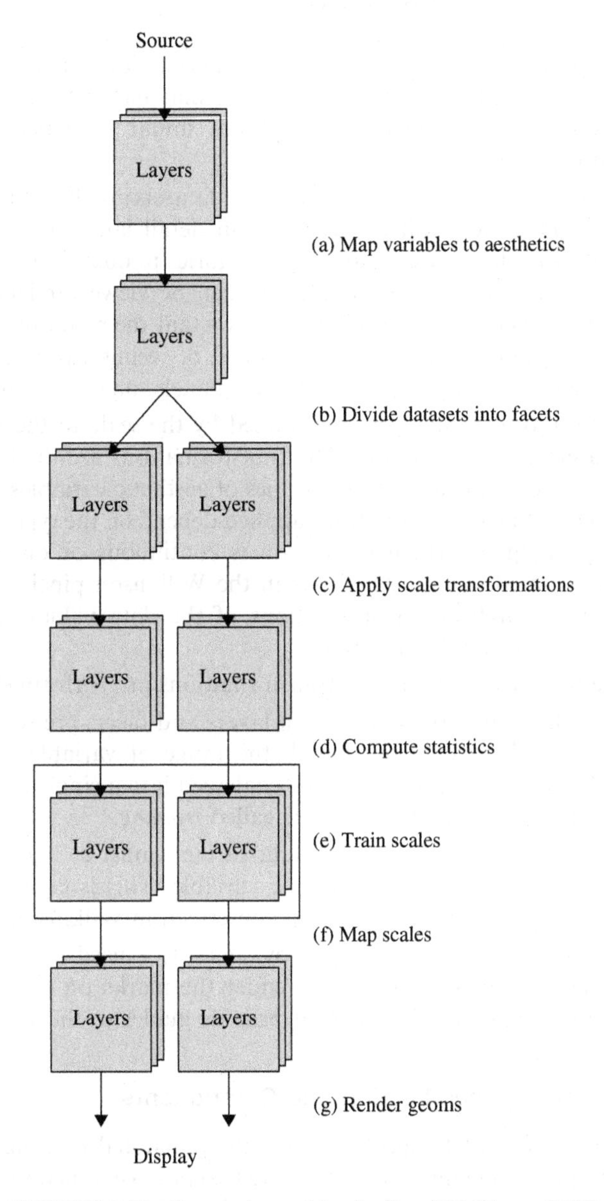


**FIGURE 3.29** Layered graphic pipeline (Wickham, 2009)

ggplot2 has no algebra for data composition. Layers contain what is required for a plot: a dataset, mappings between variables and aesthetic attributes, a function that performs statistical transformations on the dataset, a geometric object (mark), and a way to adjust the positions of marks. A single dataset may be shared across layers, or each layer may have its own.

The figure shows a plot defined to contain three layers of statistical information. The processing is done as follows:

1. In stage (a), variables in one or more datasets identified as source are mapped to aesthetic attributes. Each layer may have its own dataset and mappings. The mappings are conceptually similar to what was described in the previous section.
2. In stage (b), if faceting has been specified, datasets are divided into groups called *facets*. Facets will be described in detail later, but the effect of faceting is to split datasets, often by categories in one of the variables, so that the data related to each category can be viewed independently in separate panels of a plot. The figure shows that the result of faceting has divided the data into two groups. Instead of seeing a plot with a single panel, there will now be a plot with two panels, one for each facet.
3. In stage (c), transformations are applied by the scale to the variables in the datasets across the layers. The transformations are the first of three steps that map data values to the values of aesthetic variables (x, y, color, etc.). The specific transformations applied depend on the type of scale and vary depending on whether the data is continuous or discrete and, if discrete, ordered or unordered. As in the Wilkinson pipeline, they may include mathematical transformations of the data values such as log, square root, or power functions.
4. In stage (d), various statistical transformations are performed.
5. In stage (e), because the data within layers and facets are separate while scales across them may be shared, the range of variables across data related by layers and facets must be coalesced into a single range used by the scale that controls them. This is called *training*.
6. In stage (f), the values from each data variable must be mapped into the values of its corresponding aesthetic variable. This is equivalent to the mapping between graph and graphic described in Wilkinson's pipeline.
7. In stage (g), the dataset produced by stage (f) is used by the geometric object and coordinate systems to render the marks on the display and generate the axes and displays that provide guides to the scales.

### 3.3.2 The Graphic Specification and Components

The language of the ggplot2 specification is the functional language R and is, therefore, a programming interface. There are two primary functions: `qplot()` is designed to create plots quickly and provides a large number of defaults;

ggplot2() is designed to incrementally build a specification that can be executed to construct the plot. This section introduces the specification and components together. It illustrates the specification and capabilities of components with figures of annotated functions and the plots constructed by their execution. The intent is to show what is possible using a grammar-based approach. The "Further Reading" section will refer to references that explain R and ggplot2 in detail.

***Data Variables and Aesthetic Attributes.*** As in the Wilkinson model, mapping dataset variables to aesthetic attributes is essential to the creation of a *plot*, which is Wickham's term for a graphic. Figure 3.30 shows the specification for a bivariate scatterplot. The variable city has been mapped to the x positional attribute and displ to the y positional attribute. A geometric object specification, in this case geom_point(), identifies the mark that will be used in the plot. The resulting plot has a single layer.

By simply adding the aesthetic mapping of gears to the size attribute, an XY(Z) plot is constructed that includes two axes and a legend as guides to the three dimensions of the underlying graph as shown in Fig. 3.31.

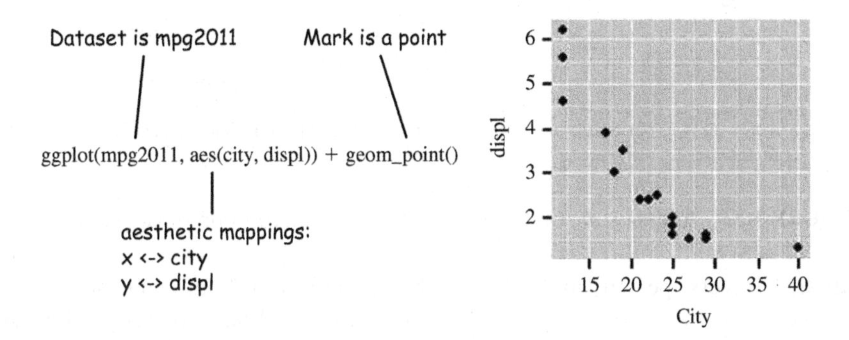


**FIGURE 3.30** A bivariate scatterplot

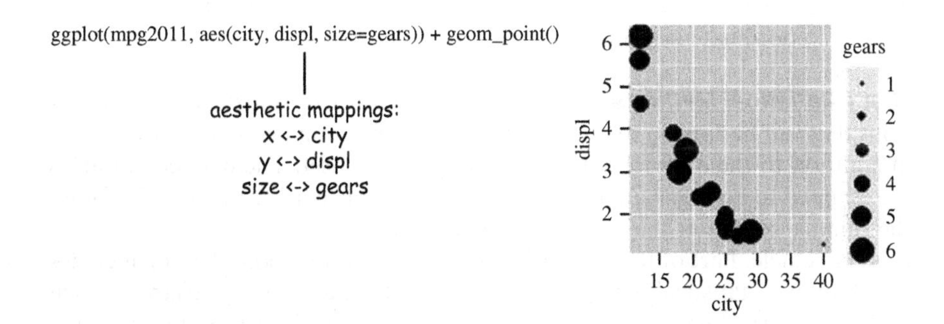


**FIGURE 3.31** An XY(Z) scatterplot with gears mapped to size

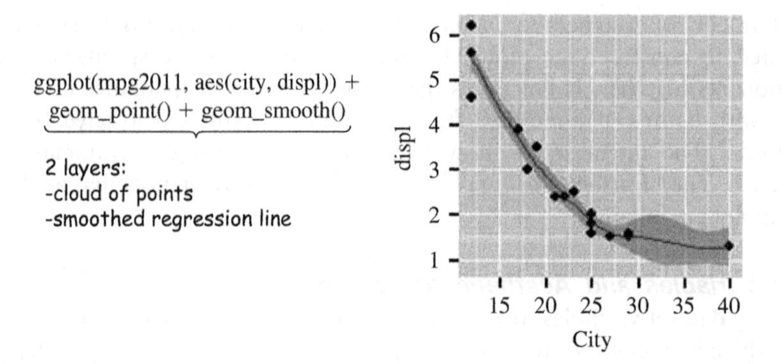


**FIGURE 3.32** A plot with two layers

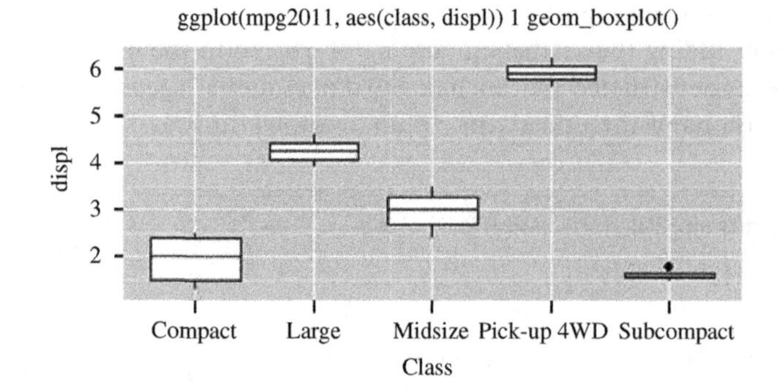


**FIGURE 3.33** Adding statistical transformations: summarization of categorical data

***Layer.*** Layers permit multiple graphs or additional information to be superimposed in the same space. The plot in Fig. 3.32 has two layers. Layers may be used for different reasons. A layer may display the raw data as shown by the points in the data area of the plot, show a statistical summary such as the regression line shown in the figure, or display annotations or other information about the raw data. In this figure, each layer of the plot panel shares the same axes and coordinate system.

***Stat (Statistical Methods).*** ggplot2 has a number of statistical transformations that can be applied to an input dataset. These compute new data variables in addition to the original data variables of the dataset. The derived variables are mapped to aesthetic attributes for use by various marks. Figure 3.33 shows the summarization of the values of displacement on the y-axis as a boxplot by class of vehicle. Part of the processing initiated by `geom_boxplot()` includes a "boxplot" statistical transformation that computes new variables, such as `lower`, `middle`, `upper`, `ymin`, and `ymax`, which are not part of the source dataset. These derived variables are used by the boxplot geometric object to render each of the five boxplots. Note that the change in specification resulted in changes to only the statistical method and the geometric object in the pipeline.

***Geoms (Marks).*** In ggplot2, geometric objects—what Wickham calls geoms—perform the rendering of the plot's surface. Wickham classifies geoms as either *individual* or *collective*. Individual geoms generate a single mark for each row in the dataset (point in a scatterplot), whereas collective geoms group multiple rows for each mark (a polygon). In Fig. 3.34, changing just the mark in the specification results in very different visible results. Only the first is useful, but the other illustrates the way in which plots can be substantially changed, or customized, by varying the statistical method, geometric object, or both.

***Coordinate Systems.*** Coordinate systems combine the x and y positional aesthetic into a coordinate, which is then transformed to determine its location on the display surface. It is the final of three transformations—the mathematical transformation during scaling and a statistical transformation being the first two—that can take place along the pipeline. ggplot2 provides ways to flip or rotate the axes as well as set limits on the axes to zoom in or out of the data.

***Facets.*** Facets are a way to subdivide a dataset into groups that can be displayed separately as a series of panels as shown in Fig. 3.35. In this example, by subdividing the dataset into groups based on the different sizes of gears, the three dimensions can be seen as a series of bivariate scatterplot panels. Each panel shows the data for only one level of gears: 1, 4, 5, or 6 as shown in the title.

## 3.4 BOSTOCK AND HEER: PROTOVIS

Wilkinson's and Wickham's work is primarily for statisticians and requires significant effort or resources to become familiar with both the environment

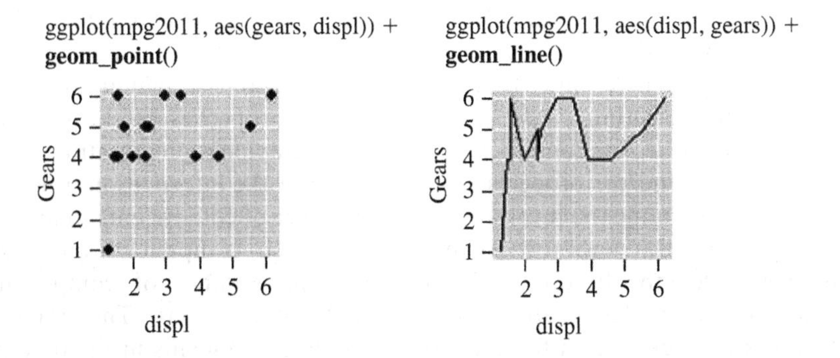


**FIGURE 3.34** The results of changing the `geom_xxx()` in a specification

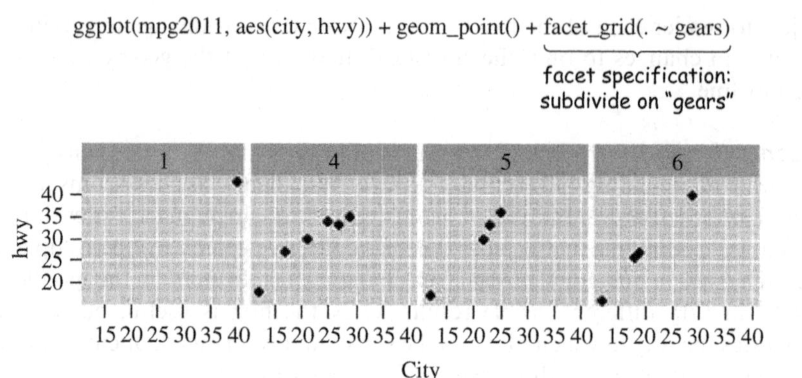


**FIGURE 3.35** The display of subsets of a dataset as facets

and the plotting software. In recent work by Michael Bostock and Jeffrey Heer, aspects of the grammar-based approach have been implemented for general use in Web browsers using JavaScript and Protovis libraries. The specification language is JavaScript, and it contains some of the same components discussed in this chapter: scales, marks, and layouts. Although it contains only a limited number of the statistical methods and support for the kind of plotting often used in statistical analysis, its use of the grammar-based approach may provide a way to efficiently design and construct a wider range of visualizations than just quantitative graphics. Chapter 5 discusses Protovis in detail.

## 3.5 SUMMARY

Humans communicate thoughts using systems of *signs*. Semiotics is a field that studies how signs and symbols become associated with meaning and the conventions or code by which signs are organized into systems to communicate and to model information about the external world. Graphics are a visual representation of data that rely on graphical elements organized in a geometric space. By seeing graphics as structured, Jacques Bertin extended the reach of semiotics into graphics systems and developed a semiotic theory for graphic communication.

Bertin outlined three things that must be understood to design graphics: the properties and structure of the information, the properties of the graphics system, and the rules for constructing an efficient graphic representation of the data. To understand the properties and structure of the information, it is first necessary to determine the invariant and the number of components to include in the graphic, the level of organization of each component, and each component's length. Bertin's view was that the number of components provided the best classification for graphical constructions. The graphics system was the sign-system for encoding the data components in the descriptions that result from an analysis of the information.

Graphics are read in three stages: identify what is external to the graphic, identify the mappings between the visual variables and components, and perceive the relevant correspondences between the marks and the subset of data that the graphic represents.

Bertin's goal was to create efficient graphics. The ideal of a good graphic was to present its information so that what was relevant for the task could be perceived at a single glance without mental effort. An efficient graphic was one with a minimum number of images.

Leland Wilkinson built on Bertin's work. He recognized that graphics, like language, has a grammar that constrains its composition, so he developed a grammar for statistical graphics. Some of the important components of Wilkinson's grammar-based system include scales, guides, statistical functions, coordinate systems, marks, and aesthetic attributes. The grammar-based approach is giving rise to new tools and toolkits such as ggplot2 and Protovis, which provide ways to generate both graphics and visualizations through specification languages rather than scripted programs that are executed sequentially.

## 3.6 FURTHER READING

It is helpful to understand the basic concepts of semiotic systems before reading Jacques Bertin. An accessible introduction to semiotics is *Semiotics: the Basics* (Chandler, 2007).

The *Semiology of Graphics* (Bertin, 1983) is a foundation for good reason. Although it requires effort to read, it provides insights on many levels and is a reference you will likely refer to repeatedly. The English translation has recently been reissued.

System R, an open-source statistical computing language and environment, was based on the S Language developed at AT&T Bell Laboratories by Rick Becker, John Chambers, and Allan Wilks. The R home page (http://www.r-project.org) contains information about the organization and instructions on how to download a copy of R for any of the major computing platforms. ggplot2 is an R library that is loaded into R. Details on how to get started with ggplot2 can be found at Hadley Wickham's site (http://had.co.az/ggplot2).

# CHAPTER 4

# DESIGNING VISUAL INTERACTIONS

## 4.1 DESIGNING FOR COMPLEXITY

Data-intensive systems are changing the scale, scope, and nature of the data to be analyzed. The tools we use to do our work are becoming increasingly sophisticated and intertwined in a mesh of data, information, and computation. Information and multidimensional data come from multiple sources and are processed using a variety of computational methods and approaches. Analyzing the data is usually a collaborative effort requiring expertise from different disciplines. Complexity is continually increasing. At the same time that users of these systems ask for simplicity, they also ask for more features and tools that extend their capabilities and allow them to explore larger datasets quickly and dynamically. Although it might seem that simplicity and complexity are opposing forces, simplicity is not the opposite of complexity (Norman, 2010). Complexity is a statement about the world; simplicity is a statement about the mind. Complexity can be tempered if a system is properly structured and organized so that it can be understood. Once understood, the effort involved in learning its structure is forgotten. A well-designed complex system, after it becomes familiar, is often described as simple or "intuitive." Simplicity is psychological. *Perceived simplicity*, which is the first impression of a system's interface based only on its look—the number of graphical and control elements that make up its visual interface—is different from *operational simplicity*. Even

*Making Sense of Data III: A Practical Guide to Designing Interactive Data Visualizations*, First Edition. Glenn J. Myatt and Wayne P. Johnson.
© 2011 John Wiley & Sons, Inc. Published 2011 by John Wiley & Sons, Inc.

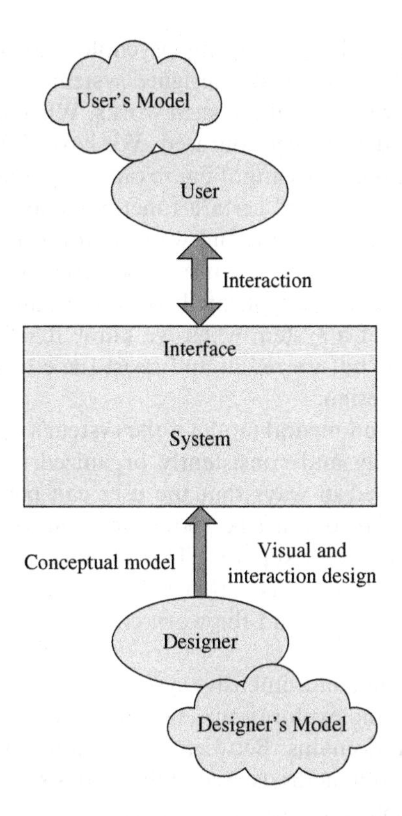


**FIGURE 4.1** Two perspectives of a system (based on Norman, 2002)

a system with a complex visual interface can be operationally simple if its structure is logical and well organized, we have taken the time to learn it, and it makes interaction efficient.

As we interact with a system or its help systems, we construct in our minds a *mental model* about how a system works. We use mental models all the time. For example, if we own a gas or propane stove, even if we have not looked inside, we likely have a simple mental model of gas flowing through pipes that lead to the burner, the igniting of the gas by a spark, and the distribution of the burning gas around the burner through many small orifices. When it doesn't light, we infer from our model that something is plugged the spark is not being generated, or the pressure is low. Our mental model of how a system functions helps us operate it without memorizing arbitrary procedures and helps us make reasonable predictions about what to do when something unexpected happens. If a system is well designed, our mental model will closely correspond to the conceptual model the designers had in mind and will be consistent with the way the system behaves or responds operationally. How the two different models relate to the system is shown in Fig. 4.1.

As human beings, we have a strong psychological need to explain. The mental model we construct of a badly designed system—we always create one—may not be consistent with how the system works. When the system's behavior doesn't match our model, we get confused. We have difficulty predicting the outcome of certain actions. We cannot figure out what combinations of buttons to push. How we engage or find certain functions may appear arbitrary. We try to commit them to memory but find we have forgotten them the next time we use the system. This leads to a sense of powerlessness, frustration, and a perception that the system is complicated. We are usually willing to study and master the complexity of a system when we know it reflects and reduces the complexity of the work, but we resent and resist the effort when complexity is the result of errors of design.

A user will form a good mental model if the system's operations support the user's tasks, are logically and consistently organized, and provide feedback about how it can be used in ways that the user can perceive. The designers' conceptual model, therefore, must be rooted in a thorough understanding of the work that the system will support. This includes the goals that must be accomplished, the activities or tasks that will be performed by individuals or groups to achieve these goals, and the work environment in which the system will operate.

For visual analytics and data-intensive systems, this means understanding the nature of data analysis and exploration. The nature of this work differs between and within application domains. Between application domains, the data to be analyzed and the tools and methods used in the analysis vary widely along many dimensions. Within application domains, there are usually layers of analysis that must be done as the raw data is successively transformed until it reaches a point where decisions can be made. Each layer of analysis may be done by individuals with different subject matter expertise. Each expert may have his or her own datasets derived from the source data, questions about the data, and computational tools and methods for analysis. The context plays an important role in understanding the work, which, in turn, affects the design of the system. The application of microarray technology in the life sciences illustrates this point.

DNA microarray data is used to measure the activity and interactions of biological genes. Within the scientific research community and industry, the uses for this data include the discovery of genes, diagnosis or prognosis of a disease, and the assessment of the toxicity of drug candidates or other chemical agents. Each of these areas have distinct scientific tasks with their own sets of questions about the data. Examples of these tasks include identifying genes with similar or different co-expression patterns, looking at gene expression patterns under various stress conditions, or mapping gene expression data to metabolic pathways (i.e., a series of chemical reactions within a biological cell). Addressing the scientific questions may involve one or more data analysis tasks with quite different analytical techniques (Berrar et al., 2003).

Designing visual interactions to support the work of analysis and exploration requires sensitivity to the context in which the tasks are being performed.

The advent of tools such as Protovis should make it possible to design systems more quickly with interactions comprised of visualizations and quantitative graphics as well as user interface controls to create tools that support specific domains. The process of design is a dialogue between designers and users. Whether the dialogue is formal or informal, a good design process contains essential elements that even small projects with just one or two individuals should consider. Of these elements, one of the most important is the *conceptual model.* The conceptual model is a high-level description of the concepts—objects, actions, and attributes—that underlie the organization, appearance, and behavior of the system being designed (Johnson, 2010). The product of design also communicates with users. Its visual interface is a system of signs and conventions—a semiotic system—that reflects a conceptual model, whether the model has been carefully designed or not. If the signs and conventions have been adopted from the workplace or are familiar, and they are visually organized to fit the tasks, the system will be more easily understood, and the interaction will flow.

The remainder of this chapter is divided into two sections. The first section discusses the critical activities of design that loosely form a process: analysis of the user and work; design of the conceptual model, the architecture of the interface, and the visual interaction; prototyping as a means for reflecting about the conceptual model and exploring alternative designs for the visual interface and interaction; and usability evaluation. Although some of these activities must initially be done before others, design is a process of continual refinement. The byproducts from each stage of design, such as the various models and prototypes, may be reexamined several times over as new insights are gained. The second section discusses the physical aspects of the design of visual interaction: the form and content of the user interface, visualizations, and graphics; the alternatives for interaction; and the implications for design from what has been learned about how our minds work.

## 4.2 THE PROCESS OF DESIGN

As stated earlier, good design begins with a thorough understanding of the users, their work, and their environment. Within the human-computer interaction (HCI) and interaction design(ID) communities, the activities that comprise the design of interactions with technological products go by different names depending on the emphasis and which activities are included. User-centered design, interaction design, usage-centered design, contextual design, activity-centered design, participatory design, and cognitive work analysis are some of the names that are used. The emphasis varies: user or use; individuals or groups; work that is service-oriented or work that is mostly in the mind; or interaction with a wide range of technological products and devices or with primarily a computer display and input devices. But regardless of emphasis, these different approaches always begin with a focus on people and what they are trying to do and accomplish. This is the foundation on which the rest of design relies.

Most of these approaches define a process. So that projects do not incur unnecessary overhead, some of these processes such as contextual design can be scaled from small individual projects to large projects that span corporate divisions. The simplest definition of a design process consists of three steps (Johnson, 2010):

1. Analyze the tasks.
2. Design a conceptual model.
3. Design the user interface and interaction from the task analysis and conceptual model.

This captures the basic design activities. Consider the commonly used spreadsheet application, which was initially designed to replace work done manually by accountants. In accounting, the spreadsheet was a "spread" of facing pages in a bound ledger book with columns that, among other things, were used to keep track of expenses by category. Invoices were itemized in the left margin, and the categories were entered in the headings of the columns. In the cell where the row and column intersected, the payment for an invoice was entered. One of the tasks the accountant performed was the summing of columns. In this description of the work, there is no mention of technology, interfaces, or the sequence and flow of the interaction.

The first step of design—the analysis of tasks—is done in the context of the application domain using language that would be familiar to the users—accounting, in this example. Each task, and the steps involved in doing it, must be identified. From the detailed description of tasks, a conceptual model can be designed.

The second step—designing the conceptual model—describes the operations that the system can perform and the concepts the user must know to perform them. In this example, concepts such as "spreadsheet," "column," "row," and "cell" were understood from the ledgers accountants used for bookkeeping, and operations such as "summing a column" was a task they performed. Incorporating these into the spreadsheet application's conceptual model as the objects that can be manipulated and the operations that manipulate them made it easier for those familiar with accounting to learn and understand the system.

In the final step, something concrete begins to emerge from design. The objects and operations in the conceptual model are mapped to actions and to visual signifiers—graphical elements, icons, or symbols—which are representations of familiar objects corresponding to digital representations in the system. In the example, the white space between two vertical lines of the spreadsheet application's user interface was a "column," and it appeared the way it would in a ledger. The terminology used in the system such as "spreadsheet," "worksheet," "row," and "column" would all have been familiar concepts. The mapping also has to specify the actions required to perform a task. In today's spreadsheet applications, two actions are needed to "sum a column." By clicking the mouse on the first cell in the column to be added and dragging the mouse over the cells to be summed, and then pointing the mouse to the $\sum$ symbol in the toolbar, the user invoked the operation that summed a column.

The preceding description is a simplification of what can be a complex process, but it provides a useful outline for design: observe the work to be supported; from what you learn about users and their work, design abstract models of work and systems that keep open the options for implementation; then design concrete alternatives—prototypes—from the abstract models that can be handled and tested by users. Much has been written about the various approaches, and references will be included in the "Further Reading" section of this chapter. The design process shown in Fig. 4.2 is a composite process that includes activities common to some of the design approaches listed earlier. In the following subsections, we provide an overview of the composite process but discuss only those aspects that are relevant to the design of visual interactions for data analysis and exploration. Although the figure shows the process as though it were linear from top to bottom, it is actually highly iterative with feedback and assessments at later stages often requiring changes at one or more earlier stages. The activities grouped as "abstract" are activities done to understand the problem space and what work the system should support. Within these activities, no commitments are made as to how the system will be implemented. The activities grouped as "concrete" are the stages that design how the system will "look and feel" and evaluate how well a proposed solution achieves established goals.

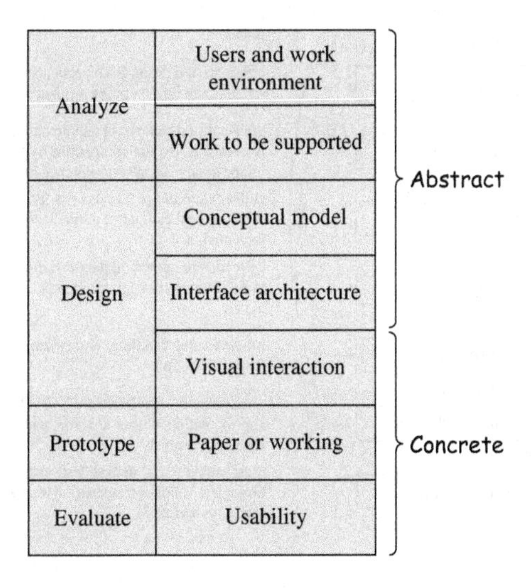


**FIGURE 4.2** A composite design process

### 4.2.1 Analyze

Assuming a preliminary study has identified a problem area that needs to be addressed, how do we design visual tools and systems to improve the ability to analyze and explore the data or see it in new ways? To begin to answer this requires data about who will use the system, for what activities and tasks, and in what context. For example, consider what is involved in the analysis of DNA microarray data. As mentioned earlier, microarray technology is used in several types of scientific studies. Scientists measure the amount of some gene product such as protein or RNA that is expressed by genes in the DNA of a cell. Knowing the level of gene expression in a cell, tissue, or organism provides useful information. Two examples of use are identifying viral infections or exploring the sensitivity or resistance to drugs used for chemotherapy in the treatment of cancer. Figure 4.3 is an overview of the data-analysis process of a microarray study. It is not necessary to understand the details of each step. What is important is to understand that user interfaces, visualizations, and graphics are part of a larger socio-technical environment based on advanced technologies in which work is practiced. Marks on a visualization or data graphic that represent genes with unusual expression patterns may become the basis for a literature search or a search against a database for other related genes. The transition to these related tasks of searching, if not the tasks themselves, is part of the work that must be supported.

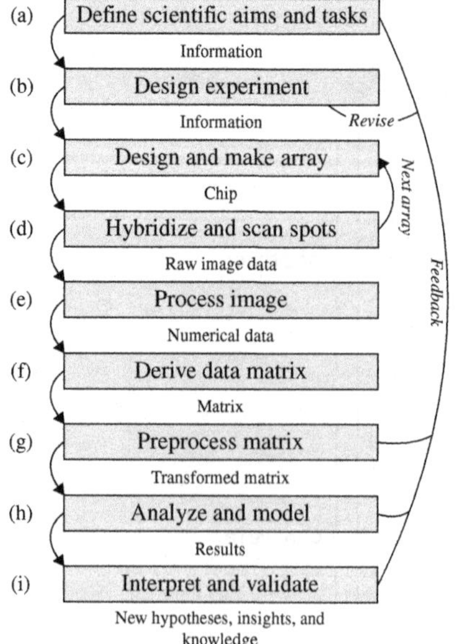


define the problem; generate a list of questions; state the hypotheses; research the literature; text mining

select factors, define conditions; determine replications; define decision criteria and analysis tasks/methods; consider statistical issues

obtain/design probe DNA sequences; arrange probes on array; obtain/track info on probe sequences

get condition/sample; obtain/prepare target RNA; obtain/prepare reference RNA; run competitive hybridization; produce digital image data; track target/reference information

collect/store images; analyze spots; derive numerical measurement estimates; normalize/standardize; track other information

collect/store numerical datasets; integrate numerical data from multiple arrays; integrate any other information; derive data matrix

missing value handling; normalization; transformation; variable/feature selection

visualization; correlation analysis; classification; regression/approximation; cluster analysis; pathway/regulatory network; modeling/analysis

cross validation; statistical test; visual inspection of results; biological validation against existing knowledge and through further experiments

**FIGURE 4.3** Overview of a DNA microarray data-analysis process (Berrar et al., 2003)

Design decisions must be based on facts and details elicited from discussions with users about their work and from observing users at work, instead of, as often happens, from speculation about users' needs. A variety of techniques exist for gathering data from work environments in a systematic way. These techniques include unstructured or structured interviews, focus groups, questionnaires, direct observation, contextual inquiry, and ethnographic interviews. Some of these techniques can provide useful information about users but limited information about how the work is actually done. The work practice has structure, but much of it is implicit, and designers must understand the work at a fine level of detail to design systems to support it. Designers need concrete data about the work as it is actually performed, not abstract summaries of it.

Of the methods previously listed, we have found variations of contextual inquiry (Holtzblatt & Beyer, 1998) to be a useful field method. The structured interviews are done on-site to directly observe the work in context as it is performed. Users, particularly experts, are so skilled at what they do that they often become unconscious of the lower-level steps they take as they perform various tasks. They may try to explain their work to the interviewer in terms they think will be understood. By engaging them in their work and asking about *how* terms, tools, or concepts are used, the details needed to develop the conceptual model will be made explicit in the interviews.

In addition to the users, the workplace (office, laboratory, or manufacturing site) also provides many clues about the work. Artifacts such as notes, documents, research papers, or even yellow sticky notes are signals that indicate the existence of information or knowledge that may be used for certain tasks. By interviewing selected individuals at work and carefully observing, the notes taken during the interviews can be interpreted and used to create models of the work environment, the work, and the resources required to do it.

***The Users and the Work Environment.*** Work is done in an environment that imposes constraints on the design. It is important to know not only the characteristics of the types of individuals that will use the visual system being designed but also other factors that might affect its design.

Some key questions to ask of those who will use the system include the following:

- What are your goals?
- What is the system expected to help you accomplish?
- What tasks are performed? How often are these done? Which are more important and which are less important?
- What is your professional and educational background?
- Does your background include an understanding of data analysis and statistics?
- How do you stay current in your field?

The following questions can be answered through a carefully planned series of *contextual inquiry* interviews:

- What are the steps of each task?
- What data or information will be used during these tasks? Where does it come from?
- What results need to be generated by the system?
- What tools or systems do you use to accomplish your work? How often are these tools used?
- What is communicated with others involved in the work, and who are those individuals?
- What is the relationship between tasks?

Questions about the cultural and physical aspects of the work environment include the following:

- How is the work divided among teams?
- What standards, practices, and policies of the organization might constrain what you do?
- What are the guidelines and standards that IT requires of systems that are deployed?

Part of the information gained by asking these questions of different users allows the designers to create a profile for each class of individuals that will use the system. These can be generalized into *user roles*. For the microarray data analysis example, three roles might emerge: a molecular biologist, a statistician, and a computational specialist who knows data-mining methods. Each will want to examine the results from different perspectives. The results from the contextual inquiries are used to construct models of the work, which are discussed later in this chapter.

***The Work to Be Supported.*** Much of data analysis and exploration is cognitive work. Cognition is what goes on in our minds as we carry out various activities throughout the day. One way to characterize cognition is as a set of processes. In an earlier chapter, we discussed three of these as they relate to vision: attention, perception, and memory. Others include recognition, learning, reading, speaking, listening, problem solving, planning, reasoning, and decision making. Donald Norman's *action theory* (Norman, 2002) is a theoretical cognitive framework that was developed to help designers understand the mental processes of users as they interacted with computational tools. Knowing about this theory helps to understand some of the language used in describing task analysis and conceptual models.

Norman described the interaction as the seven stages of action shown in Fig. 4.4. Our actions begin with a vague *goal* about doing something such as "going to the store." We translate the goal into a specific mental description—an

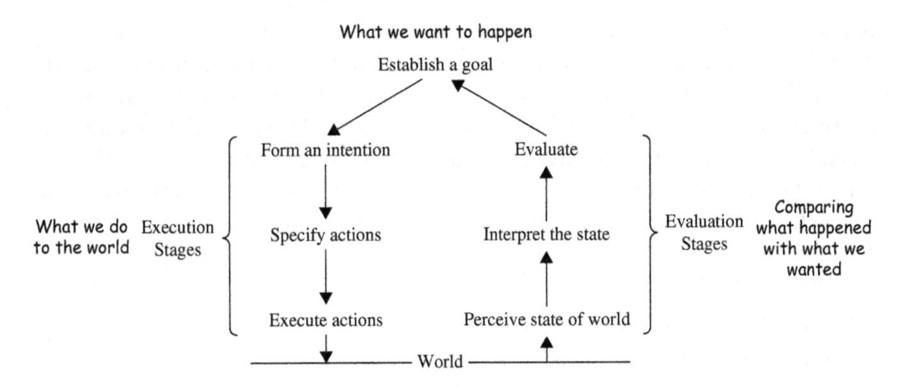


**FIGURE 4.4** Action theory (Norman, 2002)

*intention*—of what is required to achieve the goal, but not yet specific enough to perform. The intention has to be broken into an *action sequence* of physical actions such as motion, movements, or thoughts—fine-grained steps—to carry out the intention of getting to the store. The first three stages formed the *stages of execution*. The remaining stages, the *stages of evaluation*, compare what we perceive to be happening against our expectations—the intention and goal.

Norman described two things that could make a tool difficult to use. He called these the *gulf of execution* and the *gulf of evaluation*. The gulf of execution was the difference between what the tool user wanted and the actions a tool provided to carry out the intention. If the action sequence was long—in other words, it could not be directly executed—the gulf was large and would require considerable cognitive effort to translate the intention into the steps required to carry out the intention. The gulf of evaluation was a measure of how easy it was to perceive and interpret the visible feedback provided by the system in response to the actions the user took. If the representation for the state of the tool—what was visible in the interface—closely matched the user's mental model, the gulf of evaluation was small, and the tool could be easily learned and understood.

Understanding work requires an ongoing dialogue with those who do it. To design the system's operations to support the work, the tasks the user performs needs to be understood as a series of small steps. The contextual inquiry method is designed to elicit this information, but less formal methods may be used as well. Regardless of method, the notes taken during the interviews need to be interpreted and converted into models of the work.

Various models have been devised to capture different aspects of the work. In the contextual design methodology, the models include *sequence models*, which represent the detailed steps required to achieve an intent; *flow models*, which represent the coordination and communication of work between individuals; and *artifact models*, which represent physical things—documents, reports, important research papers—produced or used in the work. In the usage-centered design methodology (Constantine & Lockwood, 1999), *essential use cases* capture the detailed interaction between user and tool as a structured narrative of action and response. These are written in language the user would use and are generalized to capture only the intent of each action but not references to specific technology. This gives designers the freedom to consider implementing the actions of the task in other ways.

The starting point for the development of essential use cases is the user role. For each role, the following questions are asked of actual users:

- What is the goal?
- What must be done to achieve this goal?
- What capabilities are needed to do the task?

The fundamental question in task analysis asked of each action is "why?" Why is this being done? What is the intent? What is it accomplishing?

Note that it is often difficult to schedule interviews with domain experts because of the demands on their time. It is helpful to have someone on the design team who has an equivalent academic or professional background. This person will be able to ask focused questions relevant to the domain and explain the expert's answers later. There are also other ways to learn about the work. If the expert has some familiarity with the design methodology, he or she may write task descriptions as a starting point for further discussion. Although their task descriptions may include references to specific tools or visualizations, these experts provide insight into goals and intent. An example is shown in Table 4.1.

Another way to learn is by asking an expert to identify a set of key technical or research papers that are relevant to the work the tool will support. These can provide the background necessary to generate focused questions to ask during interviews. For scientific areas, research papers are structured to describe the problem, provide historical background, and explain the steps taken to solve the problem. Many also include a methods section that describes the work in more detail.

Analysis is iterative. The construction of models exposes gaps in understanding that lead to other questions. What was thought to be understood often requires confirmation or clarification. From the interviews and discussions, every key observation, insight, question, design idea, and breakdown—places in the workflow where problems arise—are also recorded. The results of analysis may contain the notes (or a model of these notes) and at least two key models: profiles about the types of individuals that will use the system (user role models) and the tasks they currently perform to do the work (task models).

### 4.2.2 Design

Design begins by developing an abstract conceptual model of the system after the analysis is well underway. From the conceptual model, an interface architecture is designed that groups related actions the user may perform into interaction spaces and shows how the user will navigate between these spaces.

**TABLE 4.1 A Task Description Written by an Expert**

| **Task-1: Group cells based on gene/miR expression.** |
|---|
| 1. Cells are grouped based on the correlation between the expression patterns of selected genes, microRNAs, or a combination of the two, and then visualized as a dendrogram. |
| 2. (optional). Filter genes/miRs (rows of T) so that X% of the values have expression levels greater than Y; for example, 10% have expression levels greater than 7. |
| 3. (optional). Select the *n* top genes with the highest variability in expression level, for example, the highest interquartile range or highest variance. |
| 4. Calculate the $60 \times 60$ cell–cell correlation matrix M based on expression levels of genes/miRs from step 2. Provide options for Pearson, Spearman, and Kendall correlation statistics. There is no missing data for genes/miRs, so that is not an issue. |
| 5. Using agglomerative clustering with $D = 1 - M$ as the distance matrix, construct a dendrogram of the cell lines. Provide options for complete and average linkage. |
| 6. Display the dendrogram with options for coloring the cell lines based on tissue type *as shown in Fig. 2 of [MCT07 1483]*. (Note that the italicized remark does not refer to a figure in this book, but to a figure in a paper containing a visualization the expert is familiar with. Before considering other approaches, it is necessary to understand the intent behind the use of the dendrogram in that figure. The paper referred to is an artifact that provides additional context for understanding the work.) |
| 7. Provide an interactive device for selecting cell line groups. |

The visual interface and interactions are designed from the various analysis and design models.

***Conceptual Model.*** Recalling the gulfs of execution and evaluation from action theory, the design goals should make operations available that are close to what the user intends to do and should make the state of the system visible in a visual language that can be easily understood. The graphical user interfaces of most applications communicate a world of digital objects on which actions can be performed; the interfaces embody concepts and relationships between the objects. The conceptual model is a model of the system that the designers hope the users will form as their mental model as they interact with the system. If the model reflects the users' world of work, it will be familiar. Designing the conceptual model involves asking the following questions of the task models produced in analysis (Johnson & Henderson, 2002; Johnson, 2008):

- What concepts should be presented to the users?
- What data will be manipulated, created, or viewed through the visible representations drawn on the display? What actions will be used to do this?
- What options, attributes, or parameters does the tool need to provide?

Applying this to the previous task description, Table 4.2 highlights in bold all the concepts that may need to be presented somewhere in the user interface

**TABLE 4.2 Concepts Identified in a Task Description**

| **Task-1: Group cells based on gene/miR expression.** |
|---|
| 1. **Cells** are grouped based on the **correlation** between the **expression patterns** of selected **genes, microRNAs**, or a combination of the two, and then visualized as a **dendrogram**. |
| 2. (optional). **Filter** genes/miRs (rows of T) so that X% of the values have **expression levels** greater than Y; for example, 10% have **expression levels** greater than 7. |
| 3. (optional). **Select** the *n* top genes with the highest **variability** in expression level, for example, the highest **interquartile range** or highest **variance**. |
| 4. Calculate the 60 × 60 cell–cell **correlation matrix M** based on expression levels of genes/miRs from step 2. Provide options for **Pearson, Spearman,** and **Kendall correlation statistics**. There is no **missing data** for genes/miRs, so that is not an issue. |
| 5. Using **agglomerative clustering** with D = 1 − M as the distance matrix, construct a **dendrogram** of the **cell lines**. Provide options for **complete and average linkage**. |
| 6. Display the **dendrogram** with options for coloring the **cell lines** based on **tissue type**. |
| 7. Provide an interactive device for selecting **cell line groups**. |

**TABLE 4.3 List of Concepts from a Task Description**

cells (biological), correlation, expression pattern, gene, microRNA, dendrogram, filter, expression level, select, variability, interquartile range, variance, correlation matrix, Pearson correlation statistic, Spearman correlation statistic, Kendall correlation statistic, missing data, agglomerative clustering, complete linkage, average linkage, cell line, tissue type, cell line group

when it is designed in a later phase. From the contextual inquiries done during analysis, the designers know that the datasets being manipulated in the task are one of two kinds that are generated from carefully controlled microarray experiments. One dataset contains the expression levels of genes, and the other contains the expression levels of microRNA. The task would be performed in step (h) of the data analysis process shown earlier in Fig. 4.3.

The list of concepts extracted from the description is shown in Table 4.3. Much of the terminology is from biology and statistics and is used to describe the analysis of microarray data.

The next step is to carry out an object/action analysis and ask which of these concepts to expose in the interface as *objects*, *actions* that the user can perform on the objects, *attributes* of objects (that can be set somewhere in the user interface), and various *relationships* between objects. Different kinds of relationships can exist between objects: in a *supertype/subtype* relationship, one object is a specialized type of another (vehicle/auto); in a *whole/part* relationship, one object is a part of another (car/wheel); and in a *containment* relationship, one object is inside another (document/folder).

An explanation of a full object/action analysis requires a background in data mining, statistics, biology, and the problem domain of chemical genomics,

**TABLE 4.4 Results of an Object/Action Analysis**

| | |
|---|---|
| objects | gene dataset, microRNA dataset, gene, microRNA, dendrogram, correlation matrix, cell line, tissue type, dendrogram, variability |
| actions | filter, correlate, cluster, group |
| attributes | interquartile range, variance, Pearson correlation, Spearman correlation, Kendall correlation, agglomerative, complete linkage, average linkage |
| relationships | correlation method: Pearson, Spearman, or Kendall<br>clustering method: agglomerative clustering<br>linkage: complete linkage, average linkage<br>microarray dataset: gene dataset, microRNA dataset |

which is beyond the scope of this book, but an initial analysis might generate the results shown in Table 4.4 for just one of many tasks. The design challenge is to keep the conceptual model as simple as possible. Actions should be generalized to apply to as many objects as possible. For example, in step 3 of the task, "Select the n top genes . . . ," the selection is really another way to filter, so a separate action is not needed as long as the filtering action provides a way to select the "top n genes" using some metric for variability.

The conceptual model has several functions:

- *Identify and structure the concepts of the domain and the work before considering how these will be presented.* A conceptual model is needed to design the interface architecture, the next step of the process.
- *Serve as a reminder to software developers that the interface is a way to communicate with users.* It provides a vocabulary for the user interface. By creating the conceptual model from the task models, which are derived from discussions with users, the user interface will embody the concepts of the workplace rather than concepts software developers generate when they implement the system.
- *Measure the complexity of the interface.* The more objects, actions, and attributes that are added, the more the user will have to learn and the more combinations of objects and actions there are to consider. Each new action added could be applied to any of the objects that already exist in the model; similarly a new object could be operated on by any action that already exists. The complexity grows exponentially.

Not all design methodologies include conceptual modeling. Contextual design uses *visioning* and *storyboarding* to conceptualize alternatives for new ways to do the work by looking across all abstract tasks and issues extracted from the interviews to see what they have in common or where they differ. The storyboards provide the detail needed to either implement paper prototypes that are used to test the ideas with users (for small projects) or begin the design of the interface architecture (larger projects).

***Interface Architecture.*** The objects, actions, and attributes that the system will make visible as tools and materials for doing the work have been identified in the conceptual model, but we are still not ready to begin sketching something that will be visible. None of these tools or materials have yet been organized. Just as physical work is done more efficiently when the tools and materials relevant to several tasks have been organized and laid out in different places, cognitive work is also done more efficiently if the visual representations of objects and actions are organized into *interaction spaces* where related tasks are performed. Interaction spaces in user interfaces are called windows, panels, dialog boxes, pages, views, wizards, and so on. If a related task cannot be done in the same interaction space, it distracts the user from the work and forces the user to think about an action sequence that will help navigate to a different space where the related task can be done. The organization and structure of actions and objects into interaction spaces along with the specification of links that show how to navigate between them is called the user environment (Holtzblatt & Beyer, 1998) or the interface architecture (Constantine & Lockwood, 1999). Like the architectural floor plans for a home, the interface architecture specifies the objects and actions that belong in the same space because they are frequently used together.

This is best illustrated with a commonly used application. Consider Gmail™ (also known as Google Mail). An application's interface reflects its internal structure. Three of Gmail's interaction spaces have been extracted as shown in Fig. 4.5. In each interaction space, the functions and objects needed for a set of related tasks are defined, as are the links to other spaces. For the system to be coherent, it not only must have a consistent user interface but also an orderly flow of interaction. Using an interface architecture to design the visual interactions results in a system where the interactions flow naturally with the work.

The analysis and design to this point have generated various abstract models, including two abstract design models: the conceptual model and the interface architecture. In the next stage of design, the abstract becomes concrete.

***Visual and Interaction Design.*** To make the abstract concrete, the visual interfaces and interaction must be designed and prototyped. User interfaces and visualizations are composed from graphical elements (such as points or lines) with visual properties, or higher level elements such as predesigned or customized components that include controls (radio buttons, menus, hyperlinks) and containers (panels, windows, dialog boxes, pages). Graphics are composed of marks, scales, and guides organized by coordinate systems or layout schemes as described in Chapter 3. Although the interaction style could have been one of several different types, we assume interactions will directly manipulate visual representations of digital objects. Finally, cognition and perception are a factor to consider in whatever is designed. We will cover all of these topics in more detail in the last section "Visual Interaction Design."

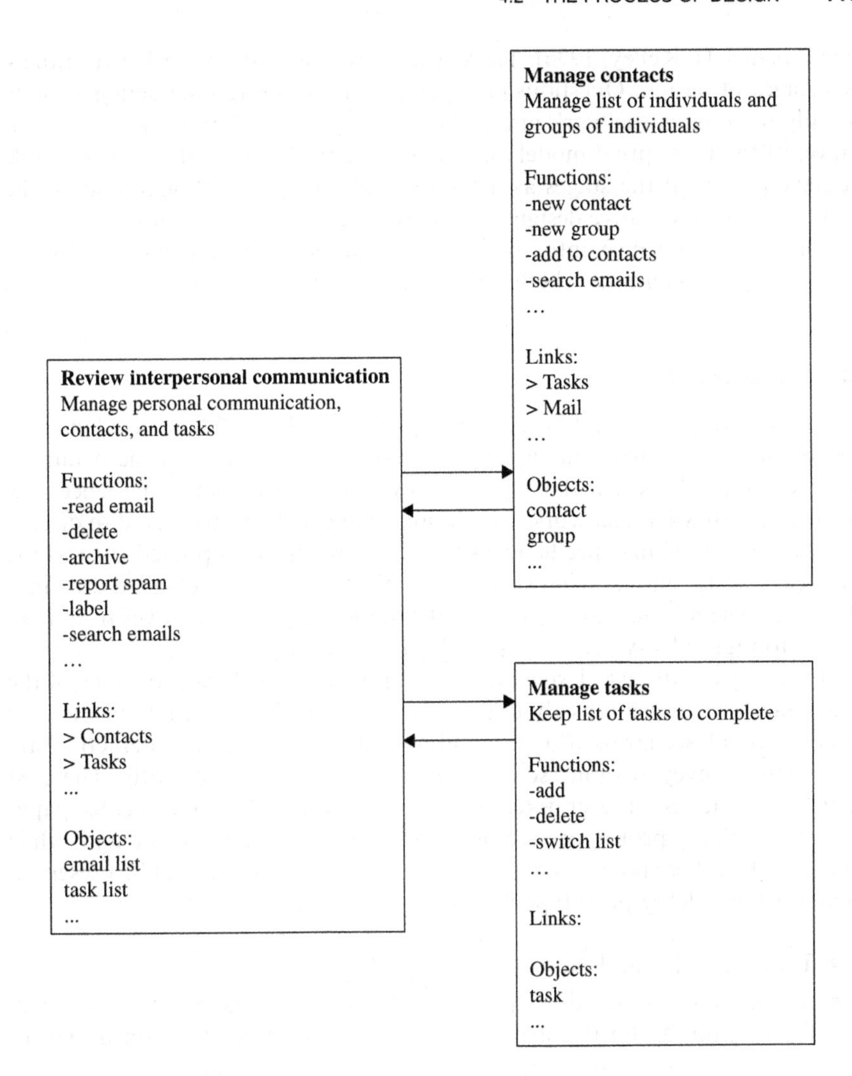


**FIGURE 4.5** Example of an interface architecture for Gmail (see the User Environment Design model in Holtzblatt & Beyer, 1998)

The various work models, the conceptual model, and the interface architecture should be seen as a specification for the physical user interface design. Using these models as a guide, different ideas could be sketched as alternatives or implemented as low-fidelity prototypes (see the next section). The models keep the physical design decisions grounded in the data gathered from users and the workplace.

Note again that there is considerable back-and-forth activity between the abstract and concrete stages of design. Design is not only problem solving. Design is also "creating beyond what the problem calls for" to find a good fit to the users' needs (Kelley, 1996). There is no hard-and-fast border between these two stages of design. Questions emerge in trying to design the details of how visually to support the work and in prototyping that will result in a reexamination of the conceptual model and the structure of the interface. Also, while we have presented the stages as if there is only one set of design models, the exploration of alternative designs might give rise to several sets of models, each of which may advance into the concrete stages of design and early-stage prototyping to determine whether they have merit.

### 4.2.3 Prototype

Prototypes are like the architect's drawings for a building. They are a language for communicating with the user about design alternatives and the details of interfaces and interaction. They provide a way to envision and experience how the new system will affect work, to explore technical alternatives, and to discuss flaws in the design and specific ideas for how it might be improved. Just as the architect's design moves from sketches of initial concepts to detailed drawings of floor plans and site plans to blueprints, the abstract design moves from low-fidelity to high-fidelity prototypes (or beta software).

Other ways to discuss design ideas with users include demonstrations of the visual interface, discussions about the written descriptions of requirements, and descriptions of scenarios. But ideas about visualizations and interaction are difficult to convey without something graphical to interact with. The first design alternatives of user interfaces are sometimes implemented as paper prototypes. Paper prototypes are one kind of *low-fidelity prototype* and, while there may be other choices, they have several qualities that should be present in whatever low-fidelity prototype is used:

- They are quick and inexpensive to build.
- Because they are hand-crafted and rough, discussions in early design will focus naturally on the structure of the system rather than on irrelevant interface issues such as the style of a particular user interface control.
- They are easy to change during the prototype interview in response to a suggestion or feedback.
- They can be used as a prop to discuss the details about what is needed, or what will or won't function in the work environment. Some issues do not surface until users need to reflect on the details of how something works.

Low-fidelity prototypes are sketches that include controls such as buttons, menu items, hyperlinks, or text fields. The user will simulate the actions of a real system by mentally pointing to controls or typing into fields. One or more interaction spaces may be mapped to a single "page," depending on the kind of user interface container selected. The sketch may be hand-drawn on a piece of paper with colored sticky notes simulating controls, or they may be digitally drawn as slides in Microsoft PowerPoint® or as artwork created by Adobe Illustrator®. The digital sketches may be saved as a file of slides, as pages in a PDF file, or as HTML pages.

*High-fidelity* prototypes are working software with interfaces and interaction. They cost more to build and are less flexible than low-fidelity prototypes. They are used to investigate technical issues that might affect design. For example, some interaction styles require the system to provide very responsive feedback. To measure this, certain operations may need to be implemented to determine whether the time criterion can be met. Designers might implement all or some of the interface and interaction to be experienced but with limited functional capabilities. In *evolutionary* prototyping, the high-fidelity prototype eventually becomes a product. A high-fidelity prototype acts as a working prototype where core sets of functionality are developed in stages. The users continue to evaluate this prototype as it evolves. The outcome of this later stage of design is a proof-of-concept prototype that can be used to demonstrate the merit of the initial design concept. In *throwaway* prototyping, the high-fidelity prototype becomes the specification for a product and is eventually discarded.

### 4.2.4 Evaluate

The goals of a usability evaluation for design differ from the goals for testing usability of preproduct applications. Design evaluation is intended to provoke discussion about better ways to structure the system, discover unnecessary tasks, or reprioritize the importance of tasks. It is iterative and must be a fundamental part of the design process from the beginning. Prototyping may be thought of as a series of interviews similar to contextual inquiry but with a focus on observing design ideas in use.

Usability testing, on the other hand, is done in the late stages of development. It measures the users' performance on a set of predefined tasks. It is intended to uncover small problems or areas that are found to be difficult to understand. The changes made are to polish the interface and refine the product by improving the interaction. If the prototyping has been done well, there should be no major surprises. Usability testing is important but beyond the scope of this book.

## 4.3 VISUAL INTERACTION DESIGN

Physical design, whether it is of something tangible such as a fountain pen or intangible such as a visual system, requires many choices and trade-offs. How these are made depends on what matters. We react to the design of a product on three levels (Norman,2004):

- At the *visceral level*, which involves the emotional system and does not involve thought or consciousness, we respond to appearance and the physical feel of the product. This is often discounted as just aesthetics, but the emotional state of mind is a factor to consider in design. Positive feelings improve creativity and broaden our thinking as well as helping us be more tolerant of minor problems in the tool; negative feelings from stressful environments make us concentrate and narrow our thinking. For complex or stressful environments, the design focuses on function and the removal of anything in the interface or interaction that is irrelevant.

- At the *behavioral level*, we respond to what the tool does, how well it performs, and whether it can be understood and learned. Designing for this level is primarily about function, followed by how to make it understandable.
- At the *reflective level*, we respond thoughtfully and consciously and reflect more deeply on past experiences, what we have learned, and the culture we live in. We consider the strengths and weaknesses, how it might be used in the work, and many other factors.

Visual tools are semiotic systems—communication-oriented tools designed for visual interaction. The graphical primitives and higher-level elements created from these primitives are deliberately designed signifiers that represent quantitative or abstract data, information, relationships, digital objects that can be manipulated, or actions that behave like tools. To understand their meaning, users must understand the connection between the signifier and what it represents. Well-designed signifiers are instantly perceived and can be easily "read." The signifiers used to represent the content and functionality of the system must be coherent—have consistent visual characteristics and style—and be understandable by the user community they are designed for. The signifiers in Fig. 4.6 are understood because they have been learned by using graphical user interfaces. A visual language is comprised of the visual properties (color, size, shape, etc.) of a set of signifiers that are related to each other by a set of rules.

Because "user interfaces," "visualizations," and "graphics" are terms loosely used in many contexts, we provide definitions for them here. We use the term "visual interface" (Mullet & Sano, 1995) instead of "(graphical) user interface". *Visual interfaces* organize content and tools so that users can efficiently do the activities or tasks that the system is being designed to support. They have *controls* that provide ways to perform actions and *containers* that provide space—regions of the display—for either content or groups of controls. To support the activities of users, the design captured in the various models produced by the analysis of the work must be translated into a physical structure of windows, pages, dialog boxes, and so on, along with ways to

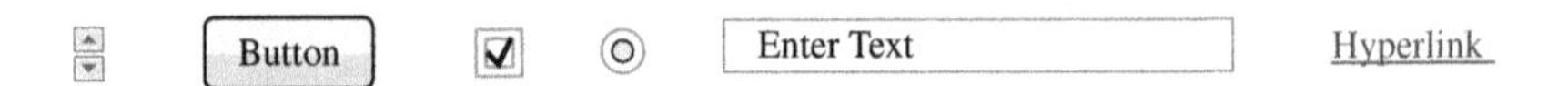


**FIGURE 4.6** The controls of a user interface are signifiers for actions

navigate between them. (Note that Web applications refer to a physical interaction space—a container of content or controls—as a "page," and desktop applications refer to these spaces with terms such as "windows" or "dialog boxes." We will refer to all kinds of containers simply as pages.) Various interface schemes for doing this are discussed in the upcoming "Visual Interfaces" section. *Visualizations* communicate information using graphical representations. Examples of visualizations include charts, diagrams, tables, guides, directories, and maps. *Graphics* (Wilkinson, 2005) are the visual representations of graphs derived from statistical data. Visualizations and graphics may both have visual interfaces that allow the user to interact with the abstract or quantitative data they represent. (Note that "graphic design" is used to describe the art of communication used to create visual messages—advertising—that educate, inform, promote, and persuade people to buy products. The principles of graphic design are used in visual interfaces and in graphical design.)

The physical design of visual systems that combine visual interfaces with visualizations and graphics must consider several dimensions to know how the users will interpret the messages communicated by the system. The first three subsections discuss the guidelines, principles, and design patterns of visual interfaces, visualizations, and graphics. Each subsection includes comments about perception and cognition that constrain the design. The final section discusses real-time constraints imposed by cognitive and perceptual processes.

### 4.3.1 Visual Interfaces

Visual interfaces organize content and tools so that users can efficiently do the activities or tasks that the system is being designed to support. When we open an application, the first thing we perceive is its visual interface. One of the first steps that must be taken is to translate the interface architecture into a physical structure of pages.

***Organizing the Application.*** Many different physical structures can be designed, but those commonly found in general applications are variations of one of the three schemes shown in Fig. 4.7. Multiple and tiled windows are found in desktop applications, while single-paged windows are found in browser-based applications.

In choosing a scheme, it is important to be aware of the limitations imposed by human attention. The brain has several mechanisms for attention and can make only a handful of items available to cognitive processes (such as problem solving). These items are indexed by our perceptual systems and whatever long-term memories have been activated by the focus of attention. After our attention shifts, this set is replaced by another, and the first set is forgotten. Tasks may require that something in one panel be referred to while content in another panel is modified or that a set of objects in one panel be compared with the objects in another. Interrupting the focus by requiring the user's attention

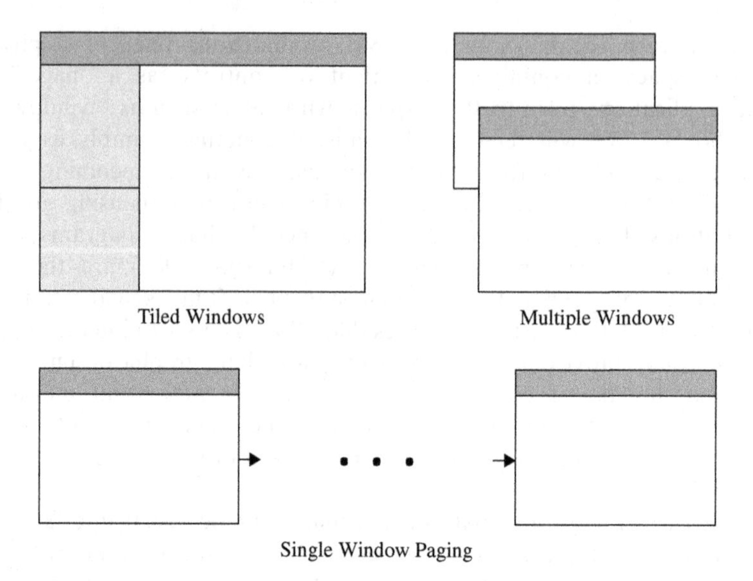


**FIGURE 4.7** Different schema for physical structures of a user interface

to move elsewhere disrupts the flow of thought. The data and the kinds of tasks to be performed determine which physical structure to use.

***Navigation.*** In complex applications, not all the functionality can be accessed from a single page. The interaction spaces of the application may be allocated to a collection of pages with various tools and views of the underlying information or data on each page as shown in Fig. 4.8. But the user must know how to get to where they need to be to perform the sequence of actions that pertain to the goal—and how to get back. This is the problem of navigation that requires organization and navigational aids to allow users to move between interaction spaces.

Navigation has a cognitive cost. Each transition to a new page is a switch in context requiring the scanning of that page to understand its structure, content, and exits. The ideal is to not require navigation—to have all the content and tools directly accessible—but the trade-off is complexity. And, in many cases, direct accessibility is not possible because there is simply too much information (or too many different actions) to design them to be all in one physical space. Other constraints result from perception and cognition.

The visual system is cued by the goals of the task. We notice things that match the goal and often don't "see"—become aware of—what is irrelevant. Further, the brain stores information in long-term memory by activating large numbers of neurons that effectively distributes what is being perceived as patterns across memory. Recognition is easier and faster than recall. When we see something familiar, it activates overlapping patterns with what has already been stored, which allows for quicker access to long-term memory. Recall

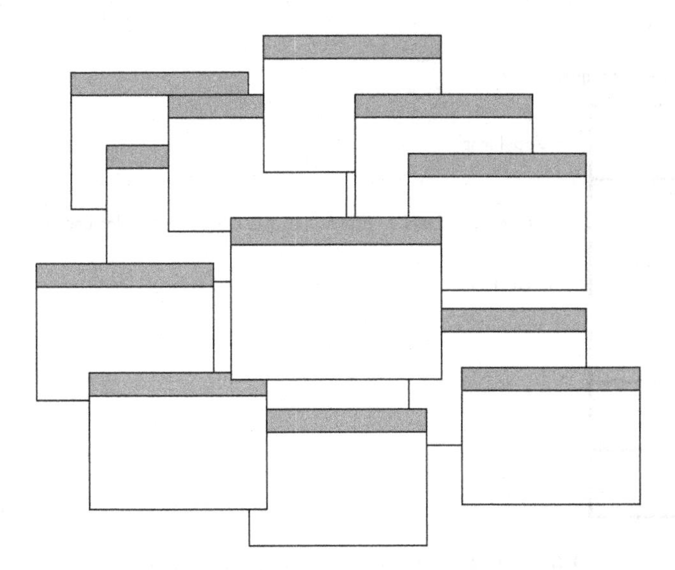

**FIGURE 4.8** A complex application structure that requires organization and navigational aids

results in a slower search through memory. We memorize and prefer familiar paths that let us accomplish a task, even if there are other ways it could be done more efficiently.

Designers use signage, maps, and clues to help users know what path takes them to the place where the next required action can be performed and to keep them from getting lost. Uniformly organized, consistently placed, clear, unambiguous navigational signifiers act like highway signage (e.g., mile markers on highways give an approximate location); and when decisions must be made, overhead signs announce major intersections, and large, clearly labeled exit signs provide the name of the town or number of the highway you will be entering. The design of the interface architecture will already have functionally structured the application so that related tasks are in the same interaction space. In the mapping of the spaces to the physical structure, the goals at each decision point must be kept in mind. Fig. 4.9 shows some of the navigational signage for Gmail. The navigational structure is uniform and remains constant throughout navigation. The global map provides clear entry points to major activity areas of the application, and the local map provides entry points to interaction spaces within an activity area. The content—but not the placement—of the local map changes when moving from "mail" to "contacts."

Because of the cost of switching contexts, designers try to minimize the steps in the path. The interaction architecture is critical in ensuring that common and frequently performed tasks do not require movement through several interaction spaces. In the local maps of the mail and contact interaction spaces of

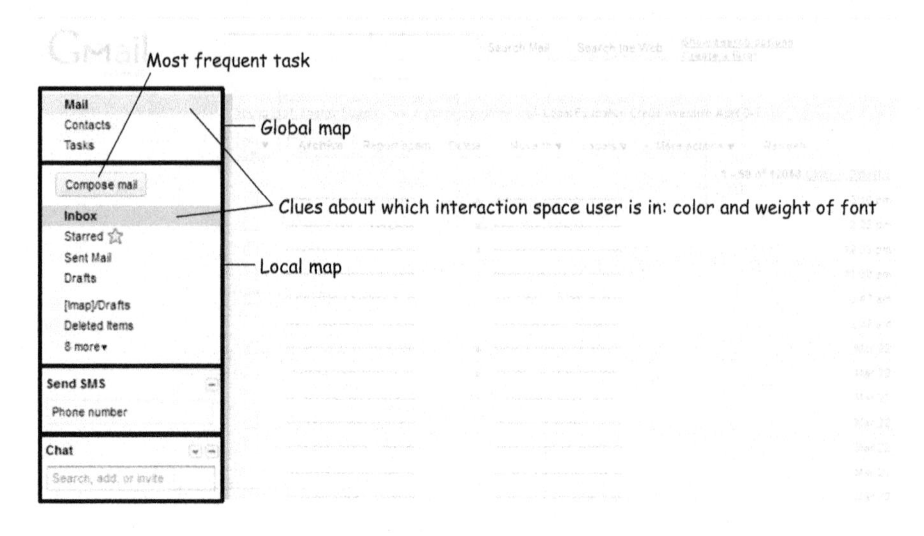


**FIGURE 4.9** Navigational signage for Gmail

Gmail, for example, a button is provided for the most frequently performed tasks: "Compose Mail" (shown in Fig. 4.9) and "New Contact" (in the contact interaction space that is not shown), respectively.

***Organizing the Page.*** As was discussed in Chapter 2, more information is in a scene—or a page in the context of an application—than can be perceived at one time. The visual system uses various search strategies to sample and scan to find what is relevant to the current goal. We perceive the fine detail of a page only through the fovea, a very small region near the center of the retinal visual field. The peripheral vision has such low resolution that its function is to primarily provide cues that guide eye movements toward what is interesting: motion, fuzzy shapes, brighter colors, or large features. The eye is optimized to see structure, and designers exploit this to convey meaning, establish a sequence for the eye to follow, and create focal points of interest. Page layout uses techniques from graphic design to create a *visual hierarchy* of the content that gives weight to what is most important and a *visual flow* that leads the user through the page.

A visual hierarchy structures the content. The user should be able to infer the visual hierarchy of a page from its layout. Titles should be apparent. The most important content should be prominent, and less important content should appear as secondary regions. Techniques that can be used to create a visual hierarchy include the following, some of which are illustrated in Fig. 4.10:

- *Upper-left or upper-right corner.* The eye will begin to scan a page in the same place that it does for reading the text of the primary language. In Western cultures, the scan will begin in the upper-left corner.

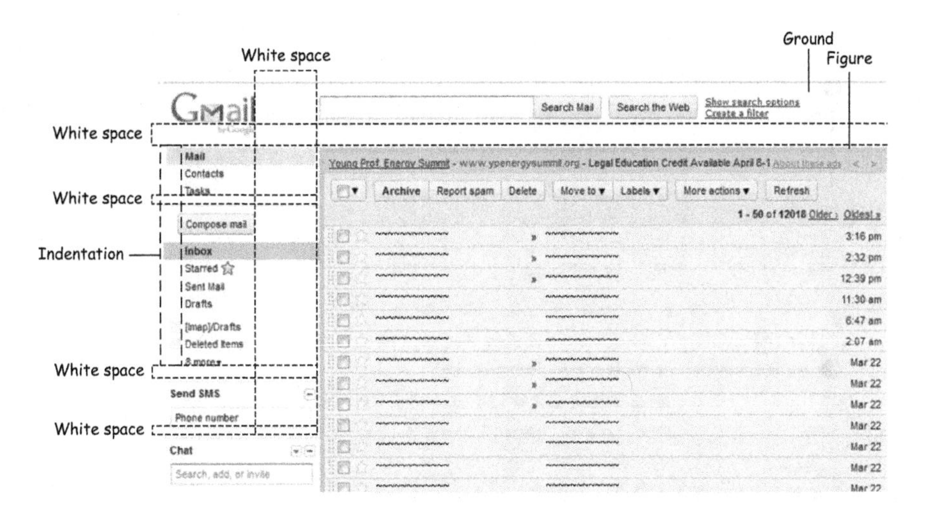


**FIGURE 4.10** Techniques for creating a visual hierarchy

- *White space.* White space is an important element in constructing a hierarchy, and its uses include separating regions or establishing hierarchical order, as shown in the example.
- *Size and weight (boldness) of fonts.* The font of more important information is given a larger size or greater weight. In the example, "Mail" and "Inbox" are emboldened to identify which application and which subset of email threads are being viewed. "Send SMS," and "Chat" are titles of major sections.
- *Contrasting colors for* figure *(the foreground) and* ground *(the background).* In the example, the white ground and the gray figure (or blue if being viewed in a Web browser) separates the navigational and search areas from the content. A darker shade of gray is the ground for the email content activity area. Lighter shades of gray separate the actions that can be performed from the email headlines that comprise the content.
- *Positioning.* In the example, the large "Gmail" lettering placed in the upper-left corner will be seen first in Western cultures. It informs the user of the application in use, and it also ensures that the branding of the application won't be missed.
- *Alignment.* Alignment is used to show a set of related items. "Mail," "Contacts," and "Tasks" are different activity areas, and "Inbox," "Sent Mail," and so on are different subsets of email threads.
- *Indentation.* Indenting text implies that it is subordinate to what is above. In Fig. 4.10, all the hyperlinks from "Mail" to "Deleted Items" are indented to show their relationship to the major activity area for email as opposed to texting ("Send SMS") or chatting ("Chat").

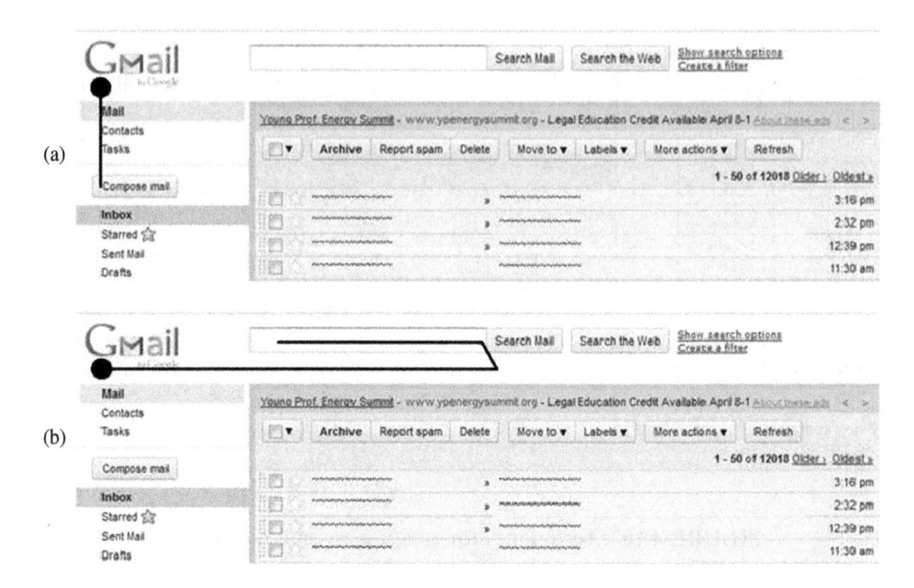


**FIGURE 4.11** Possible visual flows for composing or searching mail

Visual flow is designed to manipulate the path the eye takes as it scans the page. A well-designed visual hierarchy incorporates focal points—graphical features that emphasize important elements—that lead into secondary regions of less important content. The design of focal points relies on the way the visual system processes certain graphical images. Recall from Chapter 2 that when attention is not focused on something specific, visual features with certain properties appear to "pop out" given the right conditions. Graphic designers use visual properties—lightness, color, orientation, texture, shape, size, and motion—combined with other techniques to attract the eye. (Well-designed advertising in high-quality publications provides good examples of the techniques.) The techniques include the use of white space, contrasts of color and lightness of shapes or weight and size of fonts, and converging lines or hard edges. Your gaze follows the focal points from strongest to weakest. The focal points can be overridden by the goal of the current task, the meaning in content we see, or the natural tendency to scan a page as if we were reading text. Fig. 4.11 shows possible visual flows for composing an email (a), and entering a search query for an email (b). The large, appropriately labeled buttons provide a focal point for actions that can be performed.

Visually grouping and aligning content elements indicate that they are related to each other. As shown in Fig. 4.12, four methods based on the Gestalt principles (discussed in Chapter 2) of perceptual organization are used to convey which content elements are related: similarity, proximity, continuity, and closure. In all of the methods except the one using similarity, the use of white space provides clarity by separating the clusters of related items. These

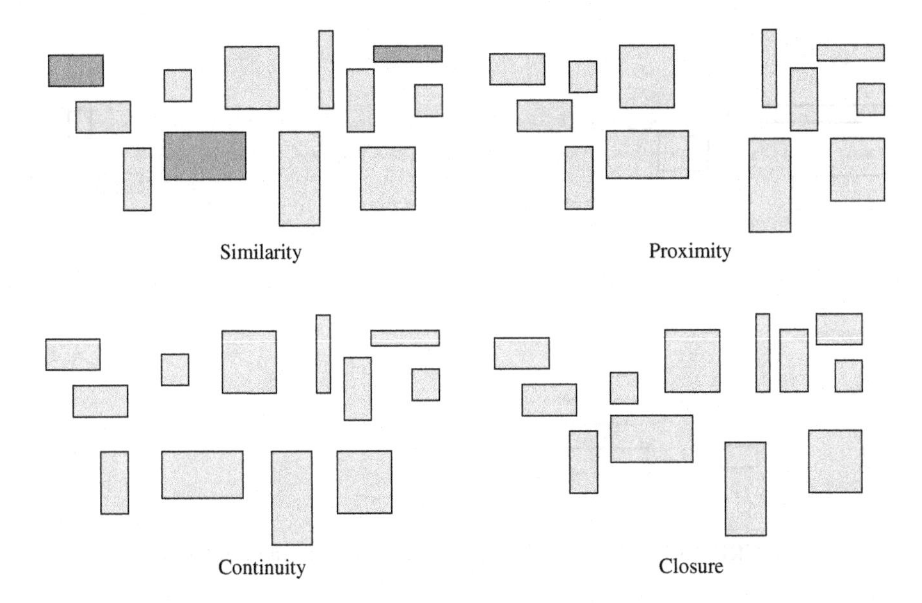


**FIGURE 4.12** The four Gestalt principles used for grouping and alignment

methods are used in Gmail, for example, to group the set of buttons that provide ways to manipulate the email threads, a set of related hyperlinks that signify folders, and the search button with its query input field.

***Organizing the Actions.*** The design of the interface so far has focused on how to create the physical structure for interaction spaces and how to present the content so that it is informative and can be quickly perceived. But to make the system fully interactive, it must be capable of taking input from the user. Although the interaction style could have been one or more of several different types (instructing, conversing, manipulating, and exploring), we will focus on what are called *direct manipulation interfaces* that are prevalent in advanced graphical user interfaces.

The first method by which we can invoke actions in the system is through visible things in the interface which we will call *visible objects*. Visible objects are controls and visual representations of digital objects that are capable of providing feedback when they are pointed at or prodded (clicked) by the mouse. The controls and visible objects are signifiers of actions that the system is capable of performing. But when the user first looks at an interface without moving the mouse, such as Gmail's interface in Fig. 4.13, how are the action signifiers discovered? And what do they do? Some of the possible action signifiers in the Gmail interface have been enclosed in black rectangles, and all but one represents an action.

Some signifiers are recognized immediately by convention and experience. Users have learned that buttons (rounded rectangles), hyperlinks (underlined

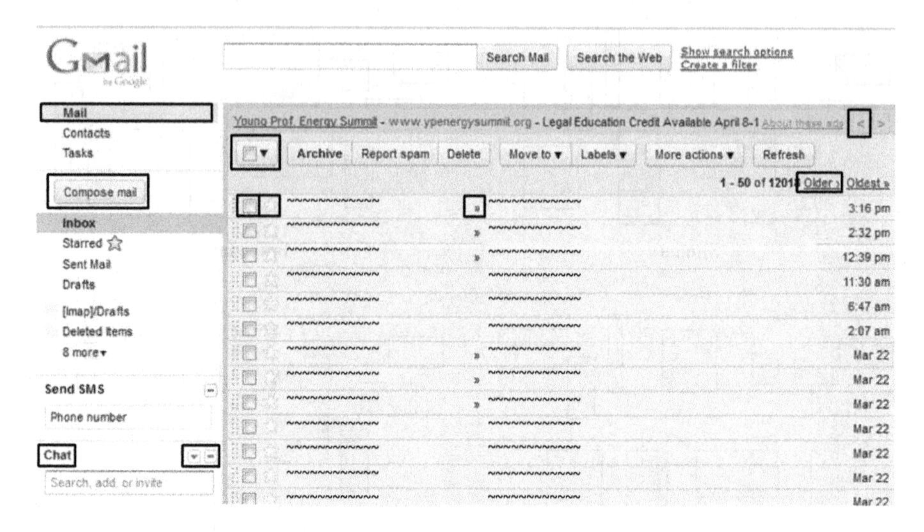


FIGURE 4.13 The possible action signifiers in a Gmail interface

text), and menus (buttons that contain upside-down triangles) are actions. These are easily identified in the Gmail interface. Style guides written for application developers for each computing platform prescribe the visual appearance and behavior of a standard set of controls. The controls in these guides should be the first source considered as signifiers of actions because the way they look and how they behave will be familiar to users.

Controls or visible objects are also designed to change appearance when the mouse rolls over the region in which they are drawn or if the state of the digital object changes when the action is invoked. In the Gmail interface, the borders of buttons are darkened, the background of the folder labels changes color, and for all controls but the checkboxes in the email, the mouse pointer symbol changes from an arrow to a hand. Selecting an email thread by clicking on the checkbox results in a change in color of the background for that email thread. The star action is an exception to these guidelines. It provides no visible feedback unless the user clicks the star. There is no way to tell it apart from the chevron (">> ") in the email thread headline which is not an action.

The visible action signifiers—buttons, hyperlinks, and menu items—provide a way to learn the actions available in the system. Users will often explore these to see what is available. There are also invisible actions that cannot be discovered: combinations of keystrokes, drag-and-drop operations, and double-clicking or right-clicking on visible objects. Users often expect these actions and have learned them from outside the interface.

A variety of organizational strategies are used to group actions. These include menu bars, tools bars, and ribbons. The controls in toolbars, such as the one from Gmail shown in Fig. 4.14a, are always visible and directly accessible. In complex systems, when the number of actions is large enough that

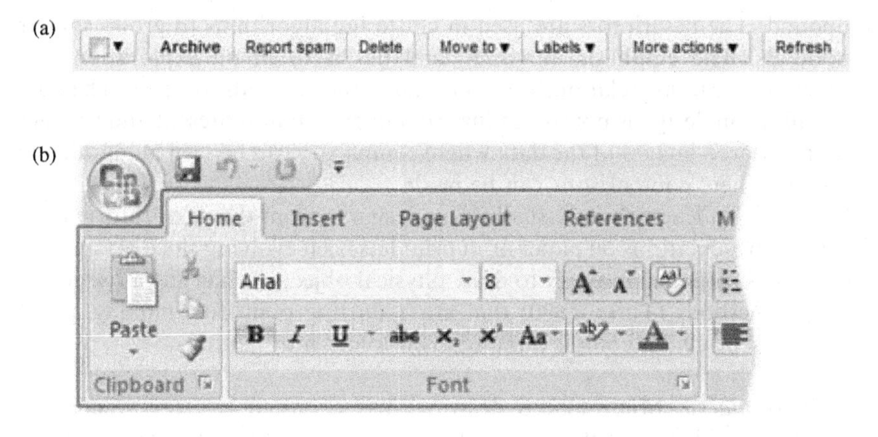


**FIGURE 4.14** A simple toolbar from Gmail and a ribbon from Microsoft Word®

menu bars become cumbersome and toolbars lack sufficient physical space to display all the controls, other organizational strategies are required. A ribbon, as shown in Fig. 4.14b, is an approach that combines aspects of the toolbar with menu bars. Groups of related actions commonly used together are available as a toolbar, but the toolbar can be switched to another for a different set of controls.

### 4.3.2 Visualizations

> "All communications between the readers of an image and the makers of an image must now take place on a two-dimensional surface. Escaping this flatland is the essential task of envisioning information—for all the interesting worlds (physical, biological, imaginary, human) that we seek to understand are inevitably and happily multivariate. Not flatlands." (Tufte, 1990)

If the first challenge of designing interactive visualizations of complex information is how to project multidimensional abstract data into a two-dimensional space without losing its richness, the second is how to visually compress or find our way through the quantity—millions of points—of data. The third challenge is to effectively link it to the growing amount of related information.

The information being visualized is *abstract*. It is often categorical or structured data that contain attributes or properties about abstract objects that have been modeled such as biological genes, chemical compounds, documents, or financial transactions. From these data objects, secondary data can be derived that is used in data mining to cluster, classify, or find associations or other relationships. For example, descriptors can be derived from chemical compounds that summarize the number of ring systems, chains of various types of atoms, or other chemical or topological features in each compound. The descriptors are used in clustering algorithms to group similar compounds. The emphasis in visual analytics is often on exploration: the discovery of patterns, relationships, clusters, outliers, trends, or gaps. The goal of visualization design is not to eliminate complexity but to present and interact with uncluttered images of the data where complexity can be seen alongside the detail and where comparisons can be made.

One place to learn about visualization design is from the past. Information design for presentations on paper or in print has been evolving since perspective drawing was invented as a way to draw physical objects. It has given rise to the following methods (Tufte, 1990) that are a starting point for thinking about how to present dense, complex, multidimensional information:

- *Micro/macro*. Micro/macro drawings use a design strategy where fine detail is added not only so it can be seen but also so it can be used to form the overall structure as shown in the 1739 Bretez-Turgot *Plan de Paris*. The detailed map of Paris, shown in Fig. 4.15, was drawn as 20 sheets. When a sheet is viewed from a distance, the detail blurs and becomes part of the texture of larger surfaces as shown in Fig. 4.16. But up close, as

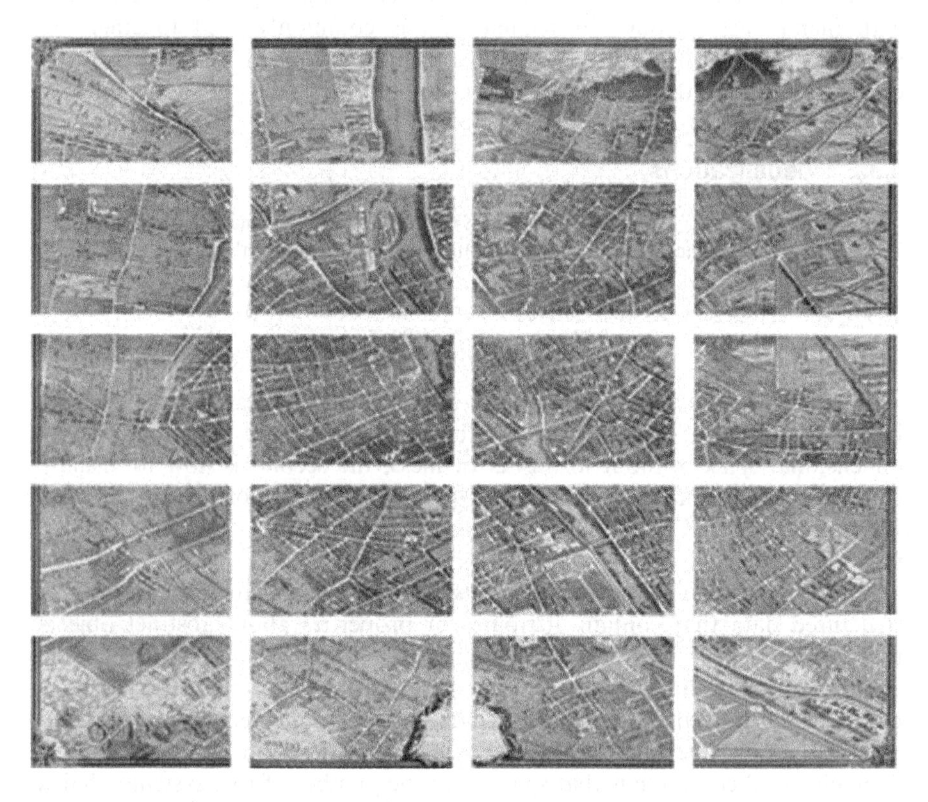

**FIGURE 4.15** The 20 sheets of the Bretez-Turgot *Plan de Paris*

**FIGURE 4.16** Sheet 15 from the Bretez-Turgot *Plan de Paris*

shown in Fig. 4.17, the details of each building and its immediate neighborhood can be seen: architecture, doors and windows, nearby buildings, and street names. The complexity is not eliminated but is controlled by organizing the information into multiple layers of context. In addition to detail becoming texture, labeling has been added to the streets, rivers, and even the roof of a hotel building to provide landmarks. Just as the design of the interface architecture brings together related tasks to avoid the cognitive costs of interrupting the flow of a task, allowing detail to be seen in its larger context avoids the cost of switching to a different image. By not stripping the complexity from the presentation, the viewer, rather than the designer, decides what information is important to see.

- *Layering and separation.* This method controls complexity by visually separating the data into strata. We have already seen one example of layer and separation through the labeling of streets, building, and rivers in the *Plan de Paris*. Another example is to use color, such as red and black, to separate annotations from data. Shape, value (light or dark), size, and color can be used to separate and layer information. Layering can be difficult to achieve because of unintended interaction between the

**FIGURE 4.17** A section of a sheet from the Bretez-Turgot *Plan de Paris*

graphical elements. The relationships must be in the right proportions and consistent with the data being represented.

- *Small multiples.* Small multiples are repetitions of the same design structure over slices of the data. Fig. 4.18 shows an example of a small multiple for unemployment rates in specific industries over 10 years. After the first element is understood, the remaining elements can be quickly read. The small multiples show all the data and make it easy to compare changes across elements without switching contexts. How much detail to include depends on the level at which the data is being viewed.

Interactive visualizations developed since the advent of high-performance computer graphics have extended the methods in the preceding list in different or new ways. These enable users to do the following:

- *See information at multiple levels of detail.* A variety of interface schemes has been developed to provide capabilities similar to the micro/macro

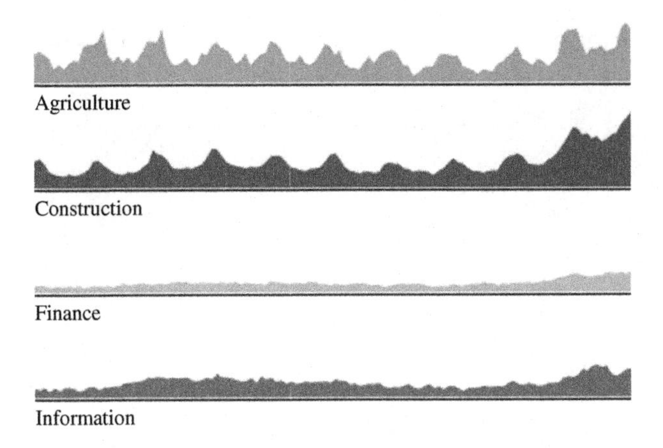


**FIGURE 4.18** Example of small multiples of unemployment rate by industry over 10 years

readings of the data. These are discussed later in this section. Interaction techniques can also be used to provide additional information. For example, datatips can provide a quick summary about a visible object as the mouse pointer rolls over the object.

- *See various relationships between the items.* Relationships can be represented using the Gestalt principles of similarity, closure, proximity, and continuity, or by connected lines. Immediate highlighting of visible objects can be used to identify individual objects among many that have been selected by some criteria.
- *Filter irrelevant information.* Dynamic queries allow visible objects being displayed to be hidden or visibly changed in response to the movement of controls—sliders or checkboxes—that change the parameters of a filter action. The immediate feedback provided within a few milliseconds allows users to rapidly sift through information. Data brushing allows visible objects selected in one view to simultaneously be highlighted in other views that are linked to that view.
- *Create subsets of the information.* Visible objects can be selected, extracted, and exported into other applications or saved as files.
- *Sort or rearrange the information.* Specific attributes may be used to sort rows or columns in a table or rearrange the layout of visible objects. Sorting alphabetically, numerically, by date or time, or categorically are common. The use of statistical and data-mining algorithms, such as clustering by different distance metrics, may also be used to rearrange the data in a table.

The abstract stages of the design process for visualizations are similar to those described previously for user interfaces: understand the users, the nature of the exploration tasks they will perform, and the data. Designing the physical

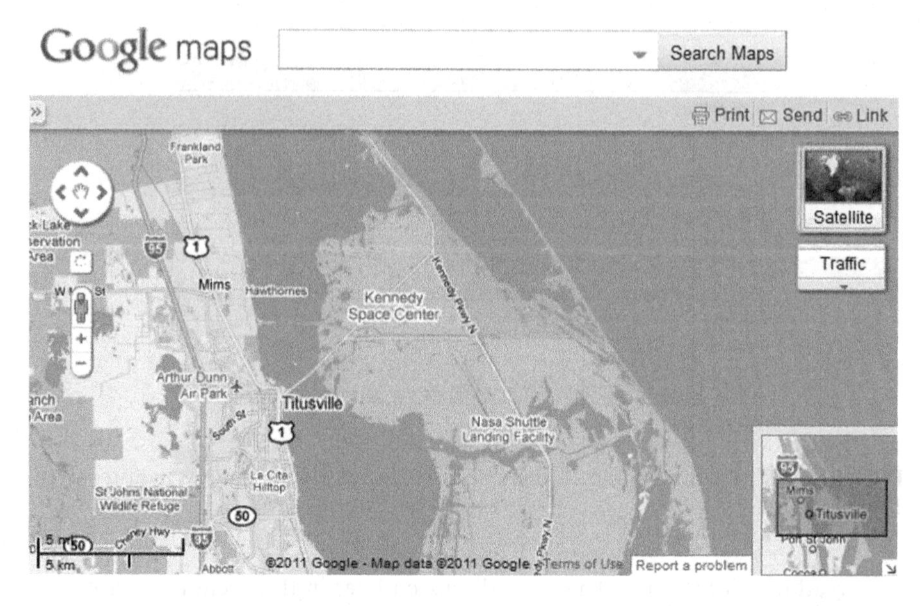


**FIGURE 4.19** Google Maps provides overview + detail and zooming interface schemes

structure of an interactive visualization is often guided by a well-known mantra: "Overview first, zoom and filter, then details on demand." (Shneiderman & Plaisant, 2010) Several interface schemes incorporate the mantra and allow users to work with and move between focused and contextual views of a dataset. Some of these separate the views spatially into panels as in the tiled windows scheme described for user interfaces. Others separate them temporally through animated transitions. These schemes include the following (Cockburn et al., 2008):

- *Overview + detail (spatial separation)*. This scheme simultaneously shows an overview and a detailed view of the underlying information space. The user interacts separately with each view, but the views are coordinated so that what changes in one is reflected in the other. The overview provides context for what is seen in the detail view. Google Maps™ is an example of this structure as shown in Fig. 4.19.
- *Zooming (temporal separation)*. This scheme uses a single window that allows the user to zoom in on the dataset. Zooming controls increase or decrease the levels of scale, and the resulting changes are done in place so the views cannot be seen simultaneously. Google Maps with the overview insert hidden is also an example of this approach.
- *Focus + context*. This scheme integrates the focus and contextual views into a single seamless view. All of the content is visible, but the area within or near the focus is distorted to provide more detail as shown in Fig. 4.20.

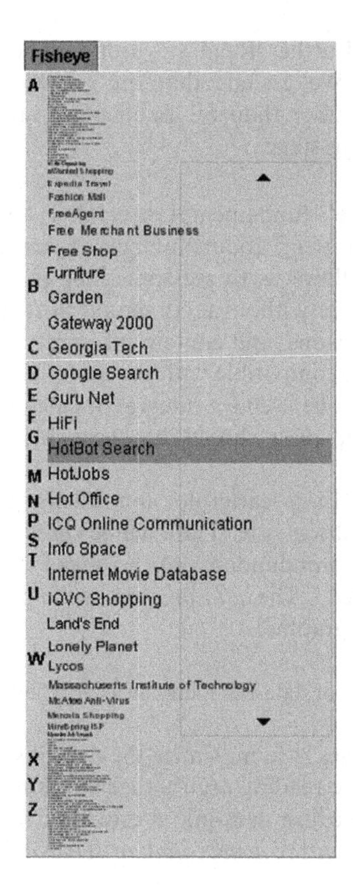


**FIGURE 4.20** Focus + context interface schemes distort to show detail in context. (Fisheye Menu courtesy of HCIL at University of Maryland: www.cs.umd.edu/hcil/fisheyemenu)

### 4.3.3 Graphics

In Chapter 3, the discussion of graphical design focused on how the choice of graphic schemas and the mappings of graph dimensions to variables affected the ability to create an image of the graphic that could be efficiently "read." Programming or scripting languages and libraries such as ggplot2 may be useful without knowing much more about the design of a graphic. Defaults provide most of what is needed to generate reasonably good graphics that can be used for exploration. However, grammar-based toolkits such as Protovis expect the designer to compose the elements of a visualization. Much more must be understood about design if the graphics are to communicate the data clearly, efficiently, and honestly. Quantitative graphics are visualizations that have been evolving since the 1700s. This section touches on only a few of the important principles and ideas that have been discussed elsewhere about graphical design. The "Further Reading" section refers to other books and articles on the subject. We assume that the graphics will be designed for interactive exploration rather than for publication and that the design may evolve in the prototypying stage.

***Displaying Data.*** Tufte's fundamental principle is "above all else, show the data." A graphic is a drawing about numbers. Even if complex, it is worth studying carefully if it allows us to see something in the data that would be harder or impossible to see without it. It should portray the data clearly and accurately, invite comparisons, and contain information that is relevant to the task. The form should be compatible with the underlying data. For example, a reference curve should not be used for integral or categorical data. As in micro/macro visualizations, the data should be accessible at several levels from overviews to fine detail.

The graphic in Fig. 4.21 is a scatterplot annotated with terminology that will be used in the following discussion. Tufte defines *data-ink* as "the nonerasable core of a graphic, the nonredundant ink arranged in response to variation in the numbers represented." The *data-ink ratio* is the proportion of data-ink to the total ink used in the graphic.

data-ink ratio = amount of data-ink / amount of total-ink

One of Tufte's principles is to maximize the data-ink ratio. (Note that more recent research on graphic reading argues that this is not always the case—see Kosslyn 2006; p. 13). Because data-ink is essential, to increase the ratio, we must erase what is not essential: *nondata-ink* or *redundant data-ink*. An example

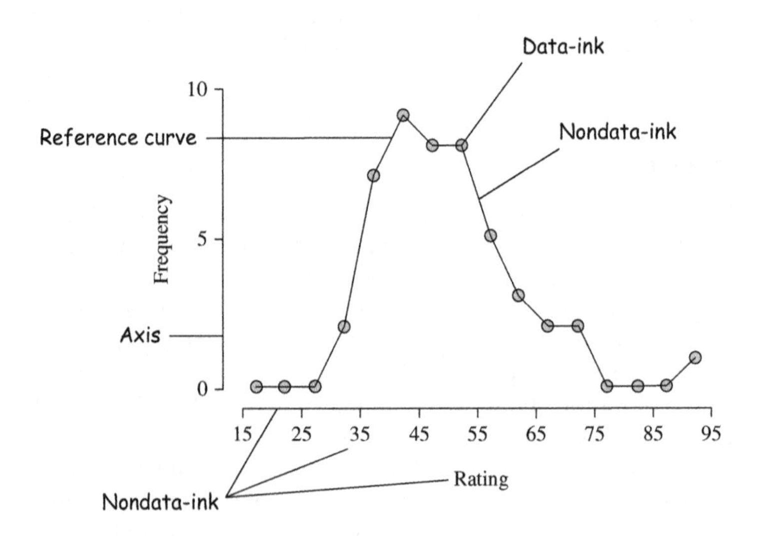


**FIGURE 4.21** A scatterplot annotated with nomenclature

of redundant data-ink is a histogram bar that has its height labeled just above the bar. The number it represents is redundantly represented in at least two ways: by the height of the bar and by the label. A designer of graphics, like an editor of text, should remove what is unnecessary. The question should be asked of whatever is drawn: does this provide new information?

For clarity, the content area should be kept free of clutter. Too much information can make it difficult to see and understand the data. Labels should not interfere with the marks. Keys and legends should be kept outside the axes, and the marks should not overlap the axes.

***Displaying Nondata: Scales and Grids.*** Grids, axes, reference curves and lines, and other accoutrements are intended to aid understanding. They are not the subject of the graphic and should be given less importance visually by being drawn in muted colors with thin lines as shown in Fig. 4.22b or, even better, erased whenever possible.

There are a number of guidelines on the use of scales, including the following:

- The minimum and maximum limits set on the scales should be chosen so that the marks fill the content area as much as possible. If multiple panels are being used to compare data, as in small multiples, both horizontal and vertical axes should be consistent. The design structure of each panel should not change. (Cleveland 1994)
- Use understandable rounded numbers for tick marks (Unwin, 2008). *Nice numbers* are familiar numbers learned in arithmetic that make mental calculations easier to do. A *nice scale* is an interval scale of numbers where the differences of the first two numbers in the scale are members of the set { . . . , .1, .2, .5, 1, 2, 5, . . . } (Wilkinson 2005). The following are all nice scales that can be used to label tick marks when appropriate:

```
{ . . . 1, 2, 3, 4, 5 . . . }
{ . . . 2, 4, 6, 8, 10 . . . }
```

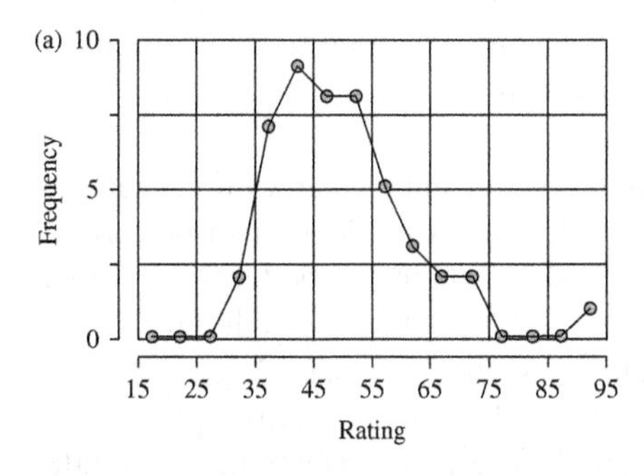


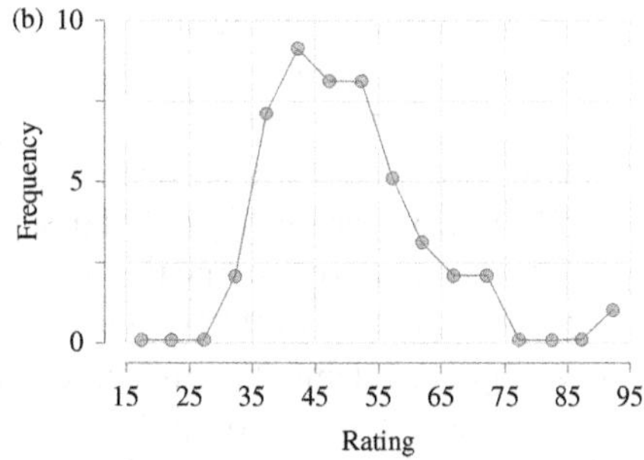


**FIGURE 4.22** Creating a visual hierarchy to emphasize the data

```
{ . . . 0, 50, 100, . . . }
{ . . . 0.001, 0.003, 0.005, 0.007 . . . }
```

- There are times when it is helpful to have different scales on top and bottom, or left and right.
- Showing data on a logarithmic scale is useful when the range of values is large, but the axis label should call attention to the use of a nonlinear scale.
- Not including zero does not necessarily distort the information (Cleveland, 1994), although this is a subject of debate.
- Avoid the scale breaks that are sometimes used in charts when there are large gaps in the data values. The mind interprets uninterrupted space as continuous, and it is difficult to perceive it otherwise. If a break is needed, a different panel should be used.

***Perceiving Graphics.*** To show the data, we must understand how we perceive it. In a graphic, the numbers and nonnumeric values have been translated into geometric objects with visual properties that have been positioned in two-dimensional space by some coordinate system or projection. Exploration involves making comparisons: How much? How much change? How much proportionally? How similar or different? The answers are inferred from comparisons made of the geometric relationships of the objects and of their visual properties: How long is this line, angle, or area? How much longer is this line, angle, or size than that one? How much lighter is this color or grayness from another? Is this point closer to this group of points or to that group? What matters is not only the actual measurements on the physical surface but also what is perceived. The following includes just a few of the findings about perception relevant to design:

- We notice large perceptible differences. We discussed earlier how focal points can be created to direct a user's attention by using brighter colors or larger features to visually emphasize certain elements. This emphasis may be used to separate the data elements from the nondata elements—such as grids, reference lines, and reference curves—and to draw attention to what is important. Visual properties, however, are relative. It's the contrasts that matter.
- We can discriminate between two values of a visual property only when the difference is proportionally large enough. For example, we can perceive a difference between a line that is 25 cm and 26 cm long no better than we can perceive a difference between a line that is 2.5 cm and 2.6 cm long.
- We group elements into units. The Gestalt principles described in a previous section also apply to the graphical elements. Spatial proximity, in particular, is a powerful way to emphasize relationships between data entities. But if grouping is not intended, the perceptual tendency to

structure the elements can make it difficult to see other patterns in the data or make it easy to see patterns that don't exist. The Gestalt principles have been effectively used in visualizations that combine a scatterplot graphic with dynamic query controls. The controls are used to filter the data. Grouping by proximity allows outliers to be seen in the data, and grouping by similarity (usually color) is used to find related items.

- When we map real numbers of a linear scale to values of a visual property (the physically measured values such as the length of a line or some value in a color scale), what we perceive (the sensation) is on a nonlinear scale. The perceived scale is the actual scale raised to some power. We estimate fairly accurately the length of a line (the power is between 0.9 and 1.1). We underestimate the area of a shape such as a square, rectangle, or circle (power is 0.6 to 0.9), and underestimate still more the volume of a solid (power is 0.5 to 0.8). (Ward et al., 2010)
- We can perceive individual marks immediately under certain conditions. This phenomenon was discussed in Chapter 2 in the section on distributed attention. For example, in a scatterplot, a few red points among a large number of gray points will be immediately noticed. The points appear to "pop out" with no cognitive effort. The degree of contrast of a specific feature in pop-out points and the other points give rise to the result. Color, brightness, orientation, size, motion, and stereoscopic depth can all be used to produce a pop-out effect. This perceptual mechanism has been effectively used in heatmaps. Heatmaps are 2-D color tables with a bipolar color scale. For each cell of the table, a lower value will be mapped to a shade of one color and a higher value to the shade of the other color. Blocks of cells of similar color and intensity immediately stand out.

***Graphical Integrity.*** For a graphic to have integrity, its visual representation must be consistent with the data. The design choices establish visual expectations for what is represented by the physical space and the marks in it. The design of each graphic should be uniform, invariant, and clearly labeled. The scales used for the content area should have regular intervals without breaks. The mappings of data variables to visual variables should remain constant for all marks in the graphic. Although perception can affect how the graphic is interpreted, what is physically measured on the surface of the graphic should at least be in direct proportion to the values in the data. More is more—a longer line means a greater magnitude of the value of some variable—and less is less. The variation of a reference curve in the graphic should correspond with a variation in the magnitude of the variables it represents. All the data that has been selected by the user should be shown so that comparisons can be made in context.

***Aesthetics.*** The elements and principles that make a graphic pleasing are elusive. Within any of the fields of functional design—graphic design, industrial design, architecture, visual interface design—certain words are used to describe the goal. For elegance and simplicity: scale, contrast, and proportion are used; for organization and visual structure: distinctiveness, integrity, comprehensiveness, and appropriateness are used. For each of these attributes, there are principles, techniques, and many examples to study. But graphics for data exploration is at its finest when it has the visual representation best suited for interacting with some specific data. The elegance of Gmail is in the way it supports the fundamental tasks it was designed to support.

***Graphics or Tables?*** Graphics are not always appropriate. Interactive data tables or spreadsheets are the best way to show exact numerical values when the datasets are not too large. Data tables or spreadsheets provide ways to filter, sort, and perform various calculations on the numbers. Tables may also be preferred when a dataset consists of highly labeled numbers.

### 4.3.4 Real-Time Constraints

Interactive systems must be responsive. Responsiveness, unlike performance, is measured on a human time scale. The time it takes to physically react, to notice a lag in the system's response, to keep our attention from drifting to something else, or to wait for a response to an email impose design constraints. A system is responsive if it provides what we need within the expected time constraint. Table 4.5 shows a generalized threshold of time constraints that are important for interaction design (Johnson, 2010). Early stage perception takes about 10 milliseconds. Above 100 milliseconds, we begin to lose the sense of cause and effect between an action we take and the reaction of the system. The longest time we expect there to be a lull in a conversation is 1 second. If the system has

**TABLE 4.5 Human Time Constraints for Interaction (Based on Johnson, 2010, p. 161)**

| Threshold | Type of Interaction |
|---|---|
| 0.01 second | feedback for stylus-based input with electronic ink on display |
| 0.1 second | feedback for hand-eye coordination (pointer movement, resizing, scrolling, drawing with mouse, etc.) |
| | feedback for click on button or link |
| | show "busy" indicators |
| | longest interval between animation frames |
| 1 second | show progress indicators |
| | finish various operations (e.g., open window, auto-save, etc.) |
| | time to wait before next visible response |
| 10 seconds | complete one action of a multisequence action (e.g., an edit in a text editor) |
| | complete user input to an operation |

not responded, we wonder why and start to become impatient. At 10 seconds, we will have completed an action in a multisequence action. To design interaction that flows, these constraints must be met as appropriate for each action the user will take.

## 4.4 SUMMARY

The complexity of data-intensive systems is continually increasing. But even a system with a complex visual interface can be operationally simple if its structure is logical and well organized, and it makes interaction efficient. To help designers better understand the mental processes of users as they interacted with computational tools, Donald Norman devised a theoretical cognitive framework of seven stages he called action theory. From a high-level goal, users form an intention, which they break into sequences of actions that they perform. As they perform each action, users evaluate and interpret the results. A system is difficult to use if it requires too many action steps or the interface is difficult to interpret.

Good design begins with a thorough understanding of the users, their work, and their environment and is done in four stages:

- *Analysis*. Design decisions must be based on facts and details elicited from discussions with the users and observations of the work. These are used to create models of the work environment, the work, and the resources required to do the work.
- *Design*. Design begins by developing an abstract conceptual model. From this, an interface architecture is designed that groups related actions into interaction spaces and shows how the user will navigate between these spaces. The visual interface and interactions are designed from the various analysis and design models.
- *Prototyping*. Prototypes provide a way to envision and experience how the new system will affect work, to explore technical alternatives, and to discuss flaws in the design and specific ideas for how it might be improved. Prototypes should be quick and inexpensive to build and relatively easy to change.
- *Evaluation*. Evaluation is intended to provoke discussion about better ways to structure the system, or to uncover tasks that aren't necessary or may be more or less important than initially thought.

Visual interfaces organize content and tools so that users can efficiently do the activities or tasks that the system is being designed to support. The following steps are required to design a visual interface:

- *Organize the application*. Complex applications require organization and navigational aids that allow users to move between interaction spaces.

Designers use signage, maps, and clues to help users know what path takes them to the place where the next required action can be performed and to keep them from getting lost.

- *Organize the page.* Page layout uses techniques to create a visual hierarchy of the content that gives weight to what is most important and a visual flow that leads the user through the page. A visual hierarchy structures the content that incorporates focal points that lead into secondary regions of less important content. Visual flow is designed to manipulate the path the eye takes as it scans the page.
- *Organize the actions.* Actions may be visible or invisible. Visible actions must provide visual feedback when they are manipulated by the user. Invisible actions are invoked through combinations of keystrokes, drag-and-drop operations, and double-clicking or right-clicking on visible objects.

Designing the physical structure of interactive visualizations is often guided by a well-known mantra: "Overview first, zoom and filter, then details on demand." Several interface schemes incorporate the mantra: overview + detail (spatial separation), zooming (temporal separation), and focus + context. Interactive techniques such as filtering, creating subsets of the information, or sorting and rearranging the information provide support for creating different views of the information.

The fundamental principle of displaying graphics is "show the data." Graphics should portray the data clearly and accurately, allow the data to be compared, and contain information that is relevant to the task.

Interactive systems must be responsive. The time it takes to physically react imposes design constraints that vary by task. The real-time constraints that affect perception and cognition range from 10 milliseconds to upward of 10 seconds. A system is responsive if it provides what we need within the expected time constraint.

## 4.5 FURTHER READING

The following are three well-known Web sites for interaction design and usability:

- The Nielsen Norman Group (www.nngroup.com/)
- Tog's First Principles of Interaction Design (www.asktog.com/)
- Jakob Nielsen on usability (www.useit.com/)

***Design Methodologies.*** The contextual design methodology is fully described along with plenty of practical advice in *Contextual Design: Defining Customer-Centered Systems* (Holtzblatt & Beyer, 1998). Variations of this methodology tailored for smaller projects with more focused goals have been further described in *Rapid Contextual Design: A How-To Guide to Key Techniques for User-Centered Design* (Holtzblatt et al., 2005). A methodology with an emphasis on *usage*-centered rather than *user*-centered design can be found in *Software for Use: A Practical Guide to the Models and Methods of Usage-Centered Design* (Constantine & Lockwood, 1999).

We have found, particularly for domains that are not well understood, that by working through a small set of core tasks and functions to be supported and combining it with low-fidelity prototyping, the designer can more quickly learn the domain. A deeper understanding of the domain makes it is easier to make decisions about how to expand the scope of the effort. Kuniavsky (2003) discusses user research methodologies used to inform the design of the user experience. Nielsen (1993) describes the importance of incorporating a usability engineering lifecycle throughout the design of software products. And, Yaffa (2007) describes the effort involved in the design and usability assessment of something as apparently straightforward as designing fonts for highway signs.

***Visual Interfaces and Interactive Design.*** *Designing Visual Interfaces* (Mullet & Sano, 1995) explains how graphic design principles and techniques are used to design visual interfaces. *Designing Interfaces* (Tidwell, 2005) discusses some of the same principles but provides more detail of interest to software developers. The book contains concrete strategies and design patterns for many aspects of visual interface design. Cockburn et al. (2008) survey the major interaction techniques for user interfaces such as overview + detail, zooming, and focus + context, while Fekete and Plaisant (2002) discuss the challenges of interaction design when applied to large datasets of more than a million data points.

***Visualization.*** There are two good places to start to learn more about visualization. The first is *Readings in Information Visualization* (Card et al., 1999), which is a collection of papers that cover a variety of topics on visualization. The second is *Envisioning Information* (Tufte, 1990), which describes general principles used in the design, editing, and analysis of data representations.

***Graphics.*** An introduction to some of the issues in graphics design is *Good Graphics?* (Unwin, 2008). Texts on quantitative graphics include works by Bertin (1983), Cleveland (1993), Kosslyn (2006), Theus & Urbanek (2008), Tufte (1983), Wainer (1997, 2005), and Wilkinson (2005). Unwin et al. (2006) discuss issues related to large datasets.

# CHAPTER 5

# HANDS-ON: CREATING INTERACTIVE VISUALIZATIONS WITH PROTOVIS

## 5.1 USING PROTOVIS

This chapter reviews the JavaScript programming language Protovis. It walks through the basic elements of the language and illustrates how to put the pieces together to generate useful interactive data visualizations to be displayed in a Web browser.

### 5.1.1 Overview

Data visualizations are generated in a number of ways. One option is to use a drawing package such as Adobe Illustrator or Microsoft® Visio to manually draw the different elements of the graphic, such as the bars to represent frequency or axes to document the graphic's scale. This option provides a flexible approach to create a single static graphic. The data underpinning the graphic must be manually mapped onto the drawing objects (such as bar or pie chart slices). This approach can be time-consuming and must be repeated for every visualization created. Simple data-analysis tools, including Microsoft Excel and Google spreadsheets, provide a series of ways to create plots directly from tabular data; however, the amount of control over the individual visualizations is limited. More sophisticated statistical analysis and charting

*Making Sense of Data III: A Practical Guide to Designing Interactive Data Visualizations*, First Edition. Glenn J. Myatt and Wayne P. Johnson.
© 2011 John Wiley & Sons, Inc. Published 2011 by John Wiley & Sons, Inc.

tools such as JMP® are available, but again, they create a fixed series of visualizations. A third option is to use a computer programming language such as Java™ that includes libraries to create graphical objects such as Java 2D. This offers a flexible approach to generating customized and interactive visualizations and, when implemented with reuse in mind, provides a solution to use over again with different datasets. The major drawback is the time necessary to learn a programming language as well as to design, implement, and test such a solution.

Because none of these approaches provide a practical option for designers of data visualizations to easily create customized and interactive plots or graphics, a team at Stanford developed the Protovis toolkit (Heer & Bostock, 2010). This JavaScript library provides control over the low-level marks of a graphic. Many types of visualizations can then be created by combining these marks. Although the toolkit is a JavaScript library that requires some understanding of programming concepts, the burden is considerably lower than with other programming languages. And because it is designed specifically for creating data visualizations, Protovis is more focused on the implementation of interactive graphics. This lower entry level means that the toolkit can be used by data-visualization designers in creating and deploying visualizations on the Web. It is possible to generate a variety of interactive graphics in Protovis, and Fig. 5.1 provides an illustration of some of the visualizations that can be implemented quickly and easily.

### 5.1.2 Getting Started

Protovis is a JavaScript library for creating visualizations to be displayed within a Web browser. To create the visualizations in this chapter, you will need to write JavaScript/Protovis code. Some of the code will be specific to Protovis, and some will be general JavaScript code; however, all the code needs to be included in an HTML file so that it can be displayed in a Web browser. Therefore, you will need the following:

- A text editor, such as NotePad or WordPad, to create and edit the code.
- A Web browser to view the visualizations, which must be capable of running HTML5 Canvas elements. Web browser options include Safari (4 and higher), Chrome, FireFox (3 and higher), Opera (9 and higher) and Internet Explorer (9 and higher).
- The Protovis library, which can be downloaded from the Protovis Web site (http://mbostock.github.com/protovis/) or this book's Web site (http://www.makingsenseofdata.com/). The book's Web site also contains instructions for downloading the library as well as the examples in this chapter. The version of the library that will be used in this chapter is 3.3.1 and is named protovis.js.

The following is a simple example of how to create a Web-based data visualization using Protovis:

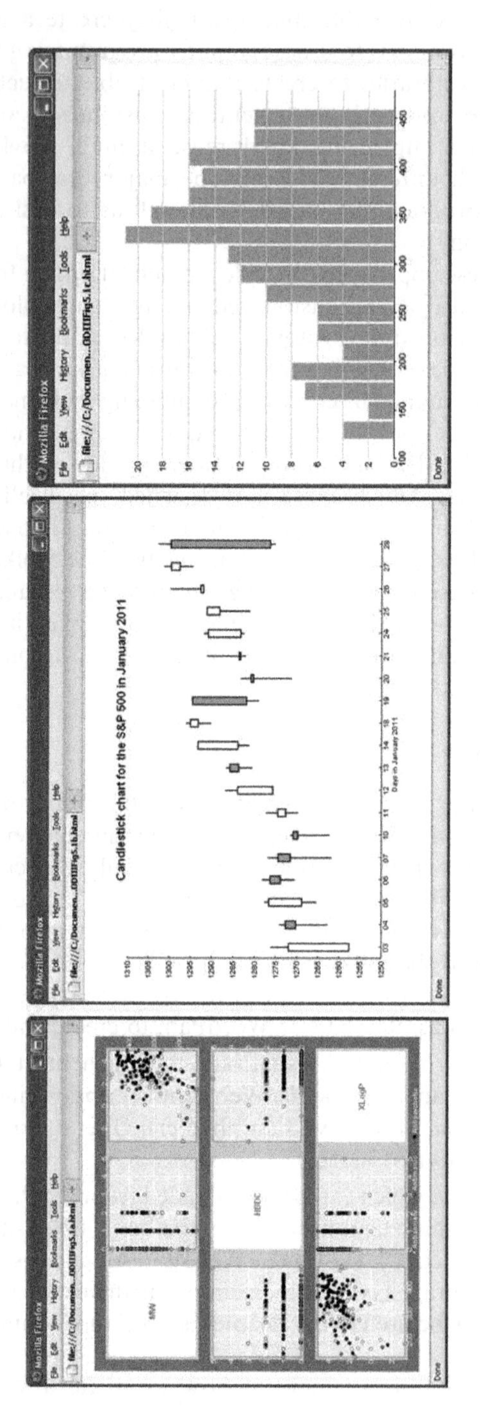


**FIGURE 5.1** Examples of different plots generated in Protovis

1. Download (and unzip) the Protovis library.
2. Create a folder in the same directory as the protovis.js file.
3. Type the following code into a text editor, and save the file as a text file called simple-bar-chart.html in the folder you just created:

```
<html>
 <head>
  <script type="text/javascript" src="../protovis.js">
    </script>
 </head>
 <body>
  <script type="text/javascript+protovis">
   new pv.Panel()
     .width(150)
     .height(100)
    .add(pv.Bar)
     .data([1.4,2.3,2.7,1.6,0.8])
     .bottom(0)
     .width(20)
     .height(function(d) d * 25)
     .left(function() this.index * 30)
    .root.render();
  </script>
 </body>
</html>
```

4. Open the file simple-bar-chart.html in any of the supported Web browsers (such as FireFox version 3 or above).

A simple bar chart should now be displayed in the Web browser, as shown in Fig. 5.2.

The HTML page we just created contains JavaScript and Protovis code to display this simple bar chart and HTML code to describe the Web page. The code contains a number of different sections that are identified using tags (phrases within angled brackets), such as `<head>` and `</head>`, which denotes the start (`<head>`) and end (`</head>`) of the header portion of the document.

There are a number of important sections within the HTML page. Between the tags `<html>` and `</html>` is the entire HTML document describing the contents of the Web page to be visualized. The tags `<head>` and `</head>` define where to find the Protovis library. In this example, the library is located in the parent directory using the ../ notation. If the library is located in a different directory or location on the Internet, the entire path or URL to the library should be provided here. Also, if a different version of the Protovis library is used that has a different name, the new library name should be reflected here as well.

Between the tags `<body>` and `</body>`, the line `<script type="text/javascript+protovis">` is used to denote that JavaScript is

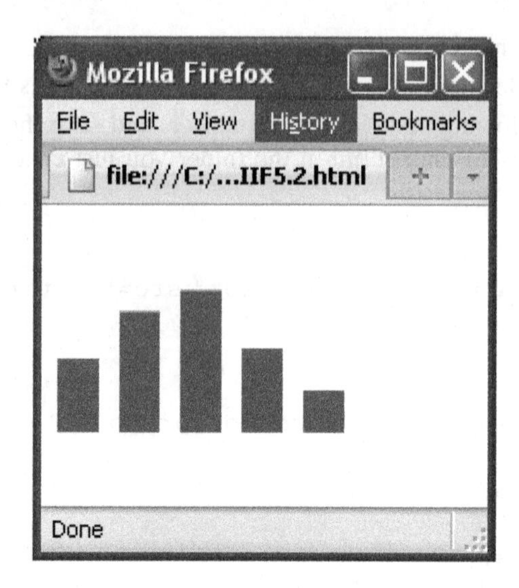


**FIGURE 5.2** Viewing a simple bar chart in a Web browser written using Protovis

extended with specific Protovis syntax. The individual visualizations to be created will be described in this part of the HTML file, which is terminated by the corresponding `</script>` tag.

The examples shown throughout this chapter generally will not be presented with the HTML wrapper that we just defined. The examples will only show the JavaScript/Protovis code contained after `<script type="text/javascript+protovis">` and before `</script>` portion of the file.

### 5.1.3 Chapter Overview

This chapter provides a description of how to use Protovis to create customized and interactive visualizations that run in a Web browser. The chapter outlines some of the basic features of Protovis as well as elements of JavaScript necessary to generate graphics, such as panels, marks, functions, and variables. The basic display elements of the Protovis language are called *marks*. These marks represent elements such as bars in a bar chart or dots in a scatterplot. The marks are described along with examples of how to generate simple plots. Marks can be coupled with techniques that customize the plots using colors and label formatting as well as annotations such as axes and grid lines. A simple plot, using a combination of these elements, will be reviewed. Commonly used graphics to support the analysis of data tables will be discussed, including frequency histograms, box-and-whisker plots, and scatterplots. The chapter also summarizes how to generate more complex visualizations that are constructed by combining elements, including the generation of stacked plots and the use of small multiple views. Layouts to create networks and hierarchies will also be summarized. Finally, the chapter briefly examines methods to interact directly with the plots, such as ways of obtaining more information on portions of the graphic.

Each section of the chapter includes step-by-step examples of how to use the elements discussed. A series of exercises are also included at the end of each section to help you learn the Protovis language, with example implementations for the exercises provided in the book's appendices.

All the code used in this chapter and Appendix A is available from www.makingsenseofdata.com. The chapter is not a comprehensive reference for writing Protovis code; however, it provides an overview of the key concepts of developing visualizations in Protovis and is best used in conjunction with the Protovis API (Application Programming Interface) documentation, found at (http://mbostock.github.com/protovis/jsdoc/).

### 5.1.4 Exercises

**5.1.4.1.** Follow the instructions in Section 5.1.2, and create the visualization as shown in Fig. 5.2.

**5.1.4.2.** Change the array values from `[1.4,2.3,2.7,1.6,0.8]` to `[1.4,2.3,2.7,1.6,0.8,1.4]`, change the width from 150 to 180, and save the file. Refresh the Web browser.

**5.1.4.3.** Change the function for the height field to `function(d) d * 20`, save the file, and refresh the Web browser.

**5.1.4.4.** Replace the word `bottom` with `top`, save the file, and refresh the Web browser.

**5.1.4.5.** Create a new label visualization by removing the block of code starting with `.add(pv.Bar)` and ending with `.left(function() this.index * 30)`, change the panel width to 100 and the panel height to 50, and add the four lines:

```
.add(pv.Label)
  .top(20)
  .left(10)
  .text("Exercise example")
```

Save the file, and refresh the Web browser.

**5.1.4.6.** Change the `left` property value to 140, save the file, and refresh the Web browser.

**5.1.4.7.** Change the `width` of the panel to 250, save the file, and refresh the Web browser.

## 5.2 CREATING CODE USING THE PROTOVIS GRAPHICAL FRAMEWORK

### 5.2.1 Overview

Protovis is designed to create flexible and interactive visualizations. These graphics are composed of different graphical elements that describe the data,

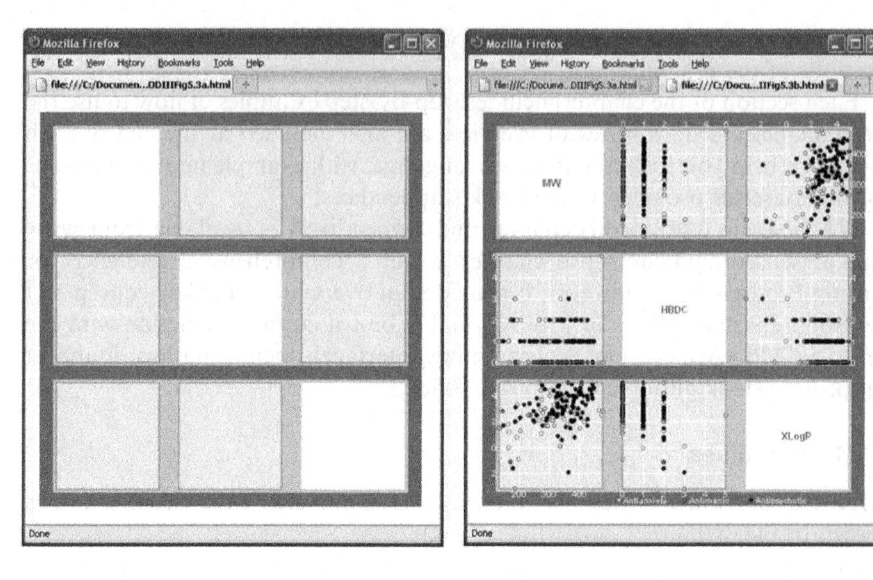


**FIGURE 5.3** Construction of a plot using different graphical elements

such as bars within a bar chart, small circles (dots) within a scatterplot, and so on. These elements provide a visualization of the underlying data, and, in Protovis, they are called *marks*. Graphics are created by combining these marks. These marks, as well as the canvas on which they are drawn (referred to as a *panel*), form the foundation of Protovis.

Figure 5.3 illustrates how a complex plot (color-coded scatterplot matrix) can be constructed using a series of low-level elements drawn on a matrix of panels. Initially a series of panels is organized on the screen as a 3-by-3 grid, shown in the left view. Those panels that will be used to display the scatterplots are colored gray with a dark gray border (those panels not on the diagonal). Axes and grid lines are drawn on selected panels as shown on the right view. Each scatterplot is created using a series of dots, with x-locations and y-locations of each of the dots calculated from the underlying data. The color of the dot is based on the specific types of observation, again using the underlying data. Axes labels are placed in the center on the diagonal panels, and a legend is added to explain the color-coding. In this example, the dots on the scatterplots are examples of Protovis marks, whereas the outer canvas and the nine inner square canvases are examples of panels.

### 5.2.2 Panels

A panel is needed to display any data visualization using Protovis, and it is defined using the expression `pv.Panel`. A *panel* is a rectangular canvas on which the graphics are drawn. The primary canvas for any visualization is referred to as the *root panel*, and data visualizations are created by adding marks to this panel or embedding other panels within the root panel. These panels are generated using the expression new `pv.Panel()`, and the height and width of the panel are usually specified when it is created.

In this example, a new panel is created with the width of the panel set at 150 pixels and the height defined as 100 pixels.

```
new pv.Panel()
  .width(150)
  .height(100)
```

A number of properties are associated with the panel object, such as width and height. These properties are set by assigning a value within the parentheses. In this example, the width property is set to 150 pixels, and the height property is set to 100 pixels. The property values can be retrieved by not including a value between the parentheses. For example, `.width()` returns the value of the panel's width. Another commonly used property is the panel's margins along all four edges of the rectangle (called margin).

Different graphical elements, such as marks or other panels, can be added to any panel using the method `.add()`. In the following code segment, a label mark (`pv.Label`) is added to the panel. The text of the label is defined as "LABEL TEXT", and the properties used to position the label on the panel (left and bottom) are both set to 50 pixels, that is, 50 pixels from the bottom of the panel and 50 pixels from the left-hand side of the panel.

```
  .add(pv.Label)
   .text("LABEL TEXT")
   .left(50)
   .bottom(50)
```

The following piece of code shows how to create and display a Protovis mark (in this case, a label) on a canvas (panel) within a browser. After all the panels and marks have been described, the final statement is to display the root panel containing the visualization. In this example, this root panel is displayed on the screen with the phrase `.root.render()`. The code segment is terminated with a semicolon (;).

```
new pv.Panel()
  .width(150)
  .height(100)

  .add(pv.Label)
   .text("LABEL TEXT")
   .left(50)
   .bottom(50)

 .root.render();
```

**FIGURE 5.4** Adding a label mark to a panel

Figure 5.4 shows the resulting Web browser launched using the code just described.

### 5.2.3 Marks

Marks are added to panels to create the graphical elements. A series of slices within a pie chart or a series of intersecting points in a line plot are examples of marks. In Protovis, there is a defined concept (referred to as a *class* in programming terminology) that describes these marks called `pv.Mark`. Although this class is not used directly, it provides many of the common fields and methods used by the individual marks that are displayed. For example, the field or property `bottom` is used to specify the vertical position of a mark from the lower margin of the panel on which it is drawn. All the individual marks (bars, dots, lines, etc.) will use the `bottom` field to specify the vertical position of the mark from the panel's lower edge. Because `pv.Mark` is not used directly, it is referred to as an *abstract class*, and all the *concrete classes* (those used directly to create a graphical element) extend the `pv.Mark` class; that is, they inherit the properties and methods from `pv.Mark`.

Each mark is usually associated with some underlying data. This data is represented as a list of values or an *array* and is stored in a field called `data`. In the following example, a bar chart is created by adding a bar mark (`pv.Bar`) to a panel using the expressions `.add(pv.Bar)`. An array of data is assigned to the bar's data field with the expression `.data([1.2,4.3,2.3,5.2])`.

```
new pv.Panel()
  .width(150)
  .height(150)
 .add(pv.Bar)
  .data([1.2,4.3,2.3,5.2])
```

The array is comprised of four real numbers, each separated by a comma, with the whole list enclosed within square brackets. Adding a bar mark to the panel and assigning this array to the data field will result in the creation of four bars, with each bar representing one of the data values in the list. Each value of the array will be linked to some visual attribute of the bar to display the bar in a meaningful manner. For example, the height of each bar could be proportional to each of the data values.

In this example, the position of the bottom of each of the bars is set to 0, which is at the bottom margin of the panel. The width of each individual bar is set to 20 pixels. The height of each bar is calculated based on a function where each value of the array is mapped onto a vertical position on the panel by multiplying the data values by 25. The horizontal position of each bar from the left margin is also set using a function based on the order of the data values in the array. The use of functions will be described in more detail in the next section.

```
new pv.Panel()
  .width(150)
  .height(150)
 .add(pv.Bar)
  .data([1.2,4.3,2.3,5.2])
  .bottom(0)
  .width(20)
  .height(function(d) d * 25)
  .left(function() this.index * 30)
 .root.render();
```

This code results in the bar chart shown in Fig. 5.5.

Other important properties of marks include `right` and `top`, which are optionally used to specify the position on the drawing canvas from the corresponding margin. If the value of any of these positioning properties is not set, Protovis assigns a sensible default value.

Other regularly used properties common to all marks include `visible` and `title`. When the visible property is set to true, the mark will be displayed on the canvas, whereas if it is set to false, the mark is not shown. This property has many uses, such as to simplify a bar chart with labels describing each bar's value. With many data values to display, the plot may become crowded, and the labels may be difficult to see. The visible property can be used to only display those values above a specific threshold. The title property can be used to

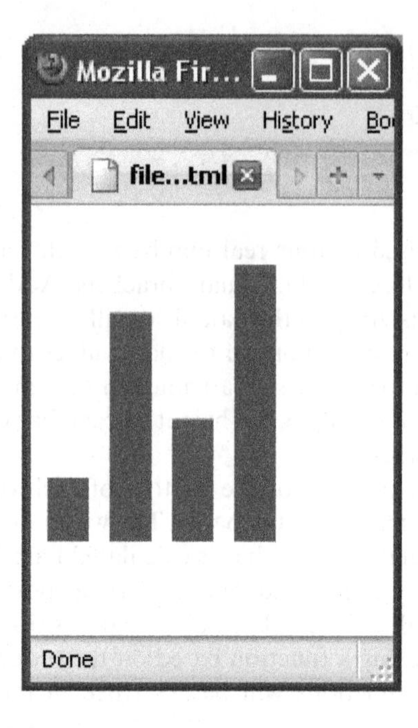


**FIGURE 5.5** Adding marks (`pv.Bar`) to a panel

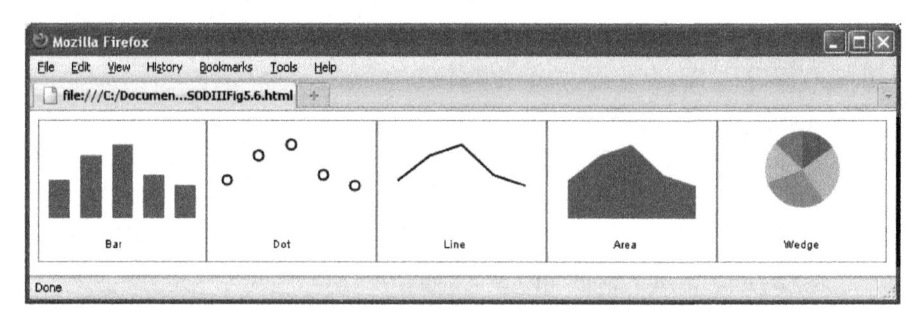


**FIGURE 5.6** Illustration of a number of marks used in Protovis

assign a value for a tooltip, to be displayed when the mouse cursor hovers over a specific element of a graphic.

All classes that are extensions of the `pv.Mark` class use these basic properties as well as properties necessary for rendering and interacting with the specific marks. For example, `pv.Bar` is a concrete class for displaying bars as shown in the preceding example, and it extends `pv.Mark`. The class reuses the eight properties associated with `pv.Mark` and also makes use of specific properties necessary for rendering and interacting with bars such as a property to specify the width of a bar (`width`).

Many concrete mark classes are used in Protovis, including bar, dot, line, area, and wedge classes, as shown in Fig. 5.6. Section 5.3 will review these marks in detail. Although it is usually desirable to set properties of these marks, such as their color, the default property settings provide a reasonable starting point. In fact, the marks displayed in Fig. 5.6 are all rendered using these default color settings.

### 5.2.4 Using Functions

The visual elements of a graphic, such as the position or color of dots in a scatterplot or the length and color of bars in a bar chart, are used to help the reader understand the underlying data. In Protovis, the underlying data associated with individual marks is mapped onto the corresponding visual representation directly using a *function*. For an assigned data array, Protovis uses a shorthand representation to create a series of visual elements, each one corresponding to a value in the list. It is not necessary to go through each element in an array explicitly to create a mark from each value. Instead, Protovis will go through each data element in turn and create a mark automatically, based on the information provided. *Functions* are used to describe the relationship between the underlying data and how it is displayed. For example, to generate a bar chart whose height changes as a function of the data, the following function can be assigned to the height property of the bar.

```
.height(function(d) d * 25)
```

Instead of assigning a static property value (such as a fixed number) to a mark's field, a function is used. The function will be called automatically over each data element, and the result will be used as the value for rendering each bar. `function(d)` is an anonymous function, that is, it does not have a name. It has one argument – `d` and will return the result of the multiplication (`d * 25`) for all values in the list. If this function is applied to the data array `[1.2,4.3,2.3,5.2]`, the first element will be 30 ($1.2 \times 25$), the second element will be 107.5 ($4.3 \times 25$), the third element will be 57.5 ($2.3 \times 25$), and the last element will be 130 ($5.2 \times 25$).

Each element in the array of data has a corresponding index that represents its position from the start of the list. The first value is at index position 0, the next value is at position 1, and so on. For example, the array used in the preceding paragraph had four values—`[1.2,4.3,2.3,5.2]`—and value 1.2 is at position 0, 4.3 is at position 1, 2.3 is at position 2, and 5.2 is at position 3. The value of this index is stored under the property index in the parent class (`this.index`). The parent class in this example is the root panel. The use of this property is helpful in functions where the value's position in the list is needed rather than its actual value. In the following example, the left position is calculated by multiplying the index of the array by 30. If this function is applied to the array with four elements, the first value will be 0 ($0 \times 30$), the second value will be 30 ($1 \times 30$), the third value will be 60 ($2 \times 30$), and the last value will be 90 ($3 \times 30$).

```
.left(function() this.index * 30)
```

This function is useful in creating individual marks at evenly spaced positions.

Functions can be complex, employing different mathematical operations. In the following example, a bar chart is created from an array of five values: `[1.2,4.3,2.3,0.9,5.2]`. A function is defined to assign a value to the visible property. In this example, the function `d > 1` is used, whereby if the data is greater than 1, the function returns a true value, and if not greater than 1, a false value is returned.

```
new pv.Panel()
  .width(200)
  .height(150)
 .add(pv.Bar)
  .data([1.2,4.3,2.3,0.9,5.2])
  .bottom(0)
  .width(20)
  .height(function(d) d * 25)
  .visible(function(d) d > 1)
  .left(function() this.index * 30)
 .root.render();
```

This result is the bar chart as shown in Fig. 5.7, with bars visible for all data values except the bar at index 3 whose value is 0.9, which is less than 1.

There are a number of commonly used operators, which include == (values are equal), != (values are not equal), > (values are greater than), >= (values are greater than or equal to), < (values are less than), and <= (values are less than or equal to). These operators can also be combined with Boolean logic such as `&&` (AND) and `||` (OR) as well as within “if . . . then . . . else” statements (the format is ‘conditional operator’?‘result if true’:‘result if false’). For example, the following code colors the individual bars using the `fillStyle` field. If the value for each bar is either less than 2 or greater than 5 (‘conditional operator’), then the color `"black"` is returned (‘result if true’). If the operator is false, then the color `"gray"` is returned (‘result if false’).

```
new pv.Panel()
  .width(200)
  .height(150)
 .add(pv.Bar)
  .data([1.2,4.3,2.3,0.9,5.2])
  .bottom(0)
  .width(20)
  .height(function(d) d * 25)
```

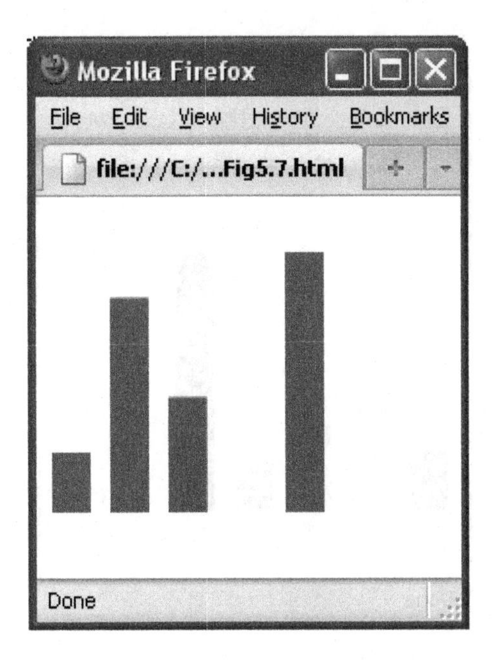


**FIGURE 5.7** Using a function to control visualization of the bars

```
    .fillStyle(function(d) (d < 2) || (d > 5) ? "black" :
      "gray")
    .left(function() this.index * 30)
  .root.render();
```

Figure 5.8 shows the resulting plot whereby the first bar is colored black because its value is less than 2, the second bar is colored gray because it is neither less than 2 nor greater than 5, and so on.

Functions are expressed in Protovis in a more concise manner than using JavaScript directly. For example, in the previous example, the function

```
.height(function(d) d * 25)
```

would be rewritten in JavaScript as

```
.height(function(d) {
return d * 25;
})
```

In JavaScript, the function explicitly defines the value to be returned, whereas the same value is returned in the Protovis expression, but it is not explicitly stated.

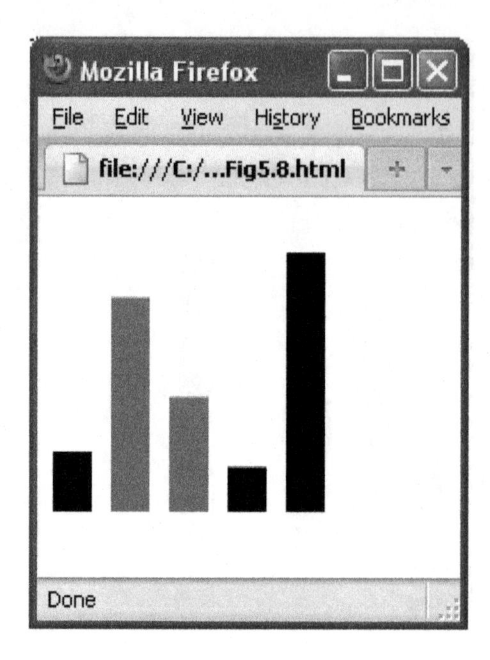


**FIGURE 5.8** Illustration of the use of a complex function to display a bar chart

### 5.2.5 Variables

*Variables* allow for the reuse of objects (such as panels, line marks, etc.) that have been defined in the code. They provide developers the opportunity to write more concise Protovis specifications. In addition, changes to the graphics at a later time can be made more easily.

In the following example, a variable is created for the root panel using the code `var barChartPanel = new pv.Panel()`. `barChartPanel` is a name to be used throughout to reference the variable just created. The name of the variable can be any single-word name or phrase, but it is helpful to make the variable's name somewhat descriptive so that the code can be reused later and easily interpreted. This variable is then reused, where `barChartPanel.add(pv.Bar)` adds a bar mark to the panel. Finally, the variable is called with the render method to display the bar chart.

```
var barChartPanel = new pv.Panel()
  .width(200)
  .height(150);

barChartPanel.add(pv.Bar)
  .data([1.2,4.3,2.3,0.9,5.2])
  .bottom(0)
```

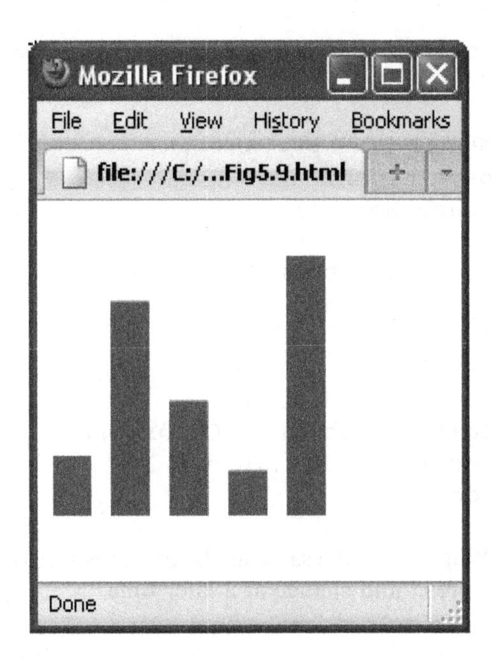


**FIGURE 5.9** Bar chart created from code using variables

```
    .width(20)
    .height(function(d) d * 25)
    .left(function() this.index * 30);

barChartPanel.render();
```

The resulting view (as shown in Fig. 5.9) is exactly the same as the previous examples where we did not use any variables; however, the code is easier to use. Semicolons are added to each section of the code to indicate that no further modification is to be made to the mark or panel.

Variables can be used to reference any object in the code. In the following example, variables that represent the panel width and height are created and assigned values (`panelWidth = 200, panelHeight = 150`) along with the width of the bar. Assigning these values upfront is helpful because it makes the code easier to modify later; that is, you do not have to search through every line to identify the specific instances where these properties are set. Variables can be used to represent other objects, such as an array of data or even functions. In this example, the variable `barChartData` is an array of values, and the variable `barColor` is a function.

```
var panelWidth = 200, panelHeight = 150, barWidth = 20;
var barChartData = [1.2,4.3,2.3,0.9,5.2];
```

```
var barColor = function(d) (d < 2) || (d > 5) ? "black" :
  "gray";

var chartPanel = new pv.Panel()
  .width(panelWidth)
  .height(panelHeight);

chartPanel.add(pv.Bar)
  .data(barChartData)
  .bottom(0)
  .width(barWidth)
  .height(function(d) d * 25)
  .fillStyle(function(d) barColor(d))
  .left(function() this.index * 30)
.root.render();
```

Again, the resulting view is the same as the previous version of the code, but the code is easier to read and change at a later time.

### 5.2.6 Exercises

5.2.6.1 Create the chart as shown in Fig. 5.10.

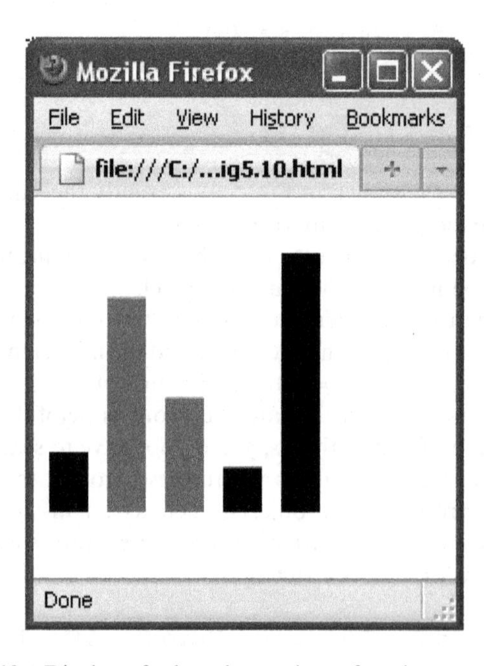


**FIGURE 5.10** Display of a bar chart using a function to color the bars

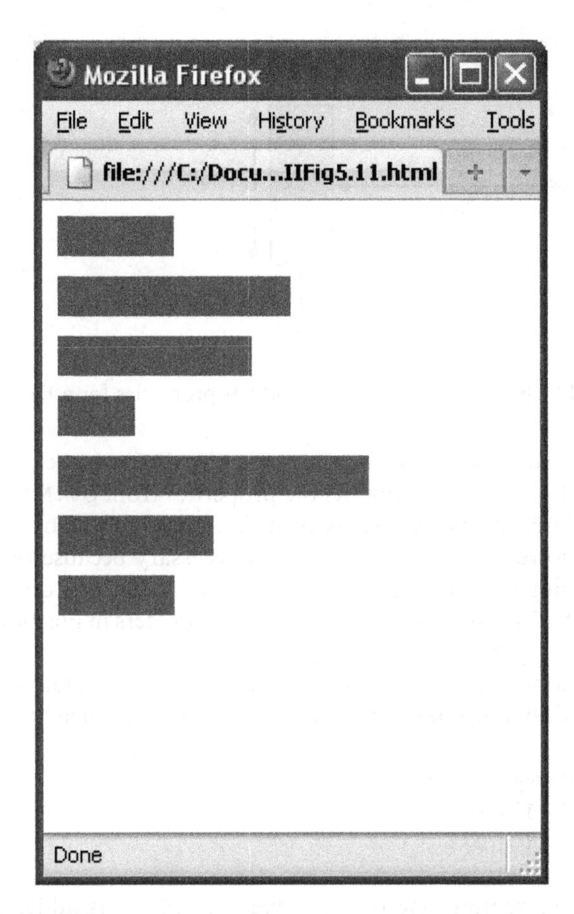


**FIGURE 5.11** Chart with bars positioned along the y-axis

5.2.6.2 Create a chart for the data `[3,6,5,2,8,4,3]` where the bars are presented as shown in Fig. 5.11.

5.2.6.3 For the chart generated in Exercise 5.2.6.2, color the bars `"orange"` for data values greater than 5; otherwise, color the bars `"lightblue"`.

5.2.6.4 For the chart created in Exercise 5.2.6.2, color even-valued bars `"lightgray"` and odd valued bars `"darkgray"`. Note that the mod function (`d%2`) will return 0 if the value is even and 1 if the value is odd.

## 5.3 BASIC PROTOVIS MARKS

### 5.3.1 Bar

Bars are used in many visualizations, including bar charts. The length or color of the bars is often used to encode information from the underlying data. For example, the length of the bars can be used to visualize the data values, or the

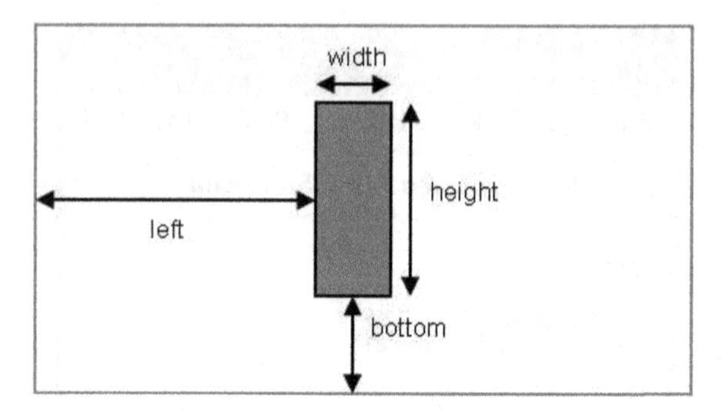


**FIGURE 5.12** bottom, width, height, and left properties for a bar on a canvas

colors can be used to highlight specific categories. Because the `pv.Bar` class is an extension of `pv.Mark`, it inherits the properties from `pv.Mark` such as the `bottom` and `left` properties we used earlier. `top` and `right` properties can also be used; however, they are not always necessary because the bars can be fully defined using the four spatial parameters: `bottom`, `width`, `height`, and `left`. Figure 5.12 illustrates the use of these parameters in positioning a bar on a canvas.

In the following example, a simple bar chart is constructed. Initially a new panel is created that is 150 pixels wide and 100 pixels in height.

```
new pv.Panel()
   .width(150)
   .height(100)
```

The class representing one or more bars (`pv.Bar`) is added to the panel using the code:

```
.add(pv.Bar)
```

Next, a list of data values or an array is added that will be used in displaying the bars:

```
.data([1.4,2.3,2.7,1.6,1.2])
```

The bar chart is positioned at the bottom of the panel, a distance of 0 pixels from the lower edge's margin:

```
.bottom(0)
```

The width of each bar is set to be 25 pixels:

```
.width(25)
```

The height of the bars will be proportional to each data value in the array. A simple function is used to calculate the height of the bar. This is calculated by multiplying each value by 30.

```
.height(function(d) d * 30)
```

The horizontal position of each individual bar is set using the `left` property. Again, this is calculated with a function based on the data array; however, the individual value is not used, just the index position. Because the width of the bars was previously set to 25, all five bars will fill up the 145 pixels of the panel width.

```
.left(function() this.index * 30)
```

The width of the bar is set to 25, yet each histogram starts at 30 pixel intervals, so there will be a horizontal space (or *gutter*) between each of the bars of 5 pixels.

The mark was added to the root panel, and the render method is called to display the bar chart in the Web browser.

The following code creates this simple bar chart, as shown in Fig. 5.13.

```
new pv.Panel()
   .width(150)
   .height(100)
  .add(pv.Bar)
```

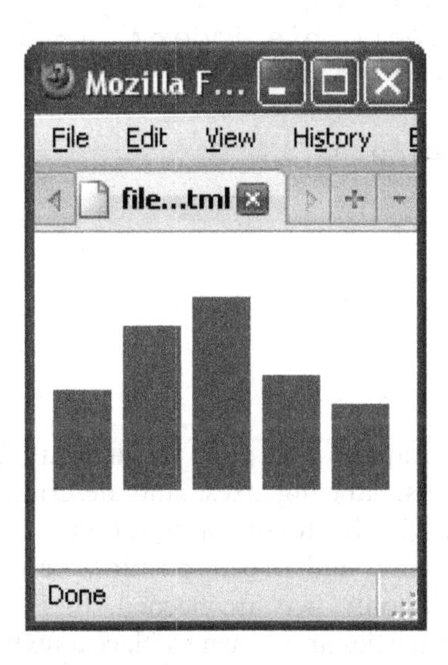


**FIGURE 5.13** Simple bar chart

```
    .data([1.4,2.3,2.7,1.6,1.2])
    .bottom(0)
    .width(25)
    .height(function(d) d * 30)
    .left(function() this.index * 30)
  .root.render();
```

A number of properties, such as `width` and `height`, are specific to the `pv.Bar` class and were also used in generating the bar chart. Other specific properties of the bar mark that control other aspects of its appearance include `fillStyle`, `lineWidth`, and `strokeStyle`. The `fillStyle` property allows you to change the color of the inner portion of the bar. The properties `lineWidth` and `strokeStyle` control the outer border of the bar; `lineWidth` controls the width of the border in pixels, and `strokeStyle` determines the color for the border. In the following example, a simple bar chart is created with `fillStyle` set to `"white"` and the border set to `"gray"` with a width of 3 pixels.

```
new pv.Panel()
    .width(150)
    .height(100)

  .add(pv.Bar)
    .data([1.4,2.3,2.7,1.6,1.2])
    .bottom(3)
    .width(25)
    .height(function(d) d * 30)
    .left(function() this.index * 30 + 3)
    .fillStyle("white")
    .lineWidth(3)
    .strokeStyle("gray")

  .root.render();
```

Figure 5.14 shows the results of this change.

### 5.3.2 Label

Text is often added to graphics and plots for annotating specific data values, providing titles for plots, labeling axes, and describing graphical elements within legends. In Protovis, this textual annotation is called a *label* mark.

The `top`, `bottom`, `left`, and `right` positions can be used to place labels onto a canvas; however, only two are needed. In Fig. 5.15, the `left` and `bottom` positioning properties are shown to place a textual annotation or label onto a canvas.

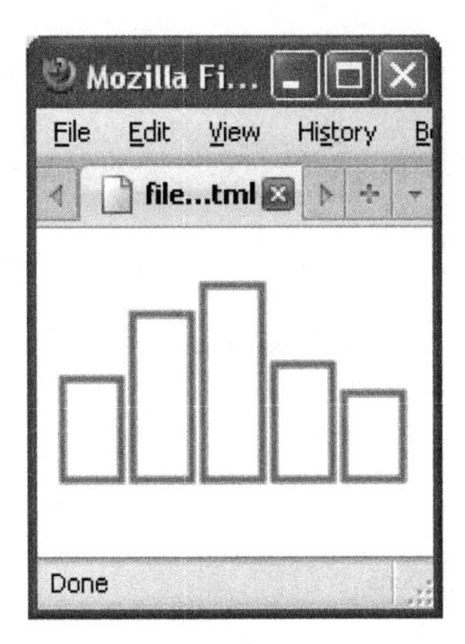


**FIGURE 5.14** Changing the bar chart's color and borders of the bars

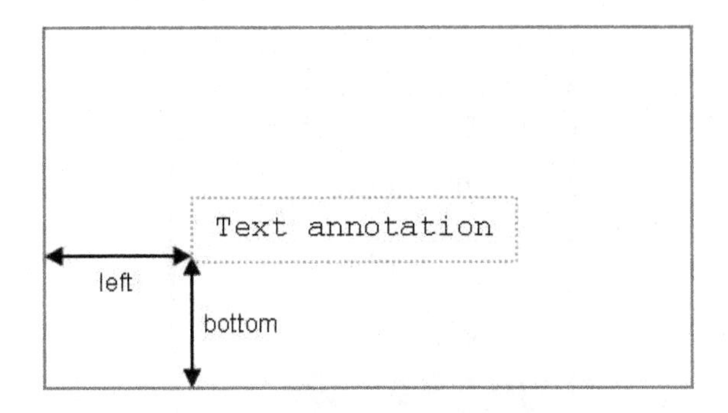


**FIGURE 5.15** Positioning labels on a canvas

In the following code, a panel is created of width 150 pixels and height 100 pixels, to which a label is added with text "`chart title`" positioned at 50 pixels from the left margin and 50 pixels from the bottom margin.

```
new pv.Panel()
  .width(150)
  .height(100)

  .add(pv.Label)
   .text("chart title")
   .left(50)
   .bottom(50)

.root.render();
```

**FIGURE 5.16** Adding a text annotation (label) to a canvas

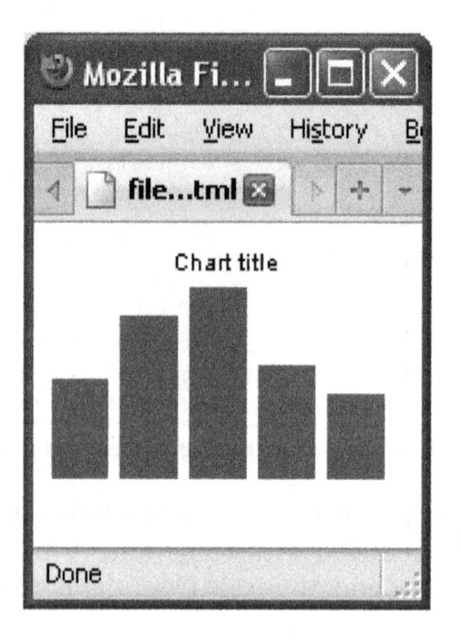


**FIGURE 5.17** Bar chart with static title

Figure 5.16 illustrates the label added to the canvas. This approach of adding a static label to a graphic is commonly used to provide titles or legend descriptions.

In the following example, the bar chart from the previous section is slightly modified to make use of variables to describe the bar chart panel. Additionally, a title “Chart title” is added to the panel that is 15 pixels from the top and 50 pixels from the left edge of the panel. The resulting plot is shown in Fig. 5.17.

```
var barChartPanel = new pv.Panel()
   .width(150)
   .height(100);

barChartPanel.add(pv.Bar)
    .data([1.4,2.3,2.7,1.6,1.2])
    .bottom(0)
    .width(25)
    .height(function(d) d * 30)
    .left(function() this.index * 30);

barChartPanel.add(pv.Label)
   .text("Chart title")
   .top(15)
   .left(50);

barChartPanel.render();
```

Using *inheritance*, labels can also be used to annotate the individual marks of a visualization such as the bars in a bar chart. This topic will be revisited later in this chapter. In the following example code, a label is added to each of the bars in the chart. The code for adding this label is `.add(pv.label)`. By adding a label to each of the bars, the label inherits the properties from the bars, such as the `bottom` and `left` property, which are used to position the label on each of the bars, as well as `data`.

```
var barChartPanel = new pv.Panel()
   .width(150)
   .height(100);

barChartPanel.add(pv.Bar)
    .data([1.4,2.3,2.7,1.6,1.2])
    .bottom(0)
    .width(25)
    .height(function(d) d * 30)
    .left(function() this.index * 30)
   .add(pv.Label).textStyle("white");

barChartPanel.add(pv.Label)
   .text("Chart title")
   .top(15)
   .left(50);

barChartPanel.render();
```

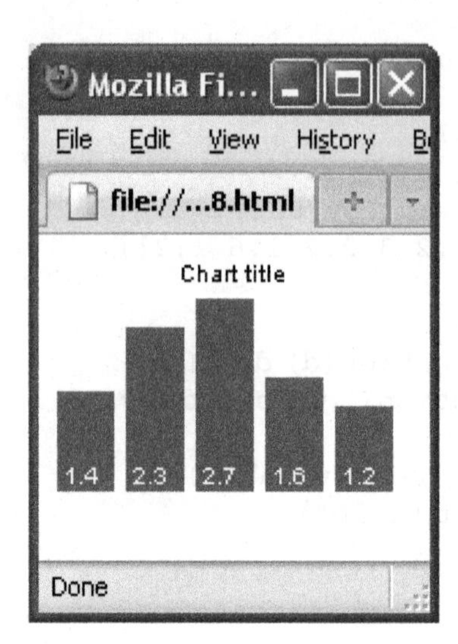


**FIGURE 5.18** Bar chart annotated with a label for the data value

In Fig. 5.18, the labels are positioned on each of the bars showing the data value for the bars. They are positioned at the bottom-left corner of the bar based on the inherited properties: `bottom` and `left`. The label values are from the `data` property.

Labels can be added to all marks, including axes or grid lines (referred to as a rule in Protovis), and other marks, such as dots in a scatterplot, lines in a line plot, and wedges in pie charts.

The way labels are rendered can be customized in Protovis using `font` (the font format), `textAlign` (horizontal alignment: `"left"`, `"center"`, `"right"`), `textAngle` (to rotate the text using radians), `textBaseline` (vertical alignment: `"top"`, `"middle"`, `"bottom"`), `textDecoration` (CSS-compliant decoration such as underline), `textMargin` (to generate a margin around the text), `textShadow` (to apply a CSS-compliant text shadow), and `textStyle` (to set the label's color) properties.

In the following code, the font of the label has been changed, as shown in Fig. 5.19.

```
var barChartPanel = new pv.Panel()
   .width(150)
   .height(100);

barChartPanel.add(pv.Bar)
    .data([1.4,2.3,2.7,1.6,1.2])
    .bottom(0)
    .width(25)
    .height(function(d) d * 30)
```

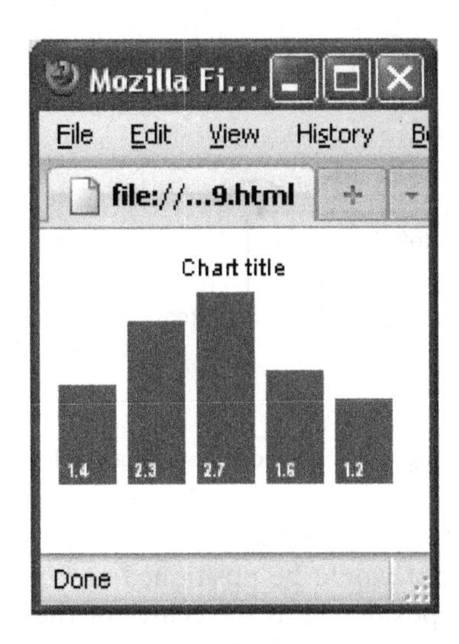


**FIGURE 5.19** Label attached to bars in white with changed fonts

```
      .left(function() this.index * 30)
    .add(pv.Label)
      .textStyle("white")
      .font("bold 8px sans-serif");

  barChartPanel.add(pv.Label)
    .text("Chart title")
    .top(15)
    .left(50);

  barChartPanel.render();
```

The placement of labels is often used in conjunction with anchors (described later in this chapter) to position the labels close to their corresponding marks.

### 5.3.3 Dot

A dot represents a point in 2-D space usually corresponding to a data value. Dots are commonly used in a number of plots, including scatterplots. The position of a dot mark is controlled by these four properties: `top`, `left`, `bottom`, and `right`. Only two are needed to position a dot on the screen, such as `left` and `bottom`, as shown in Fig. 5.20. These properties are used to set the position of the center of the dot.

In the following example, a series of dots are used to display an array of data values. The dots are vertically positioned based on the data values using the function: `d*30`. The index value of the data array is used to position the dots

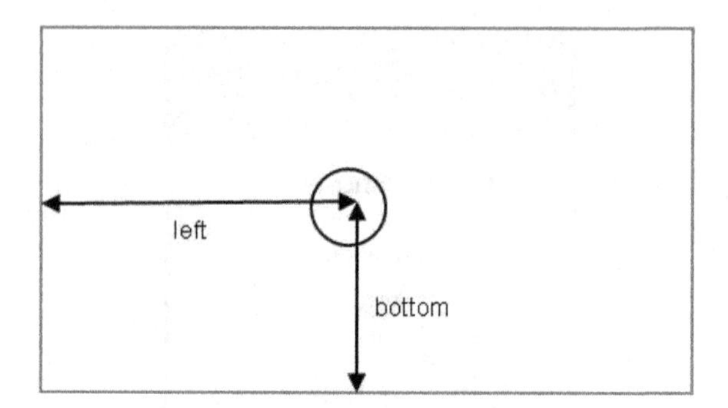


**FIGURE 5.20** Positioning a dot on a canvas using the left and bottom properties

horizontally. The first dot would be positioned at 0 pixels from the left, but because the properties' position is at the center of the dot, this dot would be cutoff. So, 5 pixels are added to the function to take this into account.

```
new pv.Panel()
   .width(150)
   .height(100)

  .add(pv.Dot)
   .data([1.4,2.3,2.7,1.6,1.2])
   .bottom(function(d) d * 30)
   .left(function() this.index * 30 + 5)

  .root.render();
```

The resulting plot of dots is shown in Fig. 5.21.

Specific properties of the dot class control its appearance, including `fillStyle`, `lineWidth`, and `strokeStyle`. `fillStyle` controls the color of the inner region of the dot. `lineWidth` and `strokeStyle` are used to control how the dot's border is displayed, with `lineWidth` controlling the thickness and `strokeStyle` the color. The size of a dot can be set with either the `radius` property (radius of the dot in pixels) or the `size` (square pixels) property. How the dot is actually drawn can also be customized using the `shape` property. A number of options are shown in Fig. 5.22: `"cross"`, `"triangle"`, `"diamond"`, `"square"`, `"circle"`, `"tick"`, and `"bar"`. The `angle` property can be used to further customize the dots by rotating them as desired.

In the following code, a series of data values are plotted as triangles set using the `.shape("triangle")` entry. The size of the dot is set to cover an area of 16 square pixels (4 × 4 pixel square). In addition, the border color is set to `"black"` and 1-pixel thick with the color of the inner region set to light gray.

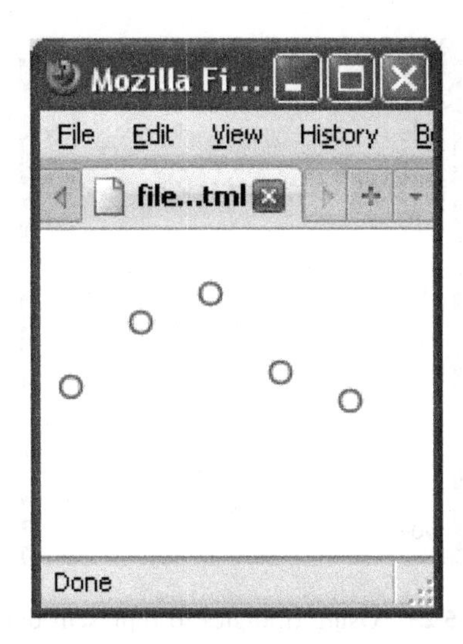


FIGURE 5.21 Simple plot with dots representing the data values of an array

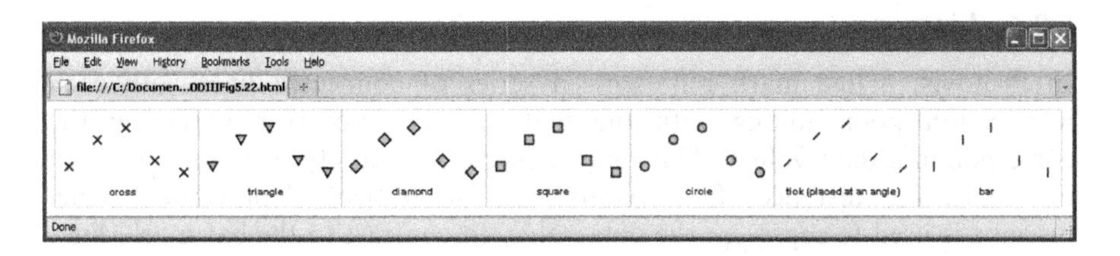


FIGURE 5.22 Different options for displaying dots

```
new pv.Panel()
   .width(150)
   .height(100)

  .add(pv.Dot)
   .data([1.4,2.3,2.7,1.6,1.2])
   .bottom(function(d) d * 30)
   .left(function() this.index * 30 + 15)
   .shape("triangle")
   .size(16)
   .strokeStyle("black")
   .lineWidth(1)
   .fillStyle("lightgray")

  .root.render();
```

Figure 5.23 shows the resulting plot.

FIGURE 5.23 Using triangles to represent data values

### 5.3.4 Line

In Protovis, a line mark represents a series of connected lines with each intersection point (along with the first and last point) representing the corresponding data values. This mark is used in a variety of plots, including line plots. The distance of the points from the horizontal or vertical panel margins is used to represent the data value. In contrast to the bar mark, lines are positioned based on the vertical distance from the bottom and the horizontal distance from the left margin; that is, there is no height property.

In the following code, a simple line plot is created. A line mark (representing a series of joined points) is added to the panel for an array of data values. The points along the line represent each data array entry, and the vertical distance is calculated using a function applied to the data values. The horizontal distance is calculated based on the array's index position, with the value at index 0 positioned at a distance of 0 ($0 \times 30$), index 1 at distance 30 ($1 \times 30$), and so on. The resulting line plot is shown in Fig. 5.24.

```
new pv.Panel()
   .width(150)
   .height(100)

  .add(pv.Line)
   .data([1.4,2.3,2.7,1.6,1.2])
   .bottom(function(d) d * 30)
   .left(function() this.index * 30)

  .root.render();
```

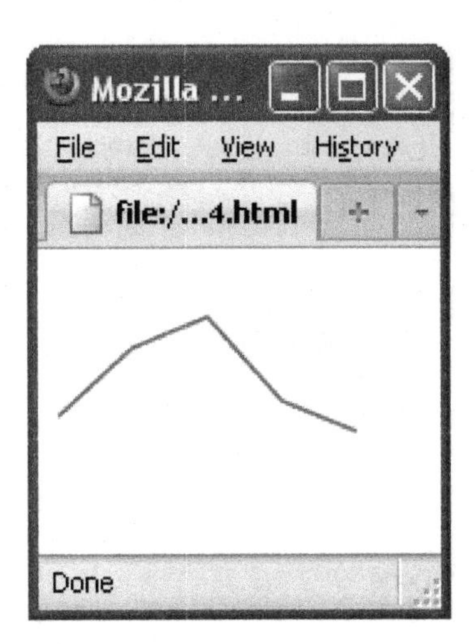


**FIGURE 5.24** Simple line plot

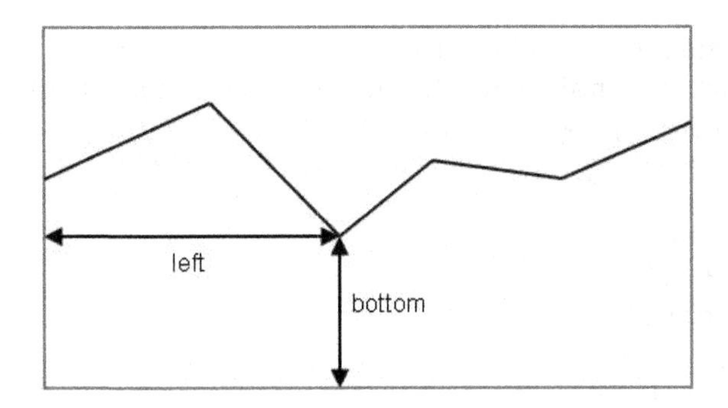


**FIGURE 5.25** Positioning of the lines

Figure 5.25 illustrates how the points along the line are positioned based on the distance from the left and the bottom of the panel.

Other specific properties of the line mark that control its appearance include `fillStyle` (for coloring the inner region of the lines), `lineWidth` (to set the thickness of the lines), and `strokeStyle` (for coloring the borders of the line). These properties are set in the same manner as the corresponding bar properties.

`Segmented` is a Boolean property that is set to true if the different line segments should be treated with different types of rendering, such as different line widths or different colors. In the following code, the segmented property is set to true. The line width and the colors for the different segments are calculated using a function. The color is defined using the RGB (red-green-blue) format that sets percentages for the red, green, and blue colors, which are combined to create the final color. This is described in more detail in the

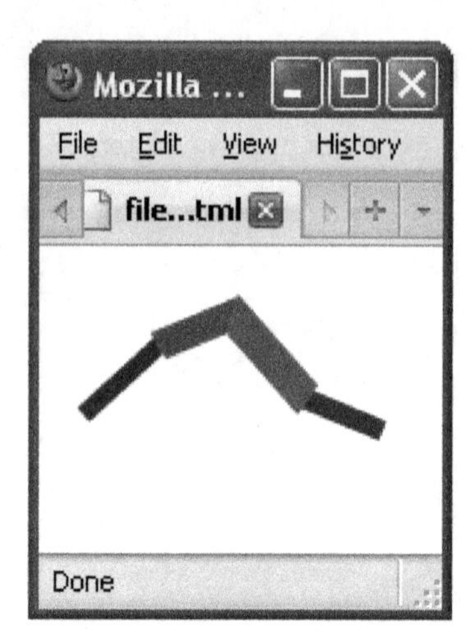


**FIGURE 5.26** Different line colors where the segmented property is set to true

following section. A final color value is created by concatenating the text `"rgb (`"with a value returned from a function to indicate the intensity of the red color, with values representing the green and blue colors (`"0%"`). For example, the first color would be represented as `"rgb(42%0%0%)"`.

```
new pv.Panel()
   .width(150)
   .height(100)

  .add(pv.Line)
   .segmented(true)
   .data([1.4,2.3,2.7,1.6,1.2])
   .bottom(function(d) d * 30)
   .left(function() this.index * 30 + 10)
  .strokeStyle(function(d) "rgb(" + (d*30) + "%,0%,0%")
   .lineWidth(function(d) d*5)
```

This generates the display as shown in Fig. 5.26.

In addition, the manner in which the lines are joined can be set using the `lineJoin` property that accepts `"bevel"`, `"round"`, and `"miter"` values (only `"miter"` can be used when `segmented` is set to true).

Finally, the property `interpolate` can be used to define how to connect the points (representing the data values). There are a number of options: `"linear"`, `"step-before"`, `"step-after"`, `"polar"`, `"polar-reverse"`, `"basis"`, and `"cardinal"`. As shown in the examples so far in this section where a straight line is drawn between the values, `"linear"` is the default setting. The other interpolate options are shown in Fig. 5.27.

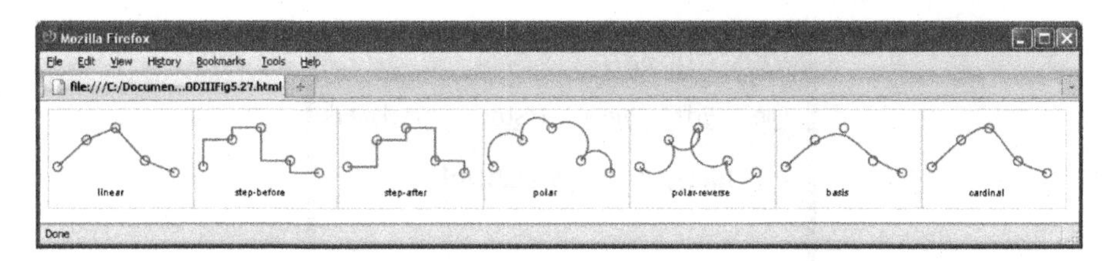


**FIGURE 5.27** Interpolate options for connecting lines

"step-before" and "step-after" connects pairs of values with horizontal and vertical lines. With "step-before", the vertical line is at the first value whereas with "step-after", the vertical line is at the second value. "polar" and "reverse-polar" join the points with an arc either clockwise ("polar") or counterclockwise ("reverse-polar"). "basis" is used to draw a b-spline fit with "cardinal" used to draw cardinal splines. The tension property is used with cardinal splines, has a value between 0 and 1 (the default is 0.7), and can be used to control the cardinal rendering. The closer to 1 the tension value is, the more linear the lines will be rendered between the values.

In the following code, a line plot is displayed where the lines are joined using the interpolate property set to "cardinal" with a tension of 0.5.

```
    new pv.Panel()
      .width(150)
      .height(100)

     .add(pv.Line)
      .data([1.4,2.3,2.7,1.6,1.2])
      .bottom(function(d) d * 30)
      .left(function() this.index * 30 + 10)
      .interpolate("cardinal")
      .tension(0.5)
      .add(pv.Dot)

.root.render();
```

This creates the plot as shown in Fig. 5.28.

### 5.3.5 Area

The line mark represents a series of connected lines or polylines. The *area* mark represents the space between two polylines. These types of marks are often used as area plots, to visualize trends for one or more properties.

Figure 5.29 illustrates the placement of the points on two polylines that make up the area mark. A number of properties can be used to control these

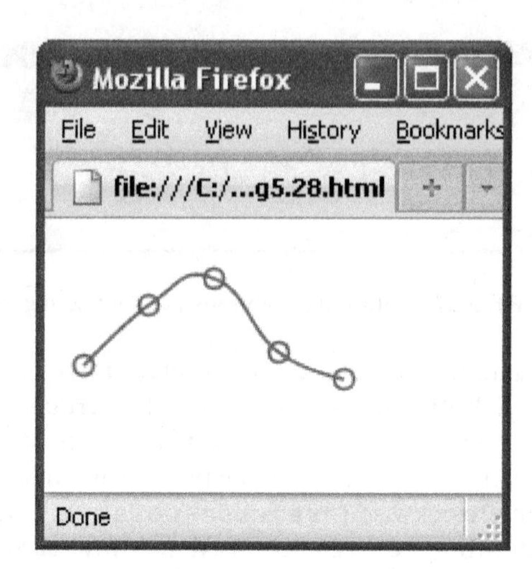


**FIGURE 5.28** Line mark using the interpolate property set to "cardinal" with a tension of 0.5

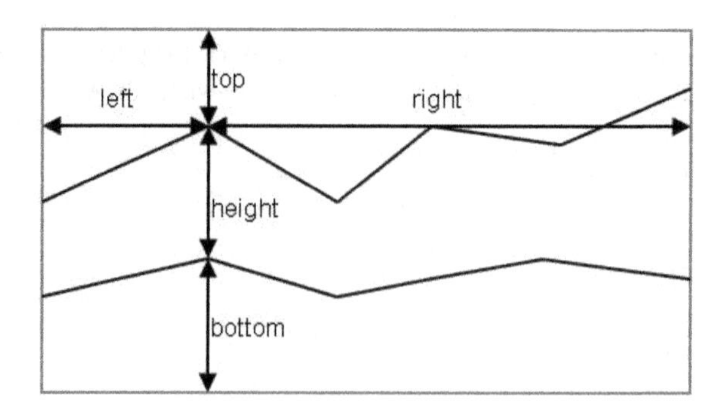


**FIGURE 5.29** Positioning properties for a horizontally aligned area mark

points: `top`, `left`, `bottom`, `right`, `width`, and `height`. For area marks, the `width` property controls the distance between the two polylines if the area mark is vertically aligned, whereas the `height` property is used to control the distance between the two polylines where the mark is horizontally aligned (as shown in Fig. 5.29). Because an area plot is either aligned horizontally or vertically, only a single property needs to be defined (`width` or `height`).

The following code displays a simple area plot. The lower polyline is drawn as a straight line at the bottom of the panel because the bottom property is set to 0. The height is used to display the top polyline based on the array of data defined. This height is based on a function `d*30` for each of the elements of the array. The horizontal position of each of the vertices of the top polyline is a function of the array index (`this.index*30`). The resulting area plot is displayed in Fig. 5.30.

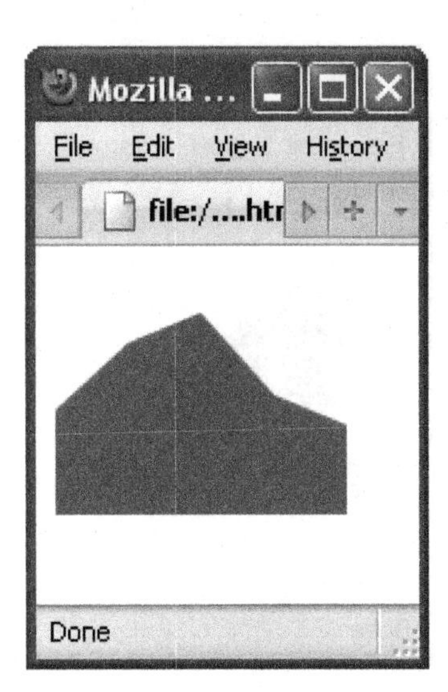


**FIGURE 5.30** Simple area plot

```
new pv.Panel()
   .width(150)
   .height(100)

  .add(pv.Area)
   .data([1.4,2.3,2.7,1.6,1.2])
   .bottom(0)
   .height(function(d) d * 30)
   .left(function() this.index * 30)

  .root.render();
```

By using different combinations of these positioning properties (`top`, `left`, `bottom`, `right`, `width`, and `height`), different types of area plots can be drawn such as those where the top and the bottom polylines vary. `fillStyle`, `lineWidth`, and `strokeStyle` can be used to alter the appearance of the area plot. Similar to the line mark, the area mark's property `segmented` can be set to true to enable different styles, such as colors, to be used for different sections of the area plot. In the following code, a vertically aligned area plot is created based on the data array specified. Two polylines are defined using the properties `left` and `width`. The `left` property, defined as a function of the data array `(d*10)` is used to position the left polyline, whereas the `width` is used to position the right polyline. The width is defined using a different function `(d*d*10)`, which represents the distance from the left polyline. Because the `segmented` property is set to true, the different areas of the area mark can be styled differently. In this code, the `fillStyle` property is used to

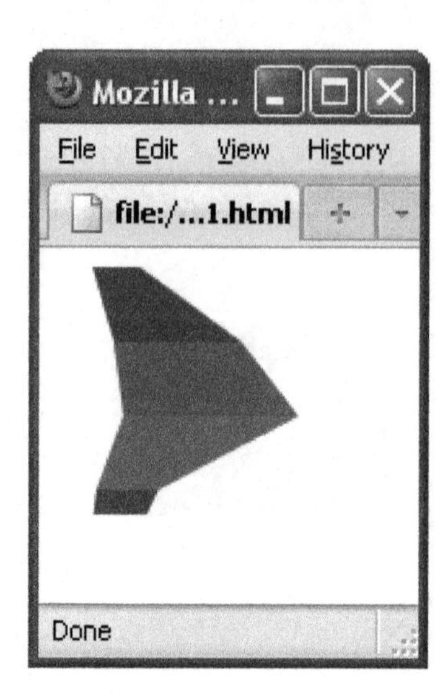


**FIGURE 5.31** Segmented area mark

color the area sections using different shades based on a function of the data array, as shown in Fig. 5.31.

```
new pv.Panel()
   .width(150)
   .height(100)

  .add(pv.Area)
   .segmented(true)
   .data([1.4,2.3,2.7,1.6,1.2])
   .left(function(d) d*10)
   .width(function(d) d*d*10)
   .top(function() this.index * 30)
   .fillStyle( function(d) "rgb(" + (d*30) + "%," + (30) +
     "%," + (30) + "%")

  .root.render();
```

The transition between the intersection points can also be controlled in a similar manner to the line mark using the `interpolate` property, which has the similar options as the line mark: `"linear"`, `"step-before"`, `"step-after"`, `"basis"`, and `"cardinal"`. Of these, `"linear"` is the default property value. In the following code, the `interpolate` property was set to `"step-after"`, and the resulting plot is shown in Fig. 5.32.

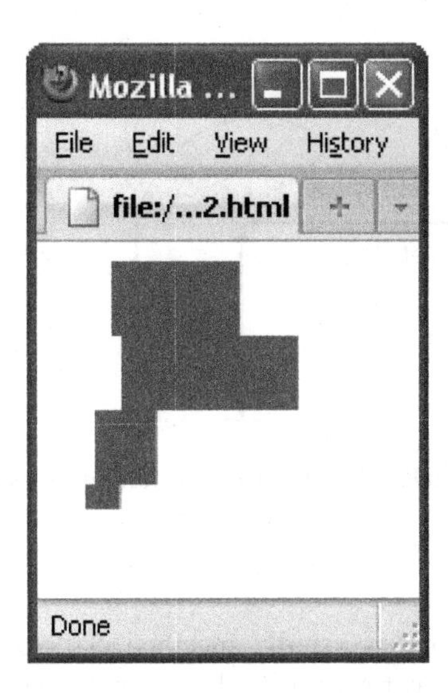


**FIGURE 5.32** Area with interpolate set to "step-after"

```
new pv.Panel()
  .width(150)
  .height(100)

 .add(pv.Area)
  .data([1.4,2.3,2.7,1.6,1.2])
  .left(function(d) d*10)
  .width(function(d) d*d*10)
  .top(function() this.index * 30)
  .interpolate("step-after")

 .root.render();
```

### 5.3.6 Wedge

A *wedge* is used to create graphics such as pie charts or donut plots. This mark uses several positioning properties to create a wedge mark on a panel: `top` (number of pixels from the top margin of the panel), `left` (number of pixels from the left margin of the panel), `bottom` (number of pixels from the bottom margin of the panel), and `right` (number of pixels from the right margin of the panel). These properties control the location of the center of the wedge mark.

Several properties specific to the wedge mark are used to construct the individual slices: `angle`, `startAngle`, `endAngle`, `innerRadius`, and `outerRadius`. The `angle` property is used to calculate the angle (in radians)

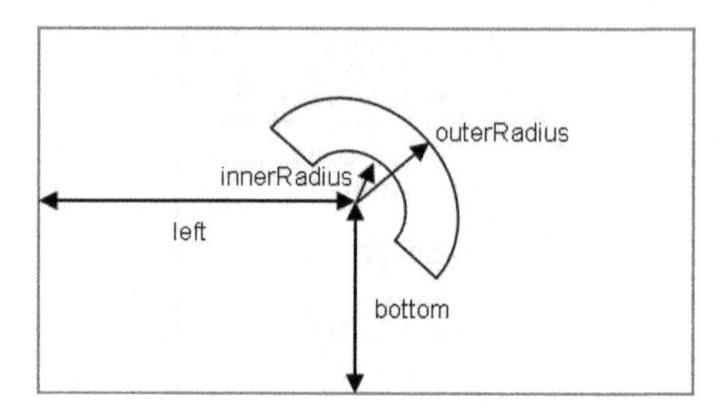


**FIGURE 5.33** Wedge-positioning properties

of the individual slices and is usually specified using a function based on the underlying data. The `startAngle` and `endAngle` properties can be used to specify a starting or ending angle. The `outerRadius` property is used to specify the radius of the outside of the pie chart, with the `innerRadius` used to specify an optional inner radius for generating donut plots (the default value for the `innerRadius` is 0). Figure 5.33 illustrates the use of `left`, `bottom`, `innerRadius`, and `outerRadius` properties to position a wedge mark.

The wedge mark is often used to create a pie chart where the entire circle represents 100% of the data, with individual slices of the chart representing a portion of the data. For example, to generate a pie chart corresponding to four data points [3,2,4,1], the entire pie chart (100%) represents the sum of all values (10), with each slice representing the corresponding proportion. The first slice (3) represents 3/10 or 30%, the second slice represents 2/10 or 20%, and so on. One approach within Protovis to calculate these proportions is to use the function `pv.normalize`, which returns a copy of the data array with each element represented as a proportion of the array. In this example, the `pv.normalize([3,2,4,1])` returns the array `[0.3,0.2,0.4,0.1]`.

The following code is used to create a simple pie chart based on an array of data. Because the pie chart requires a proportion of the data to be calculated, `pv.normalize` is used to translate the data array into proportions. Initially the chart is centered on the panel using the `bottom` and `left` properties, and the radius of the pie chart is set to 50. The angle of the individual slices of the wedge mark is computed using a function based on the normalized data array values: `2*d*Math.PI` (`Math.PI` is a JavaScript constant for $\pi$). The circumference of the circle in radians is defined as $2 \times \pi$ (or 360°); therefore, the angle for each slice is represented by the proportion of the data multiplied by $2 \times \pi$. Figure 5.34 shows the resulting pie chart. Because no styling has been defined for this pie chart, the default colors and line thickness is used to create the chart.

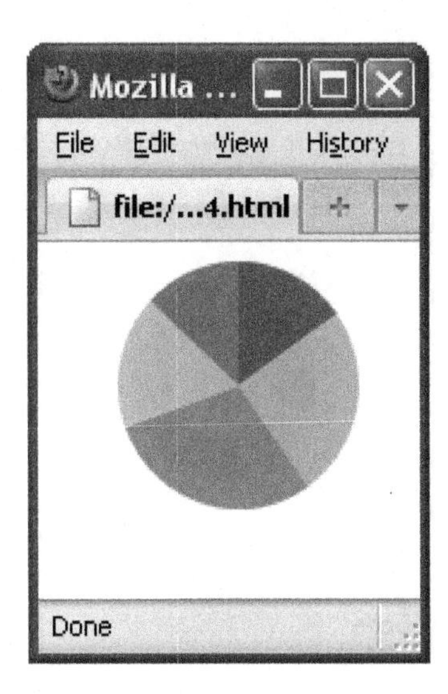


**FIGURE 5.34** Simple pie chart

```
new pv.Panel()
   .width(150)
   .height(100)

  .add(pv.Wedge)
   .data(pv.normalize([1.4,2.3,2.7,1.6,1.2]))
   .bottom(50)
   .left(75)
   .outerRadius(50)
   .angle(function(d) d * 2 * Math.PI)

  .root.render();
```

The wedge mark can be customized using several properties: `fillStyle` (controls the color of the slice), `strokeStyle` (controls the color of the boundaries for each of the slices), and `lineWidth` (controls the width of the slice boundaries).

In the following code, a simple donut plot is created from the normalized data array. The outside of the pie chart is calculated using a function based on the value of the index. If it is the first element of the array, the radius is set to 45 pixels; otherwise, it is set to 40 pixels. The `innerRadius` is set to 20 pixels to create a donut plot. The angle of the slices is set as in the previous example; however, the default colors are defined as alternating gray and black (defined with the `pv.colors` collection of colors as shown). The boundary of the slices is set to have a thickness of 4 pixels and to be rendered in white, as shown in Fig. 5.35.

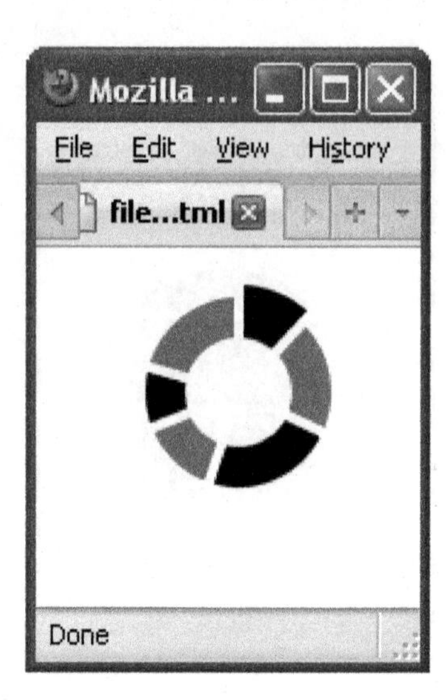


**FIGURE 5.35** Customized donut plot

```
new pv.Panel()
   .width(150)
   .height(100)

  .add(pv.Wedge)
   .data(pv.normalize([1.4,2.3,2.7,1.6,1.2,2.4]))
   .bottom(50)
   .left(75)
   .outerRadius(function() !this.index ? 45 : 40)
   .innerRadius(20)
   .angle(function(d) d * 2 * Math.PI)
   .fillStyle(pv.colors("black","gray","black","gray",
     "black","gray"))
   .strokeStyle("white")
   .lineWidth(4)

  .root.render();
```

### 5.3.7 Images

Images (such as images contained in PNG or JPEG files) can be added to a panel aligned with the bottom and left margins. Images are positioned with the `top`, `left`, `bottom`, and `right` properties. `height` and `width` properties are used to define the size of the image on the panel. Figure 5.36 illustrates the use of these properties to draw an image on a panel.

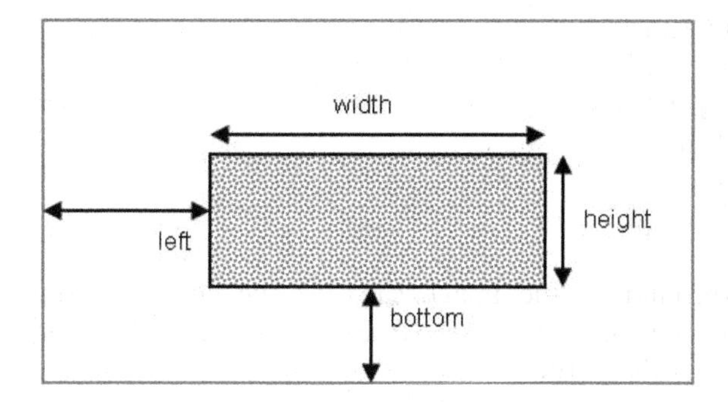


**FIGURE 5.36** Properties to position an image on-screen

**FIGURE 5.37** Display an image on the panel

The following code displays an image from a file. In this case, the image file is located in the directory where the HTML file is also located. A URL could be used to locate an image file from any URL location accessible on the Internet. The image is added to the panel, and the image file location is set using the `url` expression. The image is positioned 25 pixels from the left side of the panel and 25 pixels from the bottom of the panel. The `height` and `width` of the image are set to 50 and 151, respectively, and the results are shown in Fig. 5.37.

```
new pv.Panel()
   .width(200)
   .height(100)

  .add(pv.Image)
   .url("wiley-logo.bmp")
```

```
      .left(25)
      .bottom(25)
      .height(50)
      .width(151)

    .root.render();
```

The background to the image can also be changed using fillStyle (background color), strokeStyle (the color of the border), and lineWidth (the thickness of the border).

### 5.3.8 Exercises

5.3.8.1 Create a bar chart aligned with the x-axis from the array [3,5,6,8,9,8,11] with a "lightblue" bar, a border drawn in "blue", and data value labels.

5.3.8.2 Create a bar chart aligned with the y-axis with the same attributes as Exercise 5.3.8.1.

5.3.8.3 Create a graphic from the array [3,5,6,8,9,8,11] using "maroon" crosses at a 45° angle, with a radius of 8 pixels to represent the data values. Add a label to each of the data values.

5.3.8.4 Create a line plot from the array [4.3,5.4,7.3,6.9, 10.3,11.5] using the cardinal interpolate option, and color the line "dark-green" with a 2-pixel width.

5.3.8.5 Create five different line plots based on Exercise 5.3.8.4, with different tension values: 0, 0.25, 0.5, 0.75, 1.

5.3.8.6 Create six different line plots based on Exercise 5.3.8.4, where the interpolate property is set to "linear", "step-before", "step-after", "polar", "polar-reverse", and "basis", respectively.

5.3.8.7 Create a vertically aligned area plot from the array [43.6,54.8,47.2,34,7,58.6,34.1]where the left polyline is along the y-axis, and the variation in the data is shown in the right polyline. The inner region of the plot should be colored "steelblue" with "darkblue" border.

5.3.8.8 Create seven different plots by modifying the plot created in Exercise 5.3.8.7 such that each plot uses a different interpolate option ("linear", "step-before", "step-after", "polar", "polar-reverse", "basis", and "cardinal").

5.3.8.9 Create a donut plot corresponding to the array [4,5,7,8,2] with the colors "white", "lightgray", "gray", "darkgray", and "black", and also add a label to each of the wedges showing the data value.

5.3.8.10 Create an image using the wiley-logo.bmp file that is 302 pixels wide and 100 pixels high, with a black border of 10 pixels.

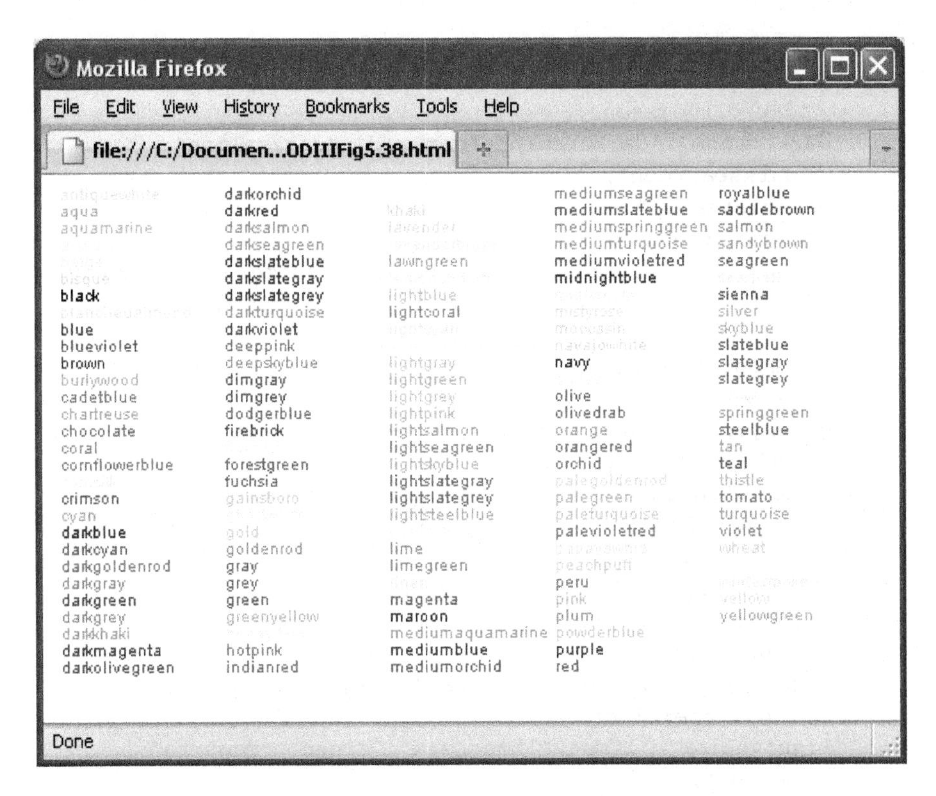


**FIGURE 5.38** Display of all named colors

## 5.4 CREATING CUSTOMIZED PLOTS

### 5.4.1 Colors

Color is often essential for visualizations to encode data or highlight elements in a graphic. Protovis makes use of standard definitions for colors (defined at www.w3.org/TR/css3-color/). This includes the use of a series of named colors, such as `"black"`, `"cyan"`, or `"lightgray"`. In the following code, the list of all named colors is provided in the variable array `namedColors`, and the code displays the name and colors each in a series of columns, as shown in Fig. 5.38.

```
var namedColors =
["aliceblue","antiquewhite","aqua","aquamarine",
    "azure","beige","bisque","black","blanchedalmond","blue",
    "blueviolet","brown","burlywood","cadetblue","chartreuse",
    "chocolate","coral","cornflowerblue","cornsilk","crimson",
```

```
"cyan","darkblue","darkcyan","darkgoldenrod","darkgray",
"darkgreen","darkgrey","darkkhaki","darkmagenta",
   "darkolivegreen",
"darkorange","darkorchid","darkred","darksalmon",
   "darkseagreen",
"darkslateblue","darkslategray","darkslategrey",
   "darkturquoise",
"darkviolet","deeppink","deepskyblue","dimgray","dimgrey",
"dodgerblue","firebrick","floralwhite","forestgreen",
   "fuchsia",
"gainsboro","ghostwhite","gold","goldenrod","gray",
   "grey","green",
"greenyellow","honeydew","hotpink","indianred",
   "indigo","ivory",
"khaki","lavender","lavenderblush","lawngreen",
  "lemonchiffon",
"lightblue","lightcoral","lightcyan",
  "lightgoldenrodyellow","lightgray",
"lightgreen","lightgrey","lightpink","lightsalmon",
   "lightseagreen",
"lightskyblue","lightslategray","lightslategrey",
   "lightsteelblue",
"lightyellow","lime","limegreen","linen","magenta",
   "maroon",
"mediumaquamarine","mediumblue","mediumorchid",
   "mediumpurple",
"mediumseagreen","mediumslateblue","mediumspringgreen",
   "mediumturquoise",
"mediumvioletred","midnightblue","mintcream",
   "mistyrose","moccasin",
"navajowhite","navy","oldlace","olive","olivedrab",
   "orange","orangered",
"orchid","palegoldenrod","palegreen",
   "paleturquoise","palevioletred",
"papayawhip","peachpuff","peru","pink","plum",
   "powderblue","purple","red",
"rosybrown","royalblue","saddlebrown","salmon",
   "sandybrown","seagreen",
"seashell","sienna","silver","skyblue","slateblue",
   "slategray","slategrey",
"snow","springgreen","steelblue","tan","teal",
   "thistle","tomato",
"turquoise","violet","wheat","white","whitesmoke",
   "yellow","yellowgreen"];
```

```
var panelWidth = 500, panelHeight = 300;
var nosRows = 30, columnWidth = 100, rowWidth = 10;

var chartPanel = new pv.Panel()
   .width(panelWidth)
   .height(panelHeight);

chartPanel.add(pv.Label)
   .data(namedColors)
   .text(function(d) d)
   .top(function() (this.index%nosRows)*rowWidth)
   .left(function() Math.floor(this.index/nosRows)
     *columnWidth)
   .textStyle(function(d) d);

chartPanel.render();
```

There are alternative and more flexible ways to define specific colors. One approach is to use the red-green-blue (RGB) format, where the strength of each of the three colors is combined to create a spectrum of colors. The RGB color can be set using hexadecimal characters whereby the symbol "#" is followed by either three characters (#rgb) or six characters (#rrggbb). For example, #fff and #ffffff denote the color white. In the following code, five dot marks are drawn using the RGB three-character color format for white, black, light dark gray, gray, and dark gray, as shown in Fig. 5.39.

```
var rgbColorExamples = ["#fff","#000","#999","#666",
  "#222"];

var panelWidth = 500, panelHeight = 200;

var chartPanel = new pv.Panel()
   .width(panelWidth)
   .height(panelHeight);

chartPanel.add(pv.Dot)
   .data(rgbColorExamples)
   .bottom(100)
   .left(function() (this.index+1)*75)
   .radius(50)
   .strokeStyle("black")
   .fillStyle(function(d) d)
```

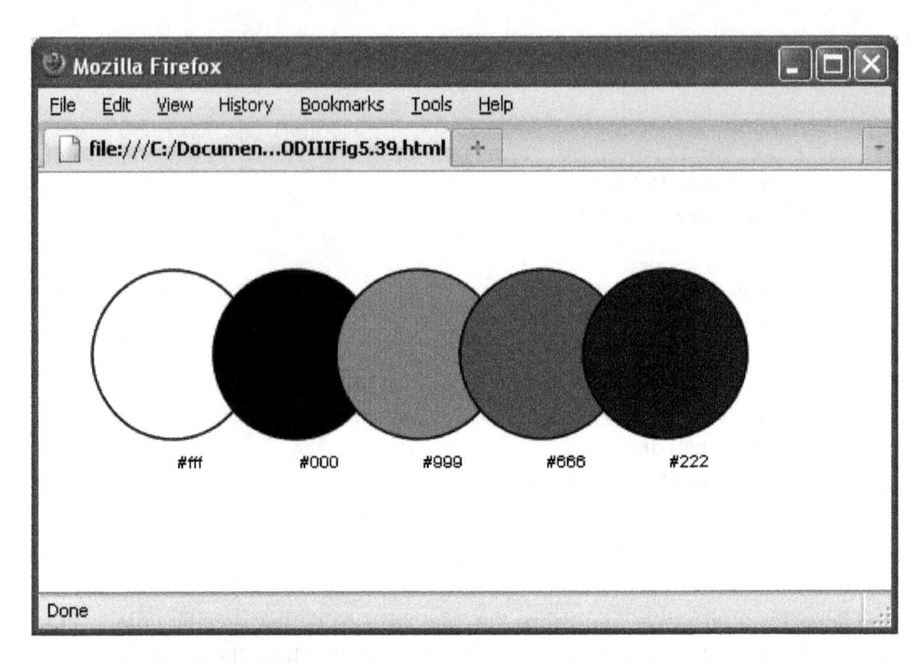


**FIGURE 5.39** Three-character RGB colors

```
    .add(pv.Label)
     .bottom(30);

  chartPanel.render();
```

Alternatively, colors can be defined with "`rgb( . . . )`" notation, where the contents of the parentheses are either three numeric values (in the range 0–255) or three percentages. For example, `rgb(255,255,255)` and `rgb (100%,100%,100%)` both denote the color white. Opacity can also be added to a color using the RGBA definitions. The "A" references an *alpha* value, which can take any number between 0 and 1 such as `rgba(255,0,0,0.5)`. The higher the value is, the more opaque is the color. Figure 5.40 illustrates its use in generating five colored dot marks, shown with the corresponding RGBA representation.

Colors can also be defined using the HSL (hue-saturation-lightness) format; for example, `hsl(120,100%,75%)` is a light green. The first number is the angle of the color and is represented as a number in the range 0–360 (red is 360, blue is 240, green is 120). Saturation and lightness are represented as percentages. Similar to the RGBA format, there is an HSLA format where an additional alpha parameter can be specified to indicate the color's opaqueness.

Protovis also contains classes to generate color objects—`pv.color.Hsl` and `pv.Color.Rgb`—that will create colors based on the HSL and RGB

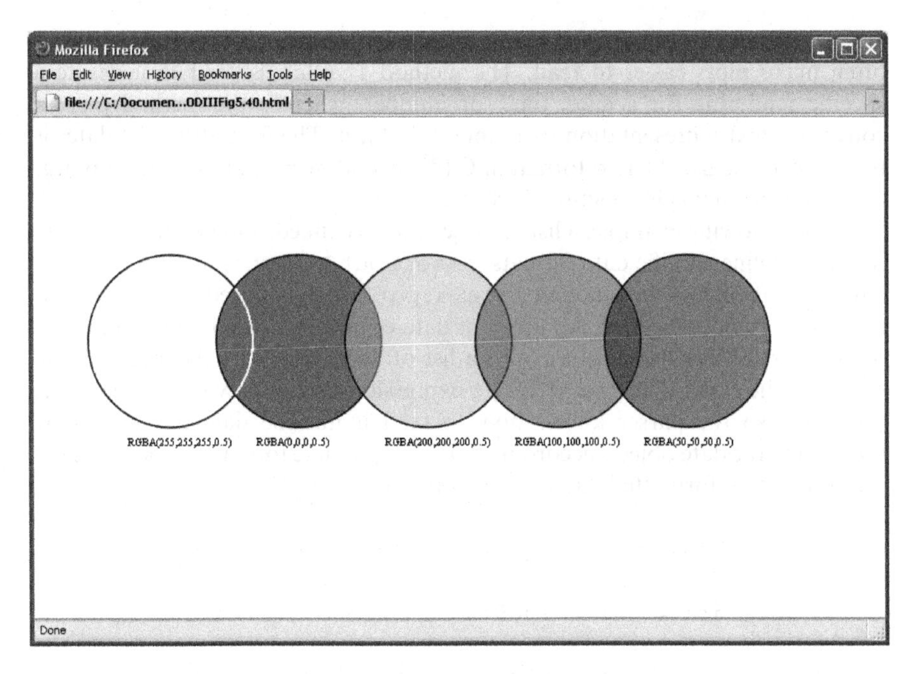


**FIGURE 5.40** Using the RGBA color format

color definitions. For `pv.color.HSL`, methods `hue`, `lightness`, and `saturation` are used to set or retrieve the color settings; for `pv.color.Rgb`, methods `red`, `green`, and `blue` will get or set the color channels. In addition, the methods `darker` and `lighter` can be used to adjust the brightness of the colors.

### 5.4.2 Formatting

The manner in which labels are presented can also be customized. The font property is set based on W3C standard font definitions (defined at: http://www.w3.org/TR/CSS2/fonts.html). For example, the following sets the font for the label to be 12 pixels, bold, sans serif.

```
chartPanel.add(pv.Label)
  .text("Temperature at Heathrow Airport")
  .top(0)
  .left(200)
  .font("bold 12px sans-serif");
```

The way numbers, times, and dates are presented can also be controlled using the `pv.Format` classes: `pv.Format.date`, `pv.Format.number`, and `pv.Format time`. Changing the way these types of objects are drawn can often make plots easier to read. The method `format` is used to generate a specific rendering for a date, time, or number, whereas the method `parse` converts a text representation to an internal object. The formatting for dates is the same as the `strftime` format in C (defined at http://pubs.opengroup.org/onlinepubs/009695399/functions/strftime.html).

In the following example, a list of three dates is defined, and two functions are set up to define specific date formats. `dateInputFormat` is used to parse the dates described, and the function `dateOutputFormat` is used to describe how the date is to be presented. An array of date objects is created by mapping the original list of dates (`dateList`) to a list of date objects by parsing the date strings in the array. The `"%m/%d/%y"` expression defines how the date string is formatted so the parser knows how to read it in. The date is redrawn by formatting the date objects according to the output date format `"%A %d %b %Y"`. The resulting reformatted dates are shown in Fig. 5.41.

```
var panelHeight = 100, panelWidth = 100;

var dateList = ["3/2/2009", "3/3/2009", "3/4/2009"];

var dateInputFormat = pv.Format.date("%m/%d/%y");
var dateOutputFormat = pv.Format.date("%A %d %b %Y");
```

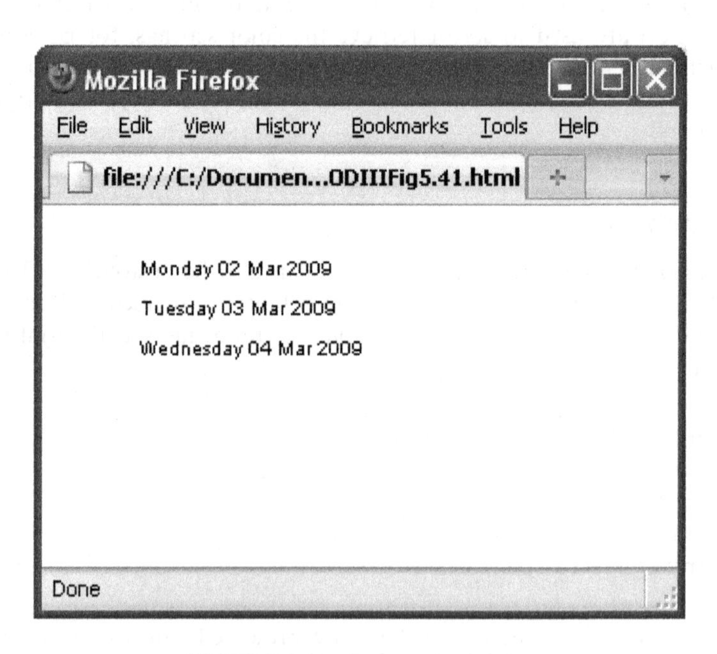


**FIGURE 5.41** Reformatted dates

```
var chartPanel = new pv.Panel()
    .width(panelWidth)
    .height(panelHeight)
    .margin(margin);

chartPanel.add(pv.Label)
.data(dateList.map(function(d) dateInputFormat.parse(d)))
 .text(function(d) dateOutputFormat(d))
 .left(10)
 .top(function() this.index*20);

 chartPanel.render();
```

Similarly, the `pv.Format.time` class will parse and format times, supporting two types: `"short"` and `"long"`. Finally, numbers can be formatted with the `pv.Format.number` class, which allows a considerable amount of flexibility in how numbers are presented. The method `fractionDigits` controls the number of digits to be presented after the decimal place, whereas the function `integerDigits` controls the number of digits to be displayed before the decimal place. For example, in the following code, a number format function is declared as a variable (`numberFormat`) with two decimal places. It is used to display the numbers in the list (`numberList`) as shown in Fig. 5.42.

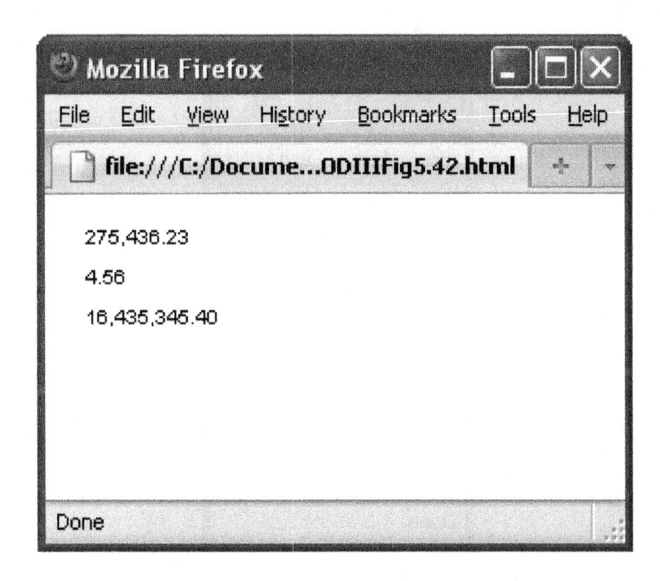


**FIGURE 5.42** Formatted numbers

```
var panelHeight = 100, panelWidth = 100;

var numberList = [275436.2343, 4.55563, 16435345.4];

var numberFormat = pv.Format.number().fractionDigits(2);

var chartPanel = new pv.Panel()
    .width(panelWidth)
    .height(panelHeight);

chartPanel.add(pv.Label)
  .data(numberList )
  .text(function(d) numberFormat(d))
  .left(10)
  .top(function() this.index*20 + 20);

 chartPanel.render();
```

### 5.4.3 Anchors

All marks have a number of defined positions close to the mark that are called *anchors*. For example, a dot mark has five defined positions: `"top"`, `"bottom"`, `"left"`, `"right"`, and `"center"`. In Fig. 5.43, label marks have been added to a dot mark in the `"top"`, `"bottom"`, `"left"`, `"right"`, and `"center"` positions.

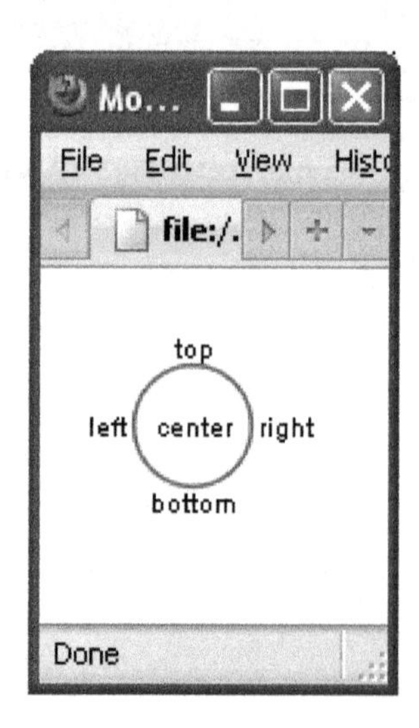


**FIGURE 5.43** Anchors used to position labels close to a dot mark

An anchor is another type of mark; however, you will never see it directly because it is used to conveniently position other marks. In the following example, a simple bar chart is constructed with labels. To display the label of the data at the top of each bar, an anchor mark is initially added with the target defined as "top". The label is then added to the anchor.

```
var barPanel = new pv.Panel()
   .width(150)
   .height(100);

var bar = barPanel.add(pv.Bar)
   .data([1.4,2.3,2.7,1.6,1.2])
   .bottom(0)
   .width(25)
   .height(function(d) d * 30)
   .left(function() this.index * 30);

bar.anchor("top").add(pv.Label);

barPanel.render();
```

Figure 5.44 shows the resulting simple bar chart with a label corresponding to the data presented at the top of each bar.

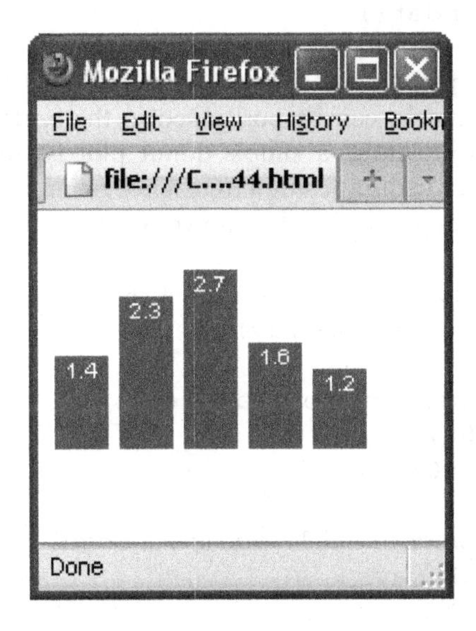


**FIGURE 5.44** Bar chart with label anchored at the top

In another example, a vertically aligned area mark is used to represent the data. Dots are used to annotate the left and right boundaries of the plot, with black diamonds used at the left and gray squares used at the right.

```
var barPanel = new pv.Panel()
  .width(100)
  .height(150);

var area = barPanel.add(pv.Area)
  .data([1.4,2.3,2.7,1.6,1.2])
  .left(10)
  .fillStyle("lightgray")
  .strokeStyle("lightgray")
  .width(function(d) d * 30)
  .bottom(function() this.index * 30 + 5);

area.anchor("left").add(pv.Dot)
  .shape("diamond")
  .fillStyle("black")
  .strokeStyle("black");

area.anchor("right").add(pv.Dot)
  .shape("square")
  .fillStyle("gray")
  .strokeStyle("gray");

barPanel.render();
```

This results in the plot shown in Fig. 5.45.

Similarly, anchor marks can be positioned in the center of marks used to display the plot. To illustrate, a simple donut plot is constructed using the wedge class. The target anchor position for each segment of the donut plot is set to "center". This results in the plot shown in Fig. 5.46.

```
var dataArray = [1.4,2.3,2.7,1.6,1.2],
  sum = pv.sum(dataArray);

var wedgePanel = new pv.Panel()
  .width(100)
  .height(100);

var wedge = wedgePanel .add(pv.Wedge)
 .data(dataArray)
 .bottom(50)
 .left(50)
```

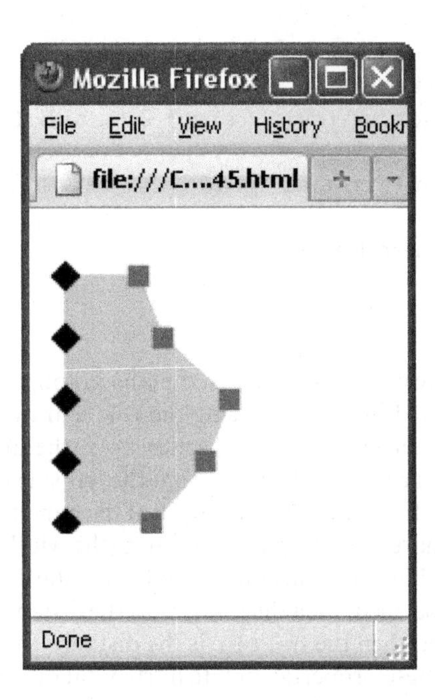


**FIGURE 5.45** Area plot with dot marks (diamonds and squares) anchored to the left and right

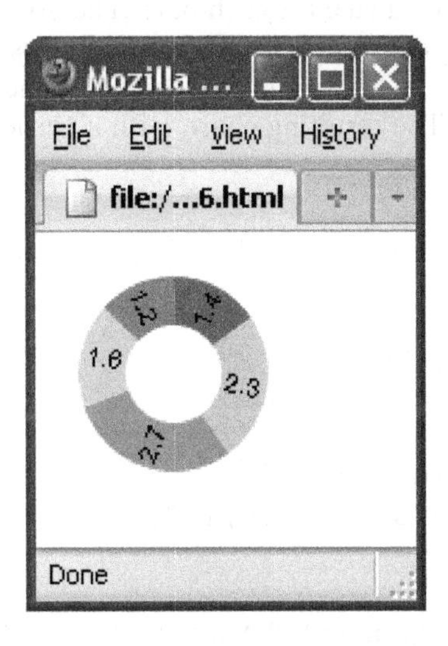


**FIGURE 5.46** Donut plot with label anchored to the center of each wedge

```
    .innerRadius(20)
    .outerRadius(40)
    .angle(function(d) d / sum * 2 * Math.PI);

  wedge.anchor("center").add(pv.Label);

  wedgePanel.render();
```

### 5.4.4 Rule

Axes and grid lines are essential elements of many graphics. These horizontal or vertical lines are created in Protovis using the rule mark (`pv.Rule`). They are positioned on the screen using the properties `top` (the distance from the top margin of the panel), `left` (the distance from the left side margin of the panel), `bottom` (the distance from the bottom margin of the panel), `right` (the distance from the right edge margin of the panel), `width` (the width of the horizontal or vertical line), and `height` (the height of the horizontal or vertical line). If the `bottom` position is specified, then the `height` is the distance from the bottom; if the `top` position is set, then the `height` is the distance from the top.

Combinations of these properties create horizontal or vertical lines. A simple approach is to set the bottom property to draw a horizontal line or set the left property to create a vertical line. More control of the position of the lines can be achieved using a combination of multiple properties; for example, you can set the `left`, `bottom`, and `top` to create a vertical line or set the `bottom`, `left`, and `right` to create a horizontal line.

In the following examples, a bar chart is created and added to a panel, and then two rule marks are added to the panel (they are not added to the bar because that would create a rule for each bar). The horizontal axis is drawn by setting the `bottom` property to 0; the vertical axis is created by setting the `left` property to 0. The length and height of the two rules is inherited from the parent panel. The resulting chart, with a vertical and horizontal axis, is shown in Fig. 5.47.

```
var barPanel = new pv.Panel()
  .width(140)
  .height(100)
  .margin(5);


barPanel.add(pv.Bar)
  .data([1.4,2.3,2.7,1.6,1.2])
  .bottom(0)
  .width(20)
  .height(function(d) d * 30)
  .left(function() this.index * 30);
```

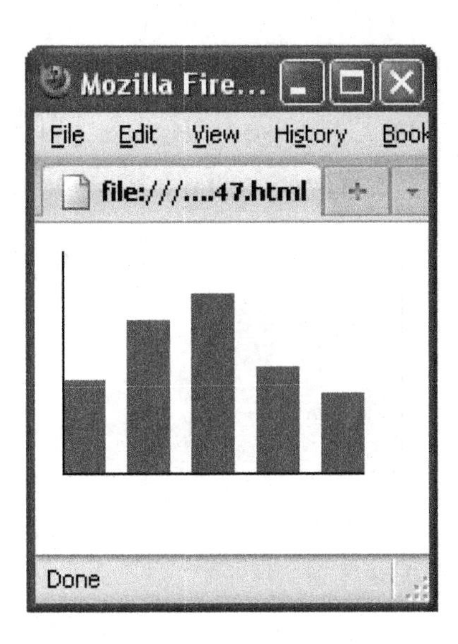


**FIGURE 5.47** Two rule marks added to the x-axis and y-axis

```
barPanel.add(pv.Rule)
  .bottom(0);

barPanel.add(pv.Rule)
  .left(0);

barPanel.render();
```

The following example generates grid lines. The code uses method `pv.range`, which creates a list of evenly spaced values starting at 0.5 and incrementing at 0.5 intervals for values less than 3.5. Adding a rule mark and assigning this data array to it creates a series of grid lines. In this example, they are colored light gray. Figure 5.48 shows the resulting chart.

```
var barPanel = new pv.Panel()
  .width(150)
  .height(100)
  .margin(5);

barPanel.add(pv.Bar)
  .data([1.4,2.3,2.7,1.6,1.2])
  .bottom(0)
  .width(20)
  .height(function(d) d * 30)
  .left(function() this.index * 30);
```

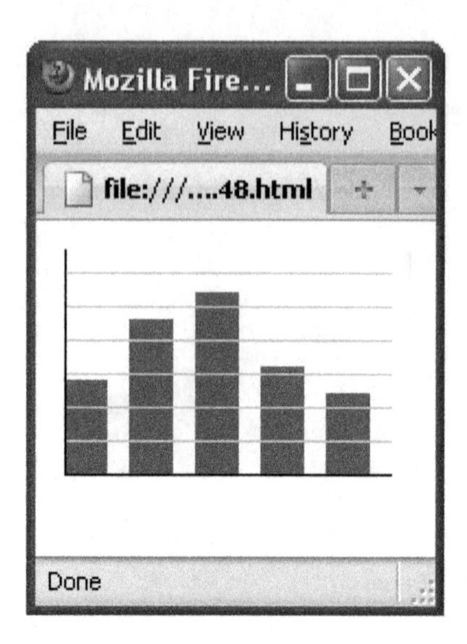


**FIGURE 5.48** Grid lines added to the chart using the rule mark

```
barPanel.add(pv.Rule)
  .data(pv.range(0.5,3.5,0.5))
  .bottom(function(d) d * 30)
  .strokeStyle("lightgray");

barPanel.add(pv.Rule)
  .bottom(0);

barPanel.add(pv.Rule)
  .left(0);

barPanel.render();
```

In a similar manner to the graphical elements discussed earlier, the `lineWidth` property is used to set the number of pixels the lines are wide, and the `strokeStyle` property is used to set the color of the line (as shown in the preceding example).

### 5.4.5 Scales

As discussed in Chapter 3, the data for many graphics is displayed using aesthetic variables such as the x- and y-positions of a mark and its color. For each data variable, its values lie in data space (the domain) and need to be mapped to values in an aesthetic space (the range). To achieve this, a function needs to perform a mapping. *Scales* are functions that map data values to aesthetic values.

In the examples discussed so far, this has been performed using a simple function, such as multiplying the data value by a constant. In Protovis, a series of classes is dedicated to these types of mappings. The `pv.Scale` abstract class is not used directly for the same reasons `pv.Mark` is not used directly. Classes that are extensions of the `pv.Scale` class provide functions to map continuous data onto aesthetic variables such as position or color (quantitative scales), continuous data onto a discrete range (quantile scale), and categorical values onto discrete colors or positions (ordinal scales). Because Protovis can assign a function to the variable, these scaling functions can be defined upfront as variables and then used throughout the code.

Quantitative scales are implemented with three methods: `pv.Scale.linear`, `pv.Scale.log`, and `pv.Scale.root`. Each of these methods is an extension of the class `pv.Scale.quantitiative`, which is an abstract class that is not used directly. In the following example, the method `pv.Scale.linear` is defined as a variable (`yMapping`). This is a function that defines a mapping from a range of data values (0 is the smallest data value, and 100 is the largest) onto a vertical pixel position (between 0 and the height of the panel). A bar mark is used to display the series of data values, and each of these values is between 0 and 100.To calculate the height of the bar, the previously defined `yMapping` variable is used. The resulting bar chart is shown in Fig. 5.49. Note that the data array can be assigned to the function instead of minimum

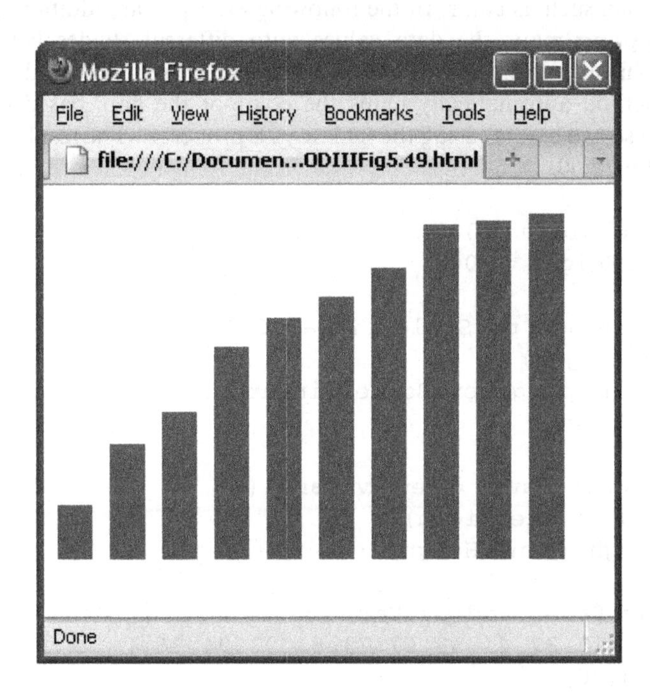


**FIGURE 5.49** Bar chart created with a linear scale function

and maximum values (along with an optional accessor function described in Section 5.5.3).

```
var panelWidth  = 300,
    panelHeight = 200;

var yMapping = pv.Scale.linear(0, 100).range(0,
   panelHeight);
var barChartPanel = new pv.Panel()
      .width(panelWidth)
      .height(panelHeight);

barChartPanel.add(pv.Bar)
  .data([15,32,41,59,67,73,81,93,94,96])
  .width(20)
  .bottom(0)
  .height(yMapping)
  .left(function() this.index * 30);

barChartPanel.render();
```

Scales can also be used to map the data domain to other aesthetic variables of the graphic, such as color. In the following example, an additional variable `colorMapping` maps the data values onto different shades between two colors. In this example, the two colors are specified as `"red"` and `"green"`, and the method automatically maps the color of the bars based on shades between these two colors (using the `fillStyle` property). The resulting chart is shown in Fig. 5.50.

```
var panelWidth  = 300,
    panelHeight = 200;

var yMapping = pv.Scale.linear(0, 100).range(0,
   panelHeight)
    colorMapping = pv.Scale.linear(0,100).range("red",
      "yellow");

var barChartPanel = new pv.Panel()
      .width(panelWidth)
      .height(panelHeight);

barChartPanel.add(pv.Bar)
  .data([15,32,41,59,67,73,81,93,94,96])
  .width(20)
  .bottom(0)
```

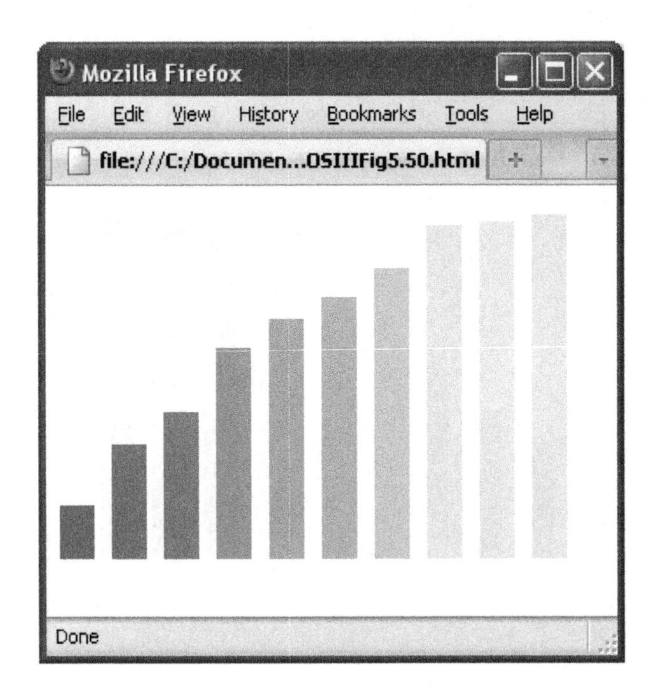


**FIGURE 5.50** Bar chart colored using the linear scale function

```
    .height(yMapping)
    .left(function() this.index * 30)
    .fillStyle(colorMapping);

  barChartPanel.render();
```

It is also possible to subdivide the range (aesthetic space) of a domain (data space). In Fig. 5.51, the bar chart was created in the same way as the previous example; however, the code that mapped the data values onto a single range defined by two colors

```
colorMapping =
pv.Scale.linear(0,100).range("red","green");
```

was replaced with code that maps the data values onto two ranges: red-to-green (for data values 0 to 75), and green-to-blue (for data values 75 to 100):

```
colorMapping =
pv.Scale.linear(0,75,100).range("red","green","blue");
```

The `ticks` method is helpful for displaying plots with defined axes. This method returns a series of evenly spaced intervals (around 5 to 10 values).

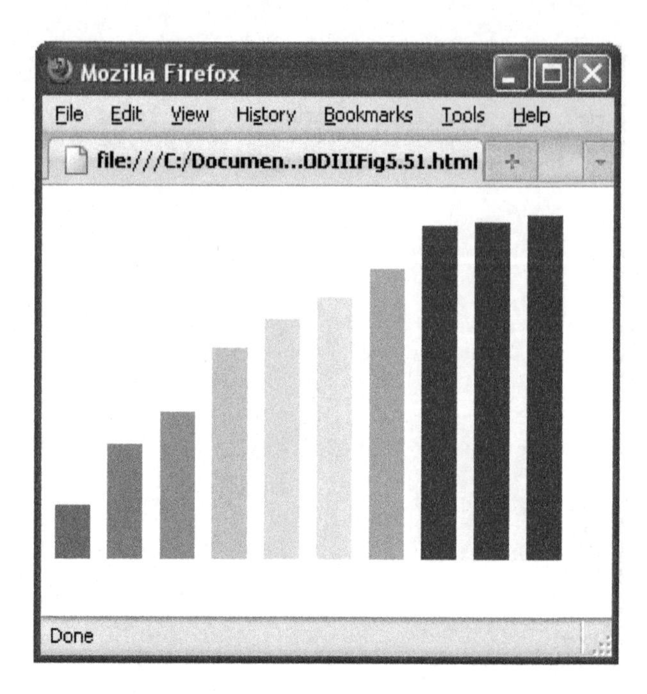


**FIGURE 5.51** Color scale using two ranges

These values can be used to annotate the plot's axes with tick marks or be used to create grid lines. In the following example, a bar chart is generated with tick marks displayed on the y-axis. The code defines a `yMapping` variable, which is a quantitative scale function to map data (in the range 0 to 100) onto the panel's vertical position (in the range 0 to the height of the panel). The y-axis is added to the chart using the rule mark. A second rule mark is used to add the ticks to the y-axis. The data property is defined as these ticks (a series of evenly spaced values) using the `ticks` methods. The ticks are positioned to start 2 pixels to the left of the y-axis mark (positioned at 4 pixels from the left). The width of the ticks is set to 2, so the ticks join with the y-axis. An anchor mark is set to `"left"` and used to display the values for the y-axis ticks using the label mark. To display these values, the method `tickformat` is called. The resulting chart is shown in Fig. 5.52.

```
var panelWidth = 304,
  panelHeight = 200;

var yMapping = pv.Scale.linear(0, 100).range(0,
 panelHeight);

var barChartPanel = new pv.Panel()
   .width(panelWidth)
```

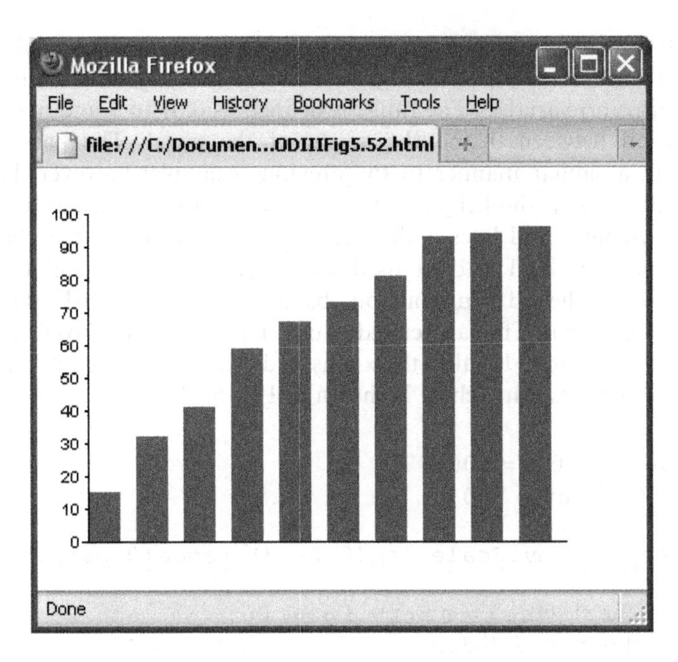


**FIGURE 5.52** Labeled y-axis using the ticks generated by the scale function

```
    .height(panelHeight)
    .margin(20)

barChartPanel.add(pv.Bar)
  .data([15,32,41,59,67,73,81,93,94,96])
  .width(20)
  .bottom(0)
  .height(yMapping)
  .left(function() this.index * 30 + 4);

barChartPanel.add(pv.Rule)
    .left(4)
    .add(pv.Rule)
     .data(yMapping.ticks())
     .bottom(yMapping)
     .left(2)
     .width(2)
    .anchor("left").add(pv.Label)
     .text(yMapping.tickFormat);

barChartPanel.add(pv.Rule)
    .bottom(0);

barChartPanel.render();
```

In the following example, a bar chart is created with the data displayed on a log scale. To set up the log mapping, the `pv.Scale.log` function is assigned to the `yMapping` variable. The range is defined as 0.01 to 7 and mapped to a pixel location between 0 and the height of the panel. The bars are then displayed in a similar manner to the previous example; however, because a log mapping is used, the height of the bars corresponds to a log function. A series of horizontal grid lines is then displayed on top of the bar chart panel using the rule mark. The data used to calculate their vertical location is computed using the `ticks` function, based on the previously defined log scale (`yMappings`). The associated chart labels are displayed using the `tickformat` method. Finally, the x-axis and y-axis are added as rule marks to the chart. The resulting chart is shown in Fig. 5.53.

```
var panelWidth = 300,
  panelHeight = 500;

var yMapping = pv.Scale.log(0.01,7).range(0,panelHeight);

var barChartPanel = new pv.Panel()
    .width(panelWidth)
    .height(panelHeight)
    .margin(30)

barChartPanel.add(pv.Bar)
   .data([0.012,0.017,0.022,0.034,0.067,0.089,1.4,2.6,
     4.6,6.4])
    .width(20)
    .bottom(0)
    .height(yMapping)
    .left(function() this.index * 30);

barChartPanel.add(pv.Rule)
    .data(yMapping.ticks())
    .strokeStyle("lightgray")
    .bottom(yMapping)
    .anchor("left").add(pv.Label)
     .font("6pt San Serif")
     .text(yMapping.tickFormat);

barChartPanel.add(pv.Rule)
    .bottom(0);

barChartPanel.add(pv.Rule)
    .left(0);

barChartPanel.render();
```

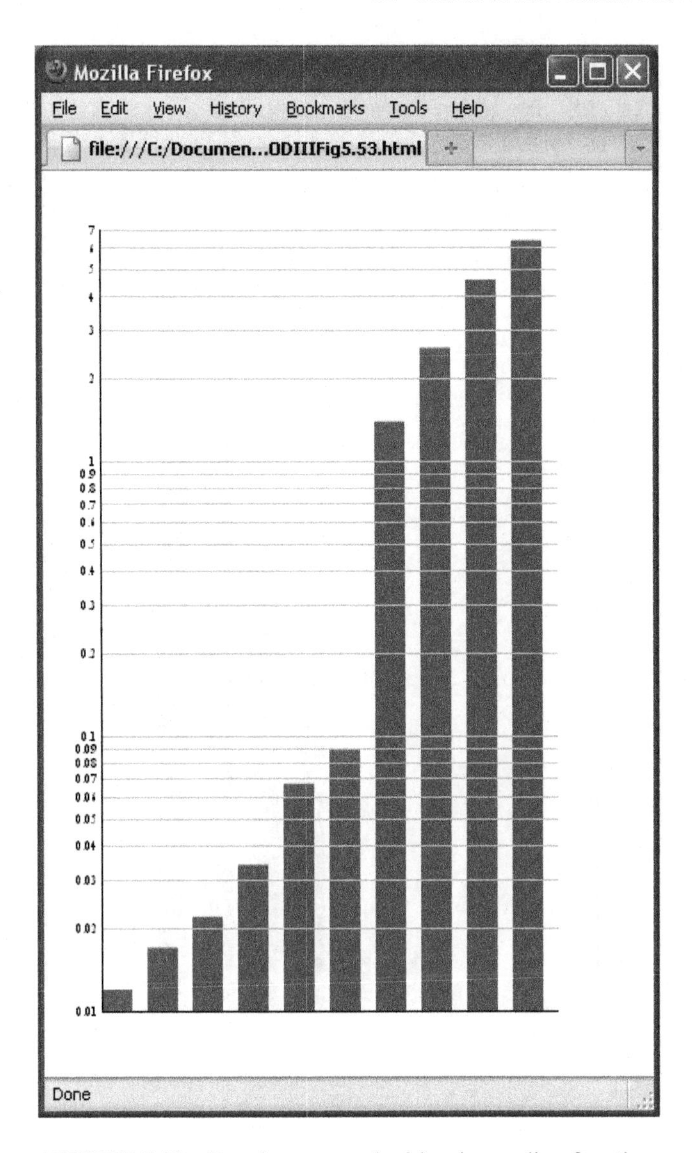


**FIGURE 5.53** Bar chart created with a log scaling function

Another quantitative mapping function is the `pv.Scale.root`, which operates in a similar manner to the `pv.Scale.log` method; however, the mapping is based on the root scale. To illustrate the use of this class, the log scale function (`pv.Scale.log`) from the previous example is replaced with the `pv.Scale.root`, and the resulting chart is shown in Fig. 5.54.

The class `pv.Scale.quantile` is an option to use when mapping values whose domain (data space) can be ordered onto quantized values. The defaults range is 0 to 1, with 0 the lowest value, and 1 the highest.

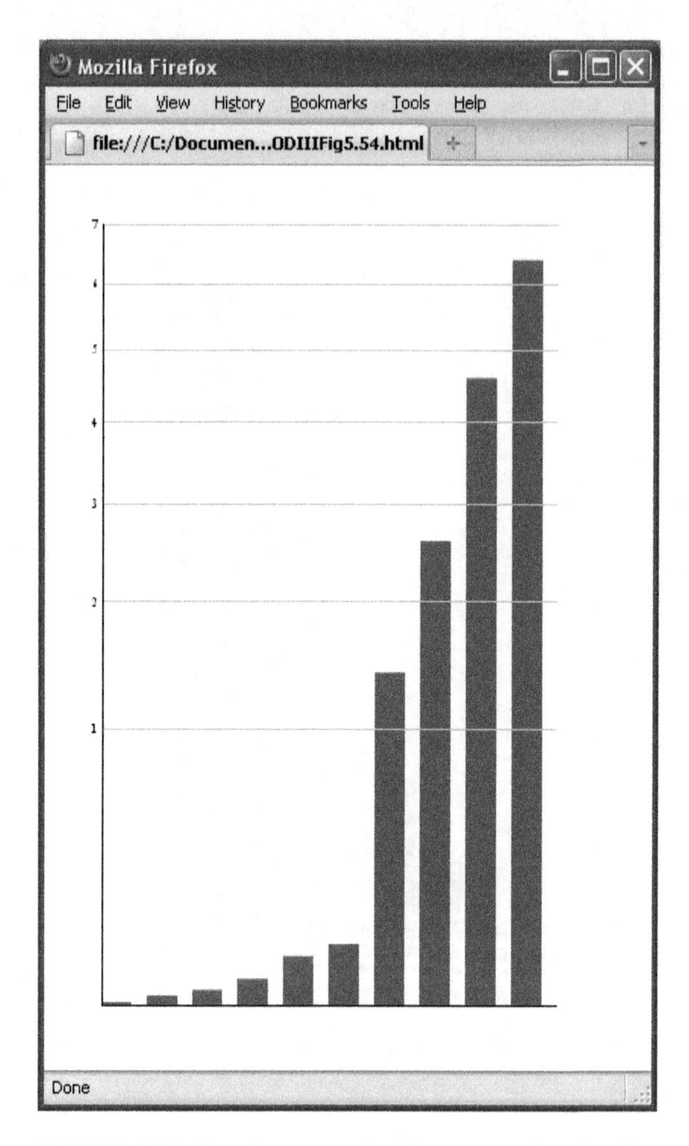


**FIGURE 5.54** Bar chart created with a root scaling function

There is often no clear mapping of categorical data onto either position or color space. The class `pv.Scale.ordinal` is used to explicitly define this mapping. This class can be called with or without an explicit list of all categories to visualize. In the following example, two `pv.Scale.ordinal` methods are assigned to the variables `ordinalYMapping` and `ordinalColorMapping`. `ordinalYMapping` maps the specified categorical values to a vertical pixel location. The class is called with the list of all domain categories as arguments ("low", "low-medium", "medium", "medium-high", "high"). Alternatively, the array can be assigned and its different values used automatically. The method split was called to define the minimum and maximum values for the mapped pixel locations, which is the distance from the bottom of the screen (0) to the height of the panel. The second ordinal mapping scale function defined is ordinalColorMapping, which maps the same categories to a list of specific colors, detailed in the range. Because the number of domain categories matches the number of colors, there is a one-to-one mapping of the categories to the colors listed. Using these scale mapping variables, a series of bar charts are generated from data whose height and color are dependent on the scales defined, as shown in Fig. 5.55. Protovis has several convenient color palettes (defined in pv.Colors) to use when coloring ordinal values: Color13, Color19, Color20 (see the Protovis API for more details).

```
var panelWidth  = 260,
    panelHeight = 200;

var ordinalYMapping = pv.Scale
  .ordinal("low","low-medium","medium","medium-high",
   "high")
  .split(0,panelHeight);
```

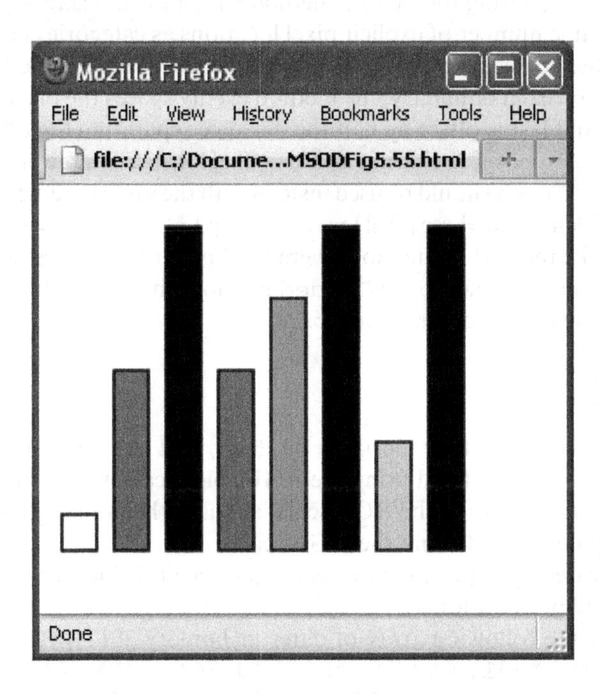


**FIGURE 5.55** Bar chart created with ordinal scaling functions

```
var ordinalColorMapping = pv.Scale
  .ordinal("low","low-medium","medium","medium-high",
    "high")
  .range("white","lightgray","gray","darkgray","black");

var barChartPanel = new pv.Panel()
  .width(panelWidth)
  .height(panelHeight);

barChartPanel.add(pv.Bar)
  .data(["low","medium","high","medium","medium-high",
    "high","low-medium","high"])
  .width(20)
  .bottom(5)
  .height(ordinalYMapping)
  .strokeStyle("black")
  .left(function() this.index * 30 + 5)
  .fillStyle(ordinalColorMapping);

barChartPanel.render();
```

The `split` method is used to calculate the pixel location for a series of categorical values. Using the `split` method with `min` and `max` arguments will result in the same number of explicit pixel locations as categories identified. The position of these categories is evenly spaced. To avoid having the first and last pixel location on each boundary, the method evenly spaces the pixel locations so that the first and last is offset by half the distance separating the other adjacent pixel locations. To position the first and last pixel location on the boundaries, the method `splitFlush` should be used instead with the same parameters. Another variation for identifying these pixel locations is `splitBanded`, which has a third parameter. Like the `split` method, there is an offset for the first and last pixel locations; however, in the `splitBanded` method, the size of this offset can be controlled using the third parameter: `band`.

### 5.4.6 Exercises

5.4.6.1 Create a bar chart for the array `[2,5,6,8,4,9]` where values above 8 are set to a blue color defined using the RGB color scale, values below 3 are set to a red color defined using the RGB scale, and all other values are defined as a light gray (again defined using the RGB scale).

5.4.6.2 Create the same bar chart as Exercise 5.4.6.1, but use the HSL color format to display the colors.

5.4.6.3 For the following array of dates in January 2011 `["2011 01 01", "2011 01 02", "2011 01 03", "2011 01 04", "2011 01 05"]`, display on the screen the full date (e.g., Monday 02 January 2011).

5.4.6.4 Create a horizontally aligned area plot (colored "lightblue"), and anchor a 3-pixel-wide green line to the top of the plot for the following array: [3,5,7,8,7,9,14].

5.4.6.5 Create a bar chart for the array [5,7,6,2,8,9,6,4,3,6,7,3,8,6,5,4,10,2,6,8,6,5,4,3,6], anchoring a label to the top of the chart where the value is greater than 8.

5.4.6.6 Create a plot with the dot mark for the same data array as in Exercise 5.4.6.5, positioning the label above the dot for those values greater than 8.

5.4.6.7 Create a bar chart aligned with the y-axis using the linear scaling functions for both axes for the array [124.5, 286.43, 134.76, 255.39, 461.38, 336.26].

5.4.6.8 Create a bar chart aligned with the y-axis chart using the log scaling functions for the array [0.95, 28.63, 1.34, 245.69, 461.38, 336.26].

## 5.5 CREATING BASIC PLOTS

### 5.5.1 Overview

When creating visualizations using Protovis, you must keep several issues in mind. First, Protovis and JavaScript are programming languages, so it is important to pay attention to case and punctuation. For example, if you define a variable myVariable, when you use it later in the code, it should be referenced with the same capitalization (if you use myvariable instead, you get an error). When defining a mark's properties, each property is set using a period (.) before the property name. Semicolons (;) are only used at the end of portions of code to signify the mark or function is completely defined. Second, it is important to note that the order in which the graphical objects are placed on the screen is important because the objects placed later will cover those placed earlier (if they overlap). Third, comments can be added to the code to describe what different portions of the code are performing. This is particularly helpful when rereading code previously written. The expression /* is used to indicate the start of a comment and */ is used to identify that the comment is complete. If the comment is contained on a single line, using // at the beginning of the line identifies that line as a comment.

### 5.5.2 Handling Arrays and Data

The examples shown so far in this chapter have made use of an array of data in a single list; for example, [1.2,3.4,2.8,3,6] is an array of four real numbers. There are a number of convenient operations that can be performed on an array, where a new array is created from the original array. filter generates a new array that is restricted to the elements that are satisfied by the defined function. For example, the following code displays the list [6,7,5,7] on the screen because these are the only elements of the list that are greater than 4.

```
var dataArray = [2,4,6,7,5,7];

var panelWidth = 200, panelHeight = 200;

var chartPanel = new pv.Panel()
      .width(panelWidth)
      .height(panelHeight);

    chartPanel.add(pv.Label)
      .data(dataArray.filter(function(d) d>4))
      .text(function(d) d)
      .top(function() this.index*15 + 15)
      .left(20);

    chartPanel.render();
```

The line containing the `filter` statement is now exchanged with the following line.

```
      .data(dataArray.map(function(d) d+4))
```

This results in the list `[6,8,10,11,9,11]` shown on the screen; that is, four is added to each element of the array.

Other helpful methods that manipulate arrays include `pv.uniq` (returns a list of unique values) and `pv.normalize` (generates an array that is normalized where all elements add up to 1, which was described earlier). The full list of methods to manipulate arrays are found in the `pv` class or the `Array` class and are documented in the Protovis API.

Arrays can also be composed of any objects, including nested arrays. For example, `[[1,3,4,2],[5,2,3,5],[3,2,8,7]]` is an array of three arrays (nested array). Each of the nested arrays is composed of four integer values. A series of methods is available to manipulate this type of data, including `pv.blend` (to create a single array by concatenating each nested array) and `pv.transpose` (to convert an $m \times n$ matrix to an $n \times m$ matrix). The Protovis API documentation outlines all methods for manipulating arrays of data.

### 5.5.3 Reading Data from Files

Data can be accessed from outside the HTML file where the Protovis code is written. The format for this external data is JSON (JavaScript Object Notation). The format uses square brackets for the array contents with curly brackets for each observation.

In the following example, a file containing data on the monthly weather at Heathrow airport is provided in a JSON format.

```
var heathrowWeather = [
 { month: "Jan", recordHigh: 63.3, averageHigh: 46.4,
   averageLow: 36.3, recordLow: 12, precipitation: 2.01},
 { month: "Feb", recordHigh: 64.8, averageHigh: 47.1,
   averageLow: 36, recordLow: 14.9, precipitation: 1.34},
 { month: "Mar", recordHigh: 73.9, averageHigh: 52.5,
   averageLow: 38.8, recordLow: 20.7, precipitation: 1.65},
 { month: "Apr", recordHigh: 84.4, averageHigh: 57.7,
   averageLow: 41.5, recordLow: 29.5, precipitation: 1.77},
 { month: "May", recordHigh: 91.8, averageHigh: 64.2,
   averageLow: 47.3, recordLow: 31.6, precipitation: 1.85},
 { month: "Jun", recordHigh: 96.8, averageHigh: 69.8,
   averageLow: 53.2, recordLow: 41.7, precipitation: 2.09},
 { month: "Jul", recordHigh: 99, averageHigh: 74.5,
   averageLow: 57.6, recordLow: 46.2, precipitation: 1.5},
 { month: "Aug", recordHigh: 101.3, averageHigh: 73.8,
   averageLow: 56.8, recordLow: 43.9, precipitation: 1.81},
 { month: "Sep", recordHigh: 92.5, averageHigh: 67.8,
   averageLow: 52.2, recordLow: 38.1, precipitation: 2.2},
 { month: "Oct", recordHigh: 84, averageHigh: 60.1,
   averageLow: 46.8, recordLow: 24.6, precipitation: 2.44},
 { month: "Nov", recordHigh: 70, averageHigh: 52,
   averageLow: 40.8, recordLow: 21.9, precipitation: 2.05},
 { month: "Dec", recordHigh: 64, averageHigh: 46.8,
   averageLow: 37, recordLow: 18.7, precipitation: 2.13}
];
```

The array uses square brackets ([ and ]) to denote the list of observations, with each individual record shown within curly braces ({ and }). Each record is composed of a consistent series of fields (in this example, `month`, `recordHigh`, `averageHigh`, `averageLow`, `recordLow`, and `precipitation`). The colon after each field indicates that the value for the field follows. Commas separate the 'FIELD':'VALUE' pairs. The value can either be a text (indicated with " . . . ") or numeric value (where the " . . . " is optional). This JSON file is equivalent to a data table where the column headers are month, recordHigh, averageHigh, averageLow, recordLow, and precipitation, and the values are the elements in each row.

The location of the file (either on your computer or a URL pointing to a location on the Web) should be specified in the header of the HTML file. For example, if this JSON file is called Heathrow-Weather.js and is located in the same directory as the HTML file containing the corresponding Protovis visualization, then the following line should be added to the header block:

```
<script type="text/javascript" src="Heathrow-Weather.js">
</script>
```

The values of the individual fields can be accessed within the Protovis code (e.g.,`function(d) d.month` accesses the month field when the `heathrowWeather` variable is added to a mark).

### 5.5.4 Worked Examples

The following example creates a plot describing the average and extreme weather conditions at Heathrow airport. The header portion of the HTML file contains code to initialize the location of the Protovis library, as well as any external data to be used. In this example, the data is contained in the file Heathrow-Weather.js, which is located in the same directory as the HTML code.

```
<html>
 <head>
  <script type="text/javascript" src=
    "../protovis.js"></script>
  <script type="text/javascript" src=
    "Heathrow Weather.js"></script>
 </head>
```

Following the header portion is the body of the code (between the `<body>` and `</body>` tags). The code starts with the script tag indicating that the code that follows is JavaScript with additional Protovis syntax.

```
<body>

<script type="text/javascript+protovis">
```

A number of variables are initially defined that indicate the dimensions of the panel as well as the location of the legend.

```
var panelHeight = 400
    panelWidth  = 760
    margin      = 30
    legendLeft  = 670
    legendTop   = 90;
```

The temperatures will be displayed on the graphic, and a scale that maps these values (from 0 to 110°F) onto the vertical position is defined because this scale will be used multiple times.

```
var temperatureScale = pv.Scale.linear(0,110).range(0,
  panelHeight);
```

Initially a new panel is created where the plot will be drawn.

```
var chartPanel = new pv.Panel()
    .width(panelWidth)
    .height(panelHeight)
    .margin(margin);
```

A label mark is added to the top of the plot for the plot title.

```
chartPanel.add(pv.Label)
  .text("Temperature at Heathrow Airport")
  .top(0)
  .left(200)
  .font("bold 14px sans-serif");
```

A series of dots are used to indicate the record-high temperatures. The `pv.Dot` mark is added to the panel, and the data array (from the external JSON file) is added (`heathrowWeather`). Each row of the data is an individual month, which means that 12 dots will be drawn on the plot. Here, the record-high temperatures are to be drawn, with the corresponding data found in the `recordHigh` field of the array (accessed via `d.recordHigh`). This statement must be within a function to access the field value. The vertical position of the dot is calculated using the `temperatureScale` function, and the horizontal position is a function of the order of the elements in the array. The dots are colored red, with a black border.

```
chartPanel.add(pv.Dot)
  .data(heathrowWeather)
  .bottom(function(d) temperatureScale(d.recordHigh))
  .left(function() this.index * 50 + 20 + 5)
  .strokeStyle("black")
  .fillStyle("red");
```

A bar is used to represent the high and low average temperatures. A bar mark is added to the plot panel, and an individual bar is generated for each element of the `heathrowWeather` array (i.e., one bar per month). The bottom of each bar maps onto the average low temperature value, and the top of the bar maps onto the average high temperature. The bottom of the bar is calculated directly by accessing the average low value and mapping it onto the vertical position using the scale function (`function(d) temperatureScale(d.averageLow)`), whereas the height is the difference between the average high and average low mapped values. The horizontal position of the bars is again based on the order in which the observations occur in the array, and the bars are colored in gray.

```
chartPanel.add(pv.Bar)
  .data(heathrowWeather)
  .bottom(function(d) temperatureScale(d.averageLow))
```

```
  .height(function(d) temperatureScale(d.averageHigh) -
    temperatureScale(d.averageLow))
  .left(function() this.index * 50 + 20)
  .width(10)
  .fillStyle("lightgray")
  .strokeStyle("lightgray");
```

Blue dots (drawn as triangles) with black borders are used to display the record-low temperatures and are mapped onto the plot in a manner similar to mapping the record-high values.

```
chartPanel.add(pv.Dot)
  .data(heathrowWeather)
  .bottom(function(d) temperatureScale(d.recordLow))
  .shape("triangle")
  .left(function() this.index * 50 + 20 + 5)
  .strokeStyle("black")
  .fillStyle("blue");
```

A y-axis is added to the plot panel at a distance of 0 pixels from the left. Because no other property is set, an axis is drawn along the entire height of the panel.

```
chartPanel.add(pv.Rule)
  .left(0);
```

A label "Temperature (Fahrenheit)" is drawn on the plot 20 pixels to the left of the plot margin. The text is rotated by 90° counterclockwise.

```
chartPanel.add(pv.Label)
  .text("Temperature (Fahrenheit)")
  .left(-20)
  .bottom(150)
  .textAngle(-Math.PI/2);
```

A series of tick-mark values are generated using the `temperatureScale` scale's `ticks` function. Each tick mark is positioned using the temperature scale function and has a width of 3 pixels. A label corresponding to each of the tick-mark values is anchored to the left of each tick.

```
chartPanel.add(pv.Rule)
   .data(temperatureScale.ticks())
   .bottom(temperatureScale)
   .left(0)
   .width(3)
    .anchor("left").add(pv.Label);
```

The x-axis is drawn using the `pv.Rule` mark starting from the bottom left-hand corner of the panel with a horizontal distance of 600 pixels.

```
chartPanel.add(pv.Rule)
  .bottom(0)
  .left(0)
  .width(600);
```

Tick marks are added to the horizontal axis corresponding to each of the observations in the array (i.e., one per month). The horizontal location of the ticks uses the same function as the position of the corresponding marks. Labels are added to each of the values and anchored at the bottom.

```
chartPanel.add(pv.Rule)
   .data(heathrowWeather)
   .bottom(0)
   .left(function() this.index * 50 + 20)
   .height(3)
     .anchor("bottom").add(pv.Label)
       .text(function(d) d.month);
```

A title for the x-axis is created and positioned in the center of the axis, 25 pixels below.

```
chartPanel.add(pv.Label)
  .text("Months")
  .bottom(-25)
  .left(300);
```

A legend is drawn to indicate the meaning of each of the graphical elements.

```
chartPanel.add(pv.Label)
   .text("Legend")
   .left(legendLeft)
   .top(legendTop)
   .font("bold 12px sans-serif")
     .add(pv.Dot)
       .left(legendLeft+4)
       .top(legendTop + 30)
       .fillStyle("red")
       .strokeStyle("black")
     .add(pv.Label)
       .text("Record high temperature")
       .left(legendLeft + 15)
```

```
      .top(legendTop + 37)
      .font("12px sans-serif")
     .add(pv.Bar)
      .top(legendTop + 50)
      .left(legendLeft)
      .height(40)
      .width(10)
      .fillStyle("gray")
      .strokeStyle("gray")
     .add(pv.Label)
      .text("Average low to high")
      .left(legendLeft + 15)
      .top(legendTop + 75)
     .add(pv.Dot)
      .fillStyle("blue")
      .strokeStyle("black")
     .shape("triangle")
      .left(legendLeft+4)
      .top(legendTop + 110)
     .add(pv.Label)
      .text("Record low")
      .left(legendLeft + 15)
      .top(legendTop + 117);

  chartPanel.render();
```

The script, body, and HTML tags are terminated to complete the code.

```
</script>
</body>
</html>
```

The entire code follows with Fig. 5.56 showing the resulting visualization.

```
<html>
 <head>
  <script type="text/javascript" src=
    "../protovis.js"></script>
  <script type="text/javascript" src=
    "Heathrow-Weather.js"></script>
 </head>

<body>

<script type="text/javascript+protovis">
```

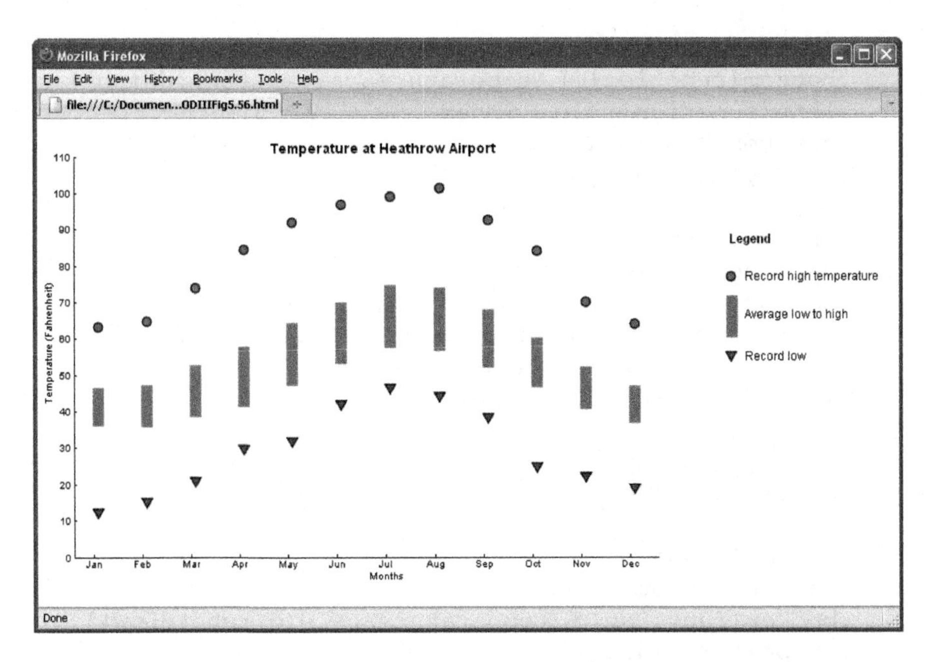


**FIGURE 5.56** Plot displaying the temperature at Heathrow airport

```
var panelHeight = 400
    panelWidth  = 800
    margin      = 30
    legendLeft  = 670
    legendTop   = 90;

 var temperatureScale = pv.Scale.linear(0,110).range(0,
   panelHeight);

var chartPanel = new pv.Panel()
    .width(panelWidth)
    .height(panelHeight)
    .margin(margin);

chartPanel.add(pv.Label)
  .text("Temperature at Heathrow Airport")
  .top(0)
  .left(200)
  .font("bold 14px sans-serif");

chartPanel.add(pv.Dot)
```

```
  .data(heathrowWeather)
  .bottom(function(d) temperatureScale(d.recordHigh))
  .left(function() this.index * 50 + 20 + 5)
  .strokeStyle("black")
  .fillStyle("red");

chartPanel.add(pv.Bar)
  .data(heathrowWeather)
  .bottom(function(d) temperatureScale(d.averageLow))
  .height(function(d) temperatureScale(d.averageHigh) -
    temperatureScale(d.averageLow))
  .left(function() this.index * 50 + 20)
  .width(10)
  .fillStyle("gray")
  .strokeStyle("gray");

chartPanel.add(pv.Dot)
  .data(heathrowWeather)
  .bottom(function(d) temperatureScale(d.recordLow))
  .shape("triangle")
  .left(function() this.index * 50 + 20 + 5)
  .strokeStyle("black")
  .fillStyle("blue");

chartPanel.add(pv.Rule)
  .left(0);

chartPanel.add(pv.Label)
  .text("Temperature (Fahrenheit)")
  .left(-20)
  .bottom(150)
  .textAngle(-Math.PI/2);

chartPanel.add(pv.Rule)
   .data(temperatureScale.ticks())
   .bottom(temperatureScale)
   .left(0)
   .width(3)
    .anchor("left").add(pv.Label);

chartPanel.add(pv.Rule)
  .bottom(0)
  .left(0)
  .width(600);
```

```
chartPanel.add(pv.Rule)
  .data(heathrowWeather)
  .bottom(0)
  .left(function() this.index * 50 + 20)
  .height(3)
   .anchor("bottom").add(pv.Label)
    .text(function(d) d.month);

chartPanel.add(pv.Label)
 .text("Months")
 .bottom(-25)
 .left(300);

chartPanel.add(pv.Label)
  .text("Legend")
  .left(legendLeft)
  .top(legendTop)
  .font("bold 12px sans-serif")
   .add(pv.Dot)
    .left(legendLeft+4)
    .top(legendTop + 30)
    .fillStyle("red")
    .strokeStyle("black")
   .add(pv.Label)
    .text("Record high temperature")
    .left(legendLeft + 15)
    .top(legendTop + 37)
    .font("12px sans-serif")
   .add(pv.Bar)
    .top(legendTop + 50)
    .left(legendLeft)
    .height(40)
    .width(10)
    .fillStyle("gray")
    .strokeStyle("gray")
   .add(pv.Label)
    .text("Average low to high")
    .left(legendLeft + 15)
    .top(legendTop + 75)
   .add(pv.Dot)
    .fillStyle("blue")
    .strokeStyle("black")
    .shape("triangle")
    .left(legendLeft+4)
```

```
      .top(legendTop + 110)
     .add(pv.Label)
      .text("Record low")
      .left(legendLeft + 15)
      .top(legendTop + 117);

  chartPanel.render();

</script>
</body>
</html>
```

### 5.5.5 Exercises

5.5.5.1 A candlestick plot is often used to represent financial data, such as the prices of stocks over a period of time. The plot presents information on the high and low values for the day (using a line) and the opening and closing values (using a bar). If the stock or fund closes higher than the opening value, the color of the bar is white, whereas if the stock closes lower, the bar is filled in (using dark gray in this example).

The following data captures information on the price for the Standard and Poor's 500 stock index for the month of January 2011. This data is located in the file CandlestickData.js.

```
var standardAndPoorsData = [
{day: "2011 01 03", open: 1257.62, close: 1271.87, high:
  1276.17, low: 1257.62},
{day: "2011 01 04", open: 1272.62, close: 1270.20, high:
  1274.12, low: 1262.66},
{day: "2011 01 05", open: 1268.78, close: 1276.56, high:
  1277.63, low: 1265.36},
{day: "2011 01 06", open: 1276.29, close: 1273.85, high:
  1278.17, low: 1270.43},
{day: "2011 01 07", open: 1274.41, close: 1271.50, high:
  1276.83, low: 1261.70},
{day: "2011 01 10", open: 1270.84, close: 1269.75, high:
  1271.52, low: 1262.18},
{day: "2011 01 11", open: 1272.58, close: 1274.48, high:
  1277.25, low: 1269.62},
{day: "2011 01 12", open: 1275.65, close: 1283.96, high:
  1286.87, low: 1275.65},
{day: "2011 01 13", open: 1285.78, close: 1283.76, high:
  1286.70, low: 1280.47},
```

```
{day: "2011 01 14", open: 1283.90, close: 1293.24, high:
  1293.24, low: 1281.24},
{day: "2011 01 18", open: 1293.22, close: 1295.02, high:
  1296.06, low: 1290.16},
{day: "2011 01 19", open: 1294.52, close: 1281.92, high:
  1294.60, low: 1278.92},
{day: "2011 01 20", open: 1280.85, close: 1280.26, high:
  1283.35, low: 1271.26},
{day: "2011 01 21", open: 1283.63, close: 1283.35, high:
  1291.21, low: 1282.07},
{day: "2011 01 24", open: 1283.29, close: 1290.84, high:
  1291.93, low: 1282.47},
{day: "2011 01 25", open: 1288.17, close: 1291.18, high:
  1291.26, low: 1281.07},
{day: "2011 01 26", open: 1291.97, close: 1292.63, high:
  1299.74, low: 1291.97},
{day: "2011 01 27", open: 1297.51, close: 1299.54, high:
  1301.29, low: 1294.41},
{day: "2011 01 28", open: 1299.63, close: 1276.34, high:
  1302.67, low: 1275.10}
];
```

Using this data, create the candlestick plot as shown in Fig. 5.57.

## 5.6 DATA GRAPHICS

### 5.6.1 Frequency Histograms

Frequency histograms are regularly used to understand the variability in a list of numbers. The numbers are divided into a series of bins that represent specific values or values within a range. The data values are examined to understand how many observations fall within each bin, and these bins are plotted using bars whose length represents the number of observations. Histograms can be used to understand the range of values as well as the type of frequency distribution, such as whether the observations follow a normal distribution.

In Protovis, two classes can be used to generate frequency histograms: `pv.histogram` and `pv.histogram.bin`. The following example generates a frequency histogram for a variable `MW` (molecular weight) for a series of tranquilizing agents, collected from the PubChem database (http://pubchem.ncbi.nlm.nih.gov/). These agents are defined under the variable `tranquilizingAgents`,which includes fields describing these chemicals as well as the type of tranquilizing agent (`"Antianxiety"`, `"Antimanic"`, and `"Antipyschotic"`). The following shows a small portion of the table.

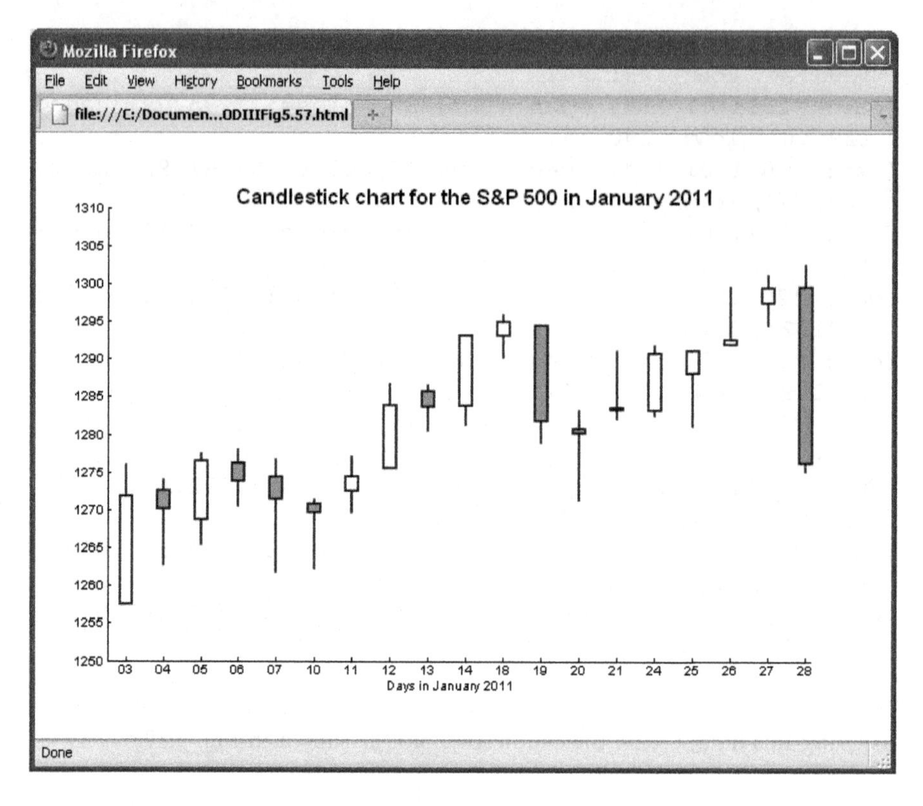


FIGURE 5.57 Candlestick data from the S&P 500 in January 2011

```
var tranquilizingAgents = [
 {
  CID: 441233,
  MW: 133.190340,
  HBDC: 1,
  HBAC: 1,
  XLogP: 1.5,
  TC: 1,
  TFC: 0,
  MF: "C9H11N",
  type: "Antianxiety",
  class: 1
 },
 {
  CID: 270840,
  MW: 334.492860 ,
  HBDC: 2 ,
```

```
  HBAC: 3 ,
  XLogP: 4.2 ,
  TC: 4 ,
  TFC: 0 ,
  MF: "C21H34O3",
  type: "Antianxiety" ,
  class: 1
 },
. . .
```

In the following code, a series of variables is initially defined outlining the height and width of the panel as well as the size of the margin.

```
var panelHeight = 300
    panelWidth  = 400
    margin      = 20
```

To understand the range of molecular weight values, the `pv.max` routine is called to calculate the highest molecular weight value.

```
maxMolWtValue = pv.max(tranquilizingAgents, function
  (d) d.molecularweight)
```

Because the ranges of molecular weights will be displayed along the x-axis, an `xMapping` scale function variable is created. This maps the domain values (from 100 to the maximum molecular weight) onto the panel's horizontal position.

```
xMapping = pv.Scale.linear(100, maxMolWtValue).range
  (0, panelWidth)
```

Next, the histogram bins are calculated by using the histogram class (`pv.histogram`) called with the data (`tranquilizingAgents`), along with specifying an *accessor* function (`function(d) d.MW`). The method is called to calculate an array of bins to be used in the frequency histogram. The `ticks` method (from the `xMapping` scale variable) is used to identify the cutoff points for the bins. An array is returned containing these bins (`pv.histogram.Bin`). `pv.histogram.Bin` has three fields: `dx` (the size of the bin's range), `x` (the start value), and `y` (the number of observations in the bin or a probability).

```
histogramBins = pv.histogram(tranquilizingAgents,
  function(d)d.MW).bins(xMapping.ticks(20))
```

The `y` value type (i.e., frequency or probability) is dependent on how the histogram was configured using the frequency method. A `yMapping` scale variable is defined that uses the maximum histogram frequency count to determine the domain range.

```
yMapping   = pv.Scale.linear(0, pv.max(histogramBins,
  function(d) d.y)).range(0, panelHeight);
```

A panel is initialized and a bar mark added whose individual bars are configured based on the histogram bins previously calculated. In addition, rule marks are used to display the x-axis and grid lines of the plot along with appropriate boundary labels.

```
var chartPanel = new pv.Panel()
    .width(panelWidth)
    .height(panelHeight)
    .margin(margin);

chartPanel.add(pv.Bar)
   .data(histogramBins)
   .bottom(0)
   .left(function(d) xMapping(d.x))
   .width(20)
   .height(function(d) yMapping(d.y))
   .fillStyle("darkgray")
   .strokeStyle("white");

chartPanel.add(pv.Rule)
   .data(yMapping.ticks(10))
   .bottom(yMapping)
   .strokeStyle("lightgray")
  .anchor("left").add(pv.Label)
   .text(yMapping.tickFormat)
   .strokeStyle("black");

chartPanel.add(pv.Rule)
   .data(xMapping.ticks())
   .left(xMapping)
   .bottom(-3)
   .height(3)
  .anchor("bottom").add(pv.Label)
   .text(xMapping.tickFormat);

chartPanel.add(pv.Rule)
  .bottom(0);

chartPanel.render();
```

Here is the entire code for displaying the histogram:

```
<html>
 <head>
  <script type="text/javascript" src="../protovis.js">
    </script>
  <script type="text/javascript" src="Tranquilizers.js">
    </script>
 </head>

<body>

<script type="text/javascript+protovis">

   var panelHeight    = 300
       panelWidth     = 400
       margin         = 20
       maxMolWtValue = pv.max(tranquilizingAgents,
         function(d)d.MW)
      xMapping        = pv.Scale.linear(100, maxMolWtValue)
        .range(0, panelWidth)
      histogramBins  = pv.histogram(tranquilizingAgents,
        function(d)d.MW).bins(xMapping.ticks(20))
      yMapping        = pv.Scale.linear(0, pv.max
        (histogramBins, function(d)d.y)).range(0,
        panelHeight);

  var chartPanel = new pv.Panel()
      .width(panelWidth)
      .height(panelHeight)
      .margin(margin);

  chartPanel.add(pv.Bar)
     .data(histogramBins)
     .bottom(0)
     .left(function(d) xMapping(d.x))
     .width(20)
     .height(function(d) yMapping(d.y))
     .fillStyle("darkgray")
     .strokeStyle("white");

  chartPanel.add(pv.Rule)
     .data(yMapping.ticks(10))
     .bottom(yMapping)
     .strokeStyle("lightgray")
    .anchor("left").add(pv.Label)
```

```
      .text(yMapping.tickFormat)
      .strokeStyle("black");

   chartPanel.add(pv.Rule)
      .data(xMapping.ticks())
      .left(xMapping)
      .bottom(-3)
      .height(3)
     .anchor("bottom").add(pv.Label)
      .text(xMapping.tickFormat);

   chartPanel.add(pv.Rule)
     .bottom(0);

    chartPanel.render();

</script>
</body>
</html>
```

The resulting frequency histogram is displayed in Fig. 5.58.

### 5.6.2 Box-and-Whisker Plots

A box-and-whisker plot is an alternative approach to understanding the frequency distribution for an array of data. In this example, the results of three experiments are initially defined as an array of nested arrays.

```
var experiments = ["Experiment 1","Experiment 2",
  "Experiment 3"];
var dataValues = [[3,2,3,2,5,6,3,8,9,14,11,13],[5,3,2,5,
  6,4,8,9,10,3,12,4],[3,6,5,8,7,3,9,10,14,12,11,3]];
```

Variables are initialized for the panel's width and height along with the width of the tail of the box-and-whisker plot, the width of the box-and-whisker bar, and the size of the dot representing the mean value, as well as the colors for the lines and boxes.

```
var width = 500, height = 200, tailWidth = 10, barWidth = 20,
  meanDotSize = 2;
var fillColor = "lightgray", lineColor = "black";
```

The experimental values are initially examined to determine the maximum and minimum values across all three experiments. This is calculated by creating a new array containing the maximum value for each of the nested arrays, from

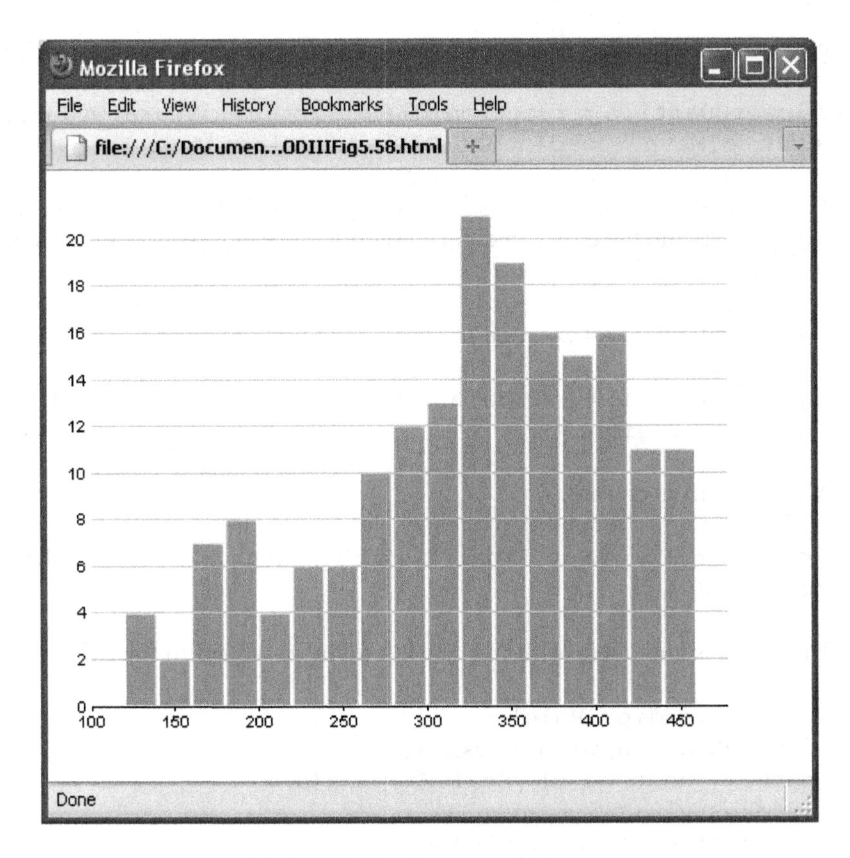


**FIGURE 5.58** Frequency histogram

which the maximum value across all three experiments is identified. Similarly, the minimum value across all three experiments is calculated.

```
var maxValue = pv.max(dataValues.map(function(d) pv.max(d)));
var minValue = pv.min(dataValues.map(function(d) pv.min(d)));
```

A variable is created for the number of observations in each experiment, using the experiment at index 0 as a prototype experiment.

```
var numberOfObservations = dataValues[0].length;
```

Two variables that reference functions are defined that will calculate the upper-quartile value and lower-quartile value for a list of observations. First, the list is ordered, and then the value in the position that is ¾ (for upper quartile) or ¼ (for lower quartile) along the array is returned. Note that that these functions only provide an approximation of upper- and lower quartiles.

```
var upperQuartile = function(d){ d.sort(pv.naturalOrder);
  return d[Math.round(numberOfObservations*3/4)]; }
var lowerQuartile = function(d){ d.sort(pv.naturalOrder);
  return d[Math.round(numberOfObservations/4)]; }
```

Two variables define functions to map the three individual experiments along the x-axis and the data values for the entire range of possible values along the y-axis.

```
var xScaleMapping = pv.Scale.ordinal(experiments).split
  (0,width),
    yScaleMapping = pv.Scale.linear(minValue,maxValue)
    .range(0,height);
```

A chart panel is created.

```
var chartPanel = new pv.Panel().width(width).height(height)
  .margin(20);
```

The y-axis is annotated with tick marks and labels added to the axis.

```
chartPanel.add(pv.Rule)
   .data(yScaleMapping.ticks())
   .bottom(function(d) yScaleMapping(d))
   .strokeStyle("lightgray")
  .anchor("left").add(pv.Label)
   .text(function(d) d);
```

An x-axis for the individual experiments is added along the bottom of the panel.

```
chartPanel.add(pv.Rule)
   .data(experiments)
   .bottom(0)
   .height(4)
   .left(function(d) xScaleMapping(d))
   .strokeStyle("lightgray")
  .anchor("bottom")
   .add(pv.Label).text(function(d) d);
```

A series of dots (drawn as horizontal ticks) are added to the chart for each experiment that represents the maximum value.

```
chartPanel.add(pv.Dot)
    .data(dataValues)
```

```
    .left(function() xScaleMapping(experiments[this.index])
      - (tailWidth/2))
    .shape("tick")
    .size(tailWidth)
    .angle(Math.PI/2)
    .strokeStyle(lineColor)
    .bottom(function(d) yScaleMapping(pv.max(d)));
```

Similarly, the minimum value tick marks are added to the chart for each experiment.

```
chartPanel.add(pv.Dot)
    .data(dataValues)
    .left(function() xScaleMapping(experiments[this.index])
      - (tailWidth/2))
    .shape("tick")
    .size(tailWidth)
    .angle(Math.PI/2)
    .strokeStyle(lineColor)
    .bottom(function(d) yScaleMapping(pv.min(d)));
```

A series of lines (using the dot mark with a `"tick"` shape) are drawn between the minimum and maximum values for each experiment.

```
chartPanel.add(pv.Dot)
    .data(dataValues)
    .left(function() xScaleMapping(experiments[this.index]))
    .shape("tick")
    .strokeStyle(lineColor)
    .size(function(d) yScaleMapping(pv.max(d)) -
      yScaleMapping(pv.min(d)))
    .bottom(function(d) yScaleMapping(pv.min(d)));
```

A box is created for each experiment; the top of the box is located at the upper quartile, and the bottom of the box is located at the lower quartile.

```
chartPanel.add(pv.Bar)
    .data(dataValues)
    .left(function() xScaleMapping(experiments[this.index])
      - (barWidth/2))
    .bottom(function(d) yScaleMapping(lowerQuartile(d)))
    .height(function(d) yScaleMapping(upperQuartile(d)) -
      yScaleMapping(lowerQuartile(d)))
    .fillStyle(fillColor)
    .strokeStyle(lineColor)
    .width(barWidth);
```

A single line (tick) is drawn for each of the experiments at the median value.

```
chartPanel.add(pv.Dot)
   .data(dataValues)
   .left(function() xScaleMapping(experiments[this.index])
     - 10)
   .shape("tick")
   .size(20)
   .strokeStyle(lineColor)
   .angle(Math.PI/2)
   .bottom(function(d) yScaleMapping(pv.median(d)));
```

A single circle (using pv.Dot) is added to each experiment's plot that represents the mean value of each experiment, and the chart is rendered.

```
chartPanel.add(pv.Dot)
   .data(dataValues)
   .left(function() xScaleMapping(experiments[this.index]))
   .size(20)
   .strokeStyle(lineColor)
   .fillStyle(lineColor)
   .angle(Math.PI/2)
   .bottom(function(d) yScaleMapping(pv.mean(d)));
chartPanel.render();
```

Figure 5.59 shows the resulting box-and-whisker plot for each of the experiments.

In this example, a series of mathematical operations were performed on the data array, including pv.min, pv.max, pv.median, and pv.mean. Many additional calculations, such as pv.deviation (estimate of the standard deviation), pv.variance (estimate of the variance), and pv.sum (summation of the values in the list), are available for use and described in the Protovis API documentation.

### 5.6.3 Scatterplots

A *scatterplot* is a helpful visualization to understand the relationship between two continuous variables. In the following example, the same tranquilizer dataset as used in Section 5.6.1 is used. The dimensions used to create the panel are initially defined.

```
var panelWidth  = 400,
    panelHeight = 400,
    margin      = 40
```

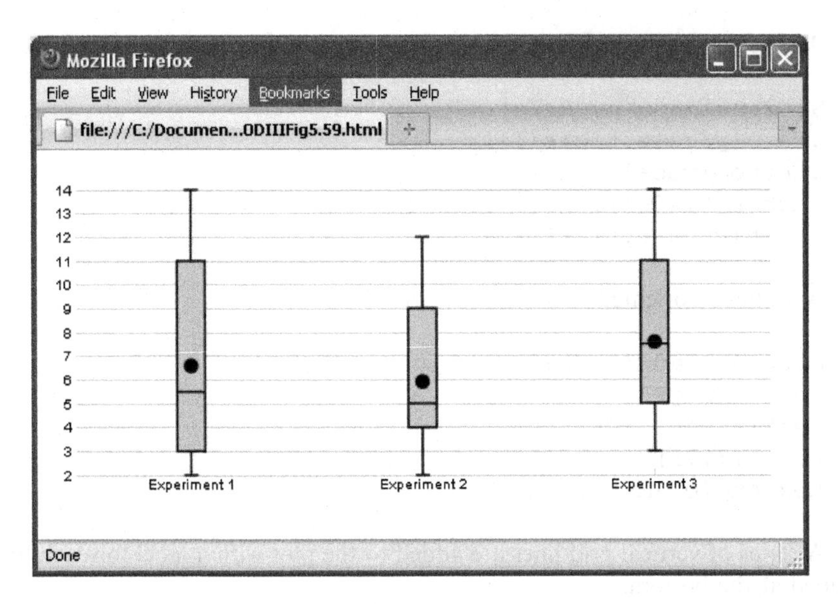


**FIGURE 5.59** Box-and-whisker plot

Molecular weight (MW) values are drawn on the x-axis, and XLogP values are displayed on the y-axis. The minimum and maximum values for each of these properties are calculated and used to generate scale functions for each property. The nice function is added to the scales to ensure that easy-to-read boundaries are produced.

```
minMolWtValue = pv.min(tranquilizingAgents, function(d)
  d.MW)
maxMolWtValue = pv.max(tranquilizingAgents, function(d)
  d.MW)
xScale = pv.Scale.linear(minMolWtValue , maxMolWtValue )
  .range(0, panelWidth).nice(),
minXLogPValue = pv.min(tranquilizingAgents, function(d)
  d.XLogP)
maxXLogPValue = pv.max(tranquilizingAgents, function(d)
  d.XLogP)
yScale = pv.Scale.linear(minXLogPValue, maxXLogPValue)
  .range(0, panelHeight).nice();
```

A chart panel is created.

```
var chartPanel = new pv.Panel()
  .width(panelWidth)
  .height(panelHeight)
  .margin(margin);
```

A series of horizontal grid lines are generated with labels placed to the left.

```
chartPanel.add(pv.Rule)
   .data(yScale.ticks())
   .bottom(yScale)
  .anchor("left").add(pv.Label)
   .text(yScale.tickFormat);
```

A vertically oriented axis label is added.

```
chartPanel.add(pv.Label)
  .text("XLogP")
  .left(-10)
  .bottom(175)
  .textAngle(-Math.PI/2);
```

A series of vertical grid lines are added to the plot with a label for each line added at the bottom.

```
chartPanel.add(pv.Rule)
   .data(xScale.ticks())
   .left(xScale)
  .anchor("bottom").add(pv.Label)
   .text(xScale.tickFormat);
```

The x-axis title is added.

```
chartPanel.add(pv.Label)
    .text("Molecular Weight")
    .bottom(-25)
    .left(150);
```

The data values are added to the plot as a series of dots (using `pv.Dot`) with the distance from the left side calculated using the `xScale` function, and the distance from the bottom of the panel calculated using the `yScale` function.

```
chartPanel.add(pv.Dot)
   .data(tranquilizingAgents)
   .left(function(d) xScale(d.MW))
   .bottom(function(d) yScale(d.XLogP))
   .fillStyle("lightgray")
   .strokeStyle("black");

chartPanel.render();
```

Figure 5.60 shows the resulting scatterplot.

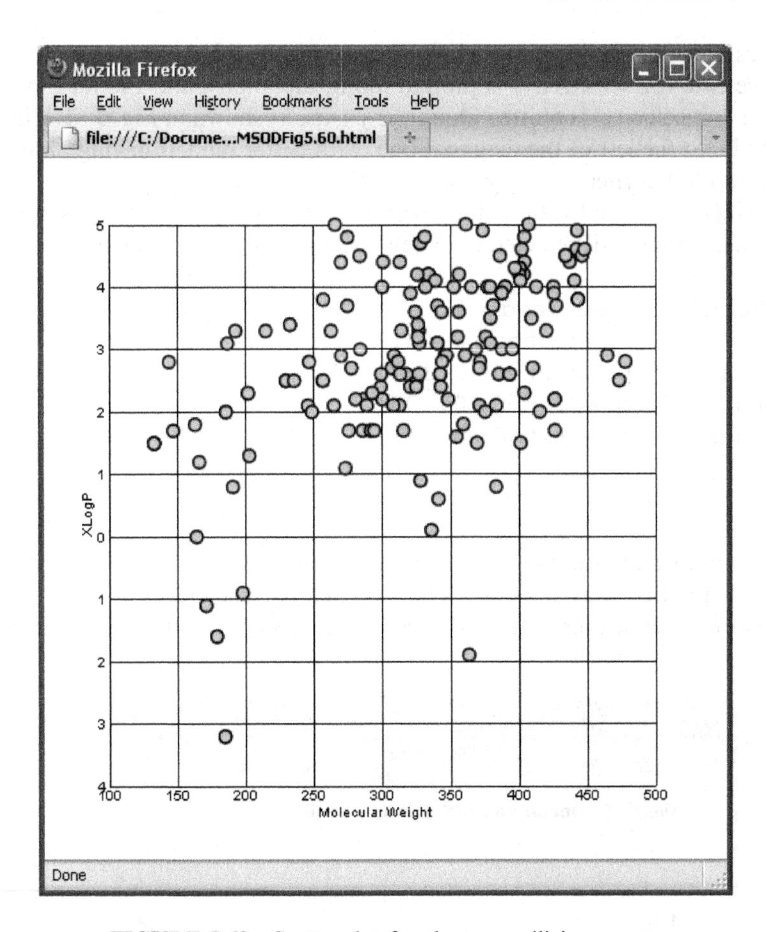


FIGURE 5.60 Scatterplot for the tranquilizing agents

## 5.6.4 Exercises

5.6.4.1 Re-create the histogram visualization from Section 5.6.1, and change the number of bins to (i) 7 and (ii) approximately 40.

5.6.4.2 Create a frequency histogram for the property `XLogP` (distribution coefficient).

5.6.4.3 Create a scatterplot using the tranquilizer data where each observations is color-coded according to the type of tranquilizer: red for antianxiety, green for antimanic, and blue for antipsychotic.

5.6.4.4 A logistic regression model (Myatt & Johnson, 2009) was generated to predict whether an observation is positive (1) or negative (0). The logistic regression model calculates a probability. A cutoff can be used to assign the calculated prediction probability as positive or negative. For example, using a 0.5 cutoff will assign values greater than or equal to 0.5 as positive and values less than 0.5 as negative. A ROC plot (Receiver Operating Characteristic) assesses the quality of a model. Different cutoff values are used, the false positive rate (or sensitivity) is plotted along the x-axis, and the false positive rate (or 1-specificity) is plotted along the y-axis, as shown in Fig. 5.61. The plot should be to the left of the diagonal line, with better models having plots close to the top-left corner.

A dataset of actual values along with the predicted probabilities is provided in the file actualAndPredicted.js. A few examples are shown here:

```
var actualVSPredicted = [
{ actual: 0 , predicted: 0.411},
{ actual: 0 , predicted: 0.305},
{ actual: 0 , predicted: 0.485},
{ actual: 0 , predicted: 0.191},
 . . .
];
```

The number of true positives (*TP*) is the number of observations that are correctly predicted as positive. The number of true negatives (*TN*) is the number of observations that are correctly predicted as negative. The number of

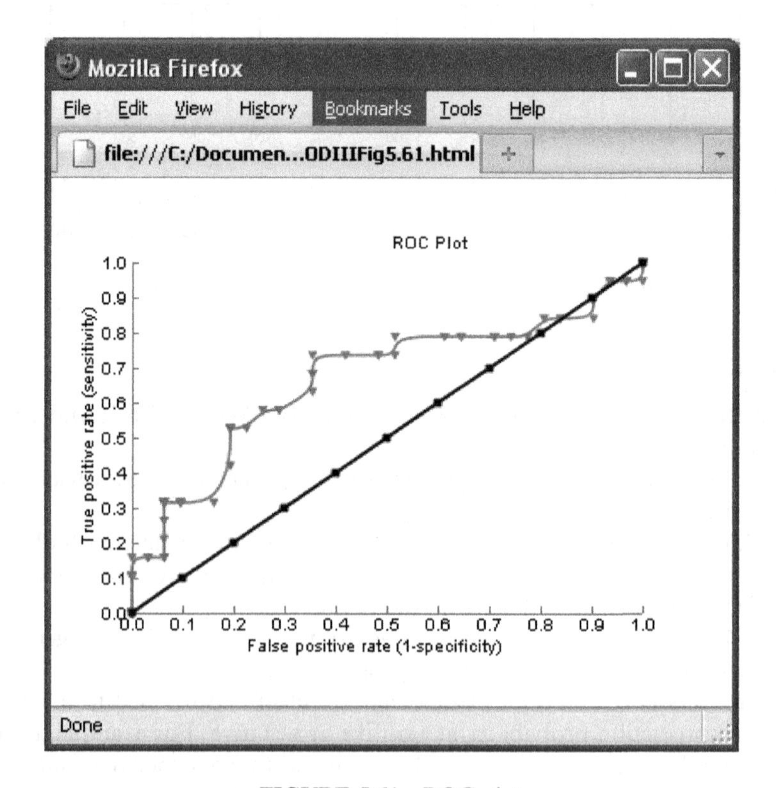


**FIGURE 5.61** ROC plot

false positives (*FP*) is the number of observations that are predicted positive but are in fact negative. The number of false negatives (*FN*) is the number of observations that are predicted as negative but are in fact positive. These numbers change with different cutoff values. Sensitivity is calculated as $TP / (TP + FN)$; specificity is calculated as $TN / (TN + FP)$. Create the ROC plot as shown in Fig. 5.61.

## 5.7 COMPOSITE PLOTS

### 5.7.1 Creating Grouped Plots Using Multiple Panels

So far in this chapter, we have used a single panel and added marks to it. More complex visualizations can be created by nesting panels within panels. In the following simple example, a `pv.Panel` container mark is added to the root panel (`chartPanel`). An array with two elements (each element a nested array) is added to the panel. This will result in two panels being added to the root panel. Each panel is positioned at a distance based on its index value. The first panel is positioned at a distance of 0 pixels from the left of the root panel ($0 \times 10$), and the second panel (at index 1) is positioned 10 pixels from the left ($1 \times 10$). A bar mark is added to each of the panels, and the `data` property is set to the corresponding nested array. The bars are positioned relative to each panel; however, because the second panel is offset by 10 pixels, the grouped histogram plot is displayed as shown in Fig. 5.62. The color of the bars is based on the index position of the parent. The bars contained in the panel at index 0 are `"lightgray"`, and the second panel bars are `"darkgray"`.

```
var chartPanel = new pv.Panel()
  .width(200)
  .height(150);

chartPanel.add(pv.Panel)
  .data([[10, 20, 25, 28, 31, 32, 34],
         [15, 22, 24, 21, 25, 31, 36]])
  .left(function() this.index * 10)
 .add(pv.Bar)
  .data(function(d) d)
  .bottom(0)
  .width(10)
  .fillStyle(function() (this.parent.index == 0)?
    "lightgray" : "darkgray")
  .height(function(d) d * 3)
  .left(function() this.index * 30);

chartPanel.render();
```

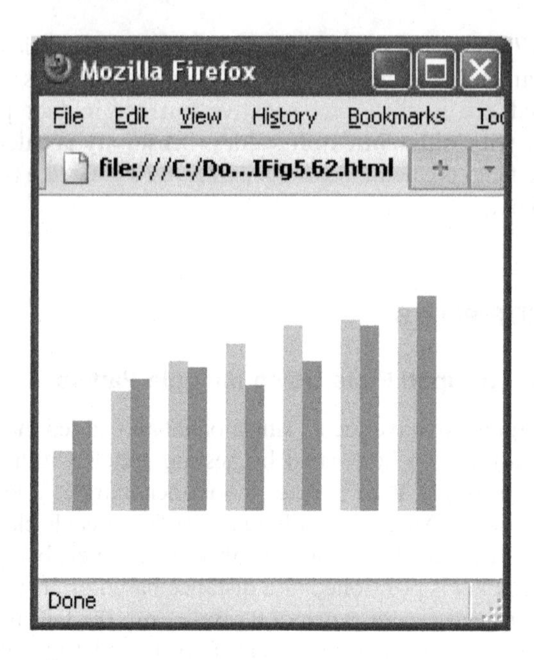


**FIGURE 5.62** Grouped histogram using multiple panels

### 5.7.2 Inheritance

An important aspect of programming in Protovis is its use of *inheritance*. This concept was reviewed earlier in this chapter. In this example, a bar mark (`pv.Bar`) is added to a panel, and appropriate properties (such as `data`, `bottom`, and `left`) are set. A second bar mark is added. This mark inherits the properties of the original bar mark (such as `bottom` and `width`). The use of inheritance simplifies the code because the second bar mark can reuse the properties assigned to the first bar mark. The properties of the new mark (such as `data` and `fillStyle`) can, in turn, be overridden to customize its appearance. The resulting grouped bar chart is shown in Fig. 5.63.

```
var chartPanel = new pv.Panel()
  .width(200)
  .height(150);

chartPanel.add(pv.Bar)
  .data([10, 20, 25, 28, 31, 32, 34])
  .left(function() this.index * 10)
  .bottom(0)
  .width(10)
  .fillStyle("darkgray")
  .height(function(d) d * 3)
```

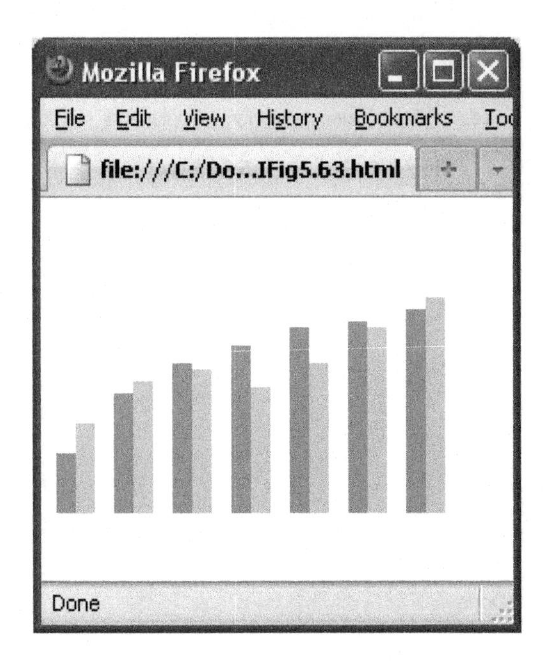


**FIGURE 5.63** Grouped bar chart using inheritance

```
  .left(function() this.index * 30)
  .add(pv.Bar)
  .data([15, 22, 24, 21, 25, 31, 36])
  .left(function() this.index * 30 + 10)
  .fillStyle("lightgray");

chartPanel.render();
```

Another approach to simplifying the code is to use *off-screen inheritance*. Here an off-screen mark is defined by creating a variable for this prototype mark and assigning values or functions to its different properties. Now, when a new mark is added to a panel, by using the `extend` expression with the off-screen mark, the new mark's properties are inherited from the off-screen mark's properties. For example, in the following code, an off-screen label mark is created (`offScreenMark`) to be used as a prototype. The `left` property is set (half the parent panel's width) along with the font type and text color. A label is now added to a chart panel, which inherits the properties from the off-screen mark using the expression `.extends(offScreenMark)`. The new label added to the panel inherits the `left`, `font`, and `textStyle` properties from `offScreenMark`. The resulting visualization is shown in Fig. 5.64.

```
var offScreenMark = new pv.Label()
  .left(function() this.parent.width() / 2)
```

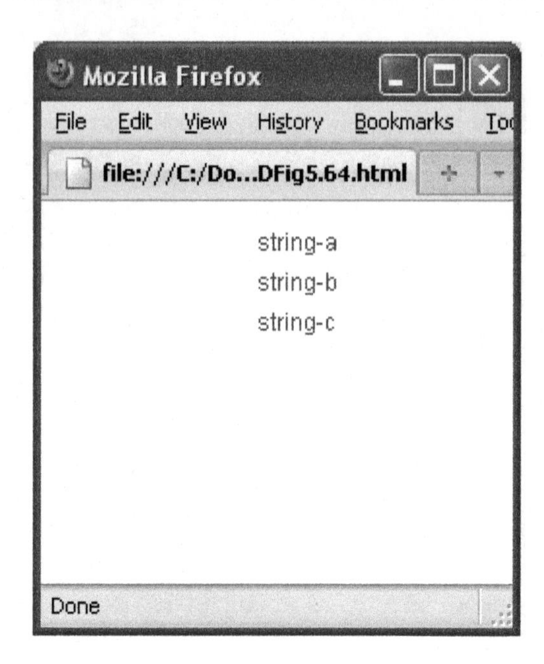


**FIGURE 5.64** Labels displayed using off-screen marks

```
.font("12px sans-serif")
  .textStyle("blue");

var chartPanel = new pv.Panel()
  .width(200)
  .height(150);

 chartPanel.add(pv.Label)
  .extend(offScreenMark)
  .data(["string-a","string-b","string-c"])
  .top(function() this.index * 20 + 20);

 chartPanel.render();
```

### 5.7.3 Property Chaining

Properties assigned to a mark can be reused with other properties or with other marks. For example, in the following code, an area mark (`pv.Area`) is added to the panel and colored black. A second area mark is added to the previous area mark. This mark inherits the properties of the first mark; however, the new area overrides a number of the properties. The second area mark then reuses the bottom and height from its parent mark to compute the new bottom for the mark (positioned at the top of the previous mark). This property reuse is

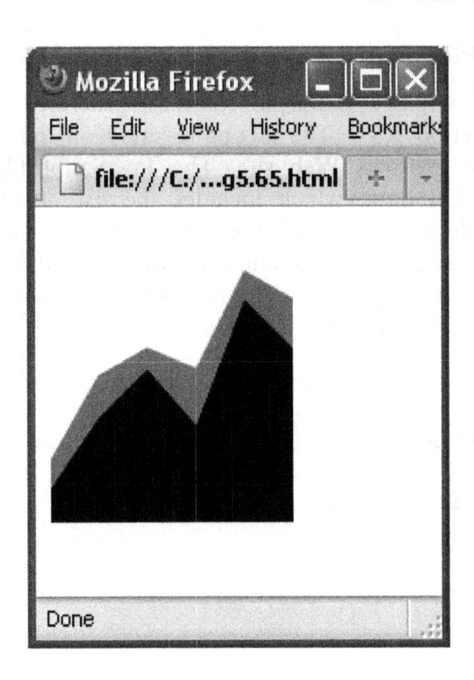


**FIGURE 5.65** Use of property chaining to create a stacked area plot

referred to as *property chaining*. The height of the new area mark inherits the previous mark's function, and it is computed using the new marks data. The resulting plot is shown in Fig. 5.65.

```
var chartPanel = new pv.Panel()
   .width(200)
   .height(150);

var area = chartPanel.add(pv.Area)
   .data([5, 15, 22, 14, 32, 25])
   .bottom(0)
   .height(function(d) d * 4)
   .left(function() this.index * 25)
   .fillStyle("black");

area.add(pv.Area)
   .data([4, 6, 3, 8, 4, 7])
   .bottom(function() area.bottom() + area.height())
   .fillStyle("gray");

chartPanel.render();
```

### 5.7.4 Creating Plot Matrices Using Multiple Panels

In the following example, a scatterplot matrix is generated using the tranquilizer data from Section 5.6.1. The layout is based on the scatterplot matrix example from the Protovis Web site. The JavaScript file Tranquilizers.js contains the external data, which includes the list of tranquilizing agent categories to consider (`agents`), a list of properties to display in the scatterplot matrix (`compoundProperties`), and the data on the different tranquilizing agents (`tranquilizingAgents`). Two example entries are displayed for the `tranquilizingAgents` array.

```
var agents = ["Antianxiety", "Antimanic", "Antipsychotic"];

var compoundProperties = ["MW", "HBDC", "XLogP"];

var tranquilizingAgents = [
 {
  CID: 441233,
  MW: 133.190340,
  HBDC: 1,
  HBAC: 1,
  XLogP: 1.5,
  TC: 1,
  TFC: 0,
  MF: "C9H11N",
  type: "Antianxiety",
  class: 1
 },
 {
  CID: 270840,
  MW: 334.492860,
  HBDC: 2,
  HBAC: 3,
  XLogP: 4.2,
  TC: 4,
  TFC: 0,
  MF: "C21H34O3",
  type: "Antianxiety",   class: 1
 },
 ...
```

To make the code easier to change at a later time, a number of variables were initially created that define the size of the scatterplot matrix elements to be visualized. A $3 \times 3$ grid of panels will be created by initially adding three panels vertically that cover the majority of the width of the root panel, and then to each of these panels three additional panels will be added. The inner panels will all have identical width and height values. The size of the scatterplots is set to 125 pixels, with the main space between the cells defined as 16 pixels and a small margin around each scatterplot panel of 4 pixels. The height and width of the entire panel is determined by summing up the individual plot sizes as well as taking into account the margins and spaces between the cells. The initial three vertically placed rectangular panels have a height set according to the height of the scatterplot panel with a small margin on either side (`rowCellHeight`).

```
var scatterplotCellSize = 125, spaceBetweenCells = 16,
cellMargin = 4;
var numberOfProperties = compoundProperties.length;
var panelSize = (numberOfProperties * scatterplotCellSize) +
        ((numberOfProperties+1) * spaceBetweenCells) +
        ((2*numberOfProperties) * cellMargin);
var rowCellHeight = scatterplotCellSize + (2 * cellMargin);
```

The scatterplots will be encoding the three different types of tranquilizers in the dataset using three colors: `"white"`, `"gray"`, and `"black"`.

```
var tranquilizerColor = pv.colors("white", "gray", "black");
```

The three properties to be used in the scatterplot matrix are listed in the `compoundProperties` array. To easily identify the corresponding scaling function for each of the properties, a mapping function is defined using `pv.dict`. Here, the scaling functions are indexed using the specific property name as a key. When this function is later called with this key (i.e., the property name), the corresponding linear scaling function is returned. The domain range of the scaling function is set using the entire list of tranquilizing agents along with an accessor function to identify the specific property.

```
var mappedPropertyScale = pv.dict(compoundProperties,
  function(property)
    pv.Scale.linear(tranquilizingAgents, function(agent)
      agent[property])
     .range(0, scatterplotCellSize));
```

The root panel is initially created based on the dimensions previously calculated and is colored gray as shown in Fig. 5.66.

```
var scatterplotMatrixPanel = new pv.Panel()
   .width(panelSize)
   .height(panelSize)
   .left(10)
   .top(10)
   .fillStyle("gray");
```

**FIGURE 5.66** Root panel of the scatterplot matrix

Into the root panel, a series of panels (three in this example) are added. These panels are positioned to fill most of the horizontal space and are evenly spaced from the top of the screen, as shown in Fig. 5.67, and colored in light gray.

```
var cellRows = scatterplotMatrixPanel.add(pv.Panel)
   .data(compoundProperties)
   .top(function() (this.index * (rowCellHeight +
      spaceBetweenCells)) + spaceBetweenCells)
   .left(spaceBetweenCells)
   .right(spaceBetweenCells)
   .height(rowCellHeight )
   .fillStyle("lightgray");
```

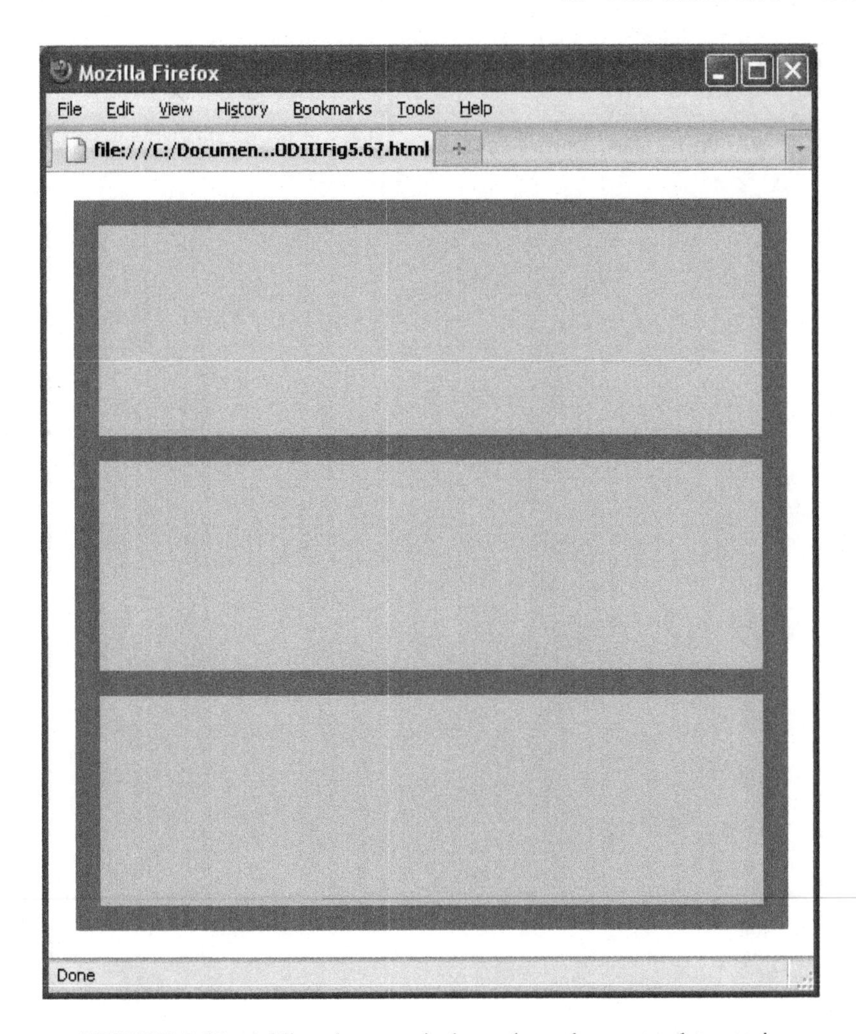


**FIGURE 5.67** Adding three vertical panels to the scatterplot matrix

Next, the individual panels making up the grid are added (nine panels in this example). For each of the row panels (*r* is used to indicate each row panel), a new array of data is created. This array contains an element for each of the values in the `compoundProperty` array. This new array is created through the array-mapping function (`map`), which creates a new array with the same number of elements but whose values are the result of the function. In this example, a new array of data values is generated where each element is a unique pair of values representing combinations of properties (referred to as `parentColumnProperty` and `parentRowProperty`). Each of the possible scatterplot cells are colored in white and shown in Fig. 5.68.

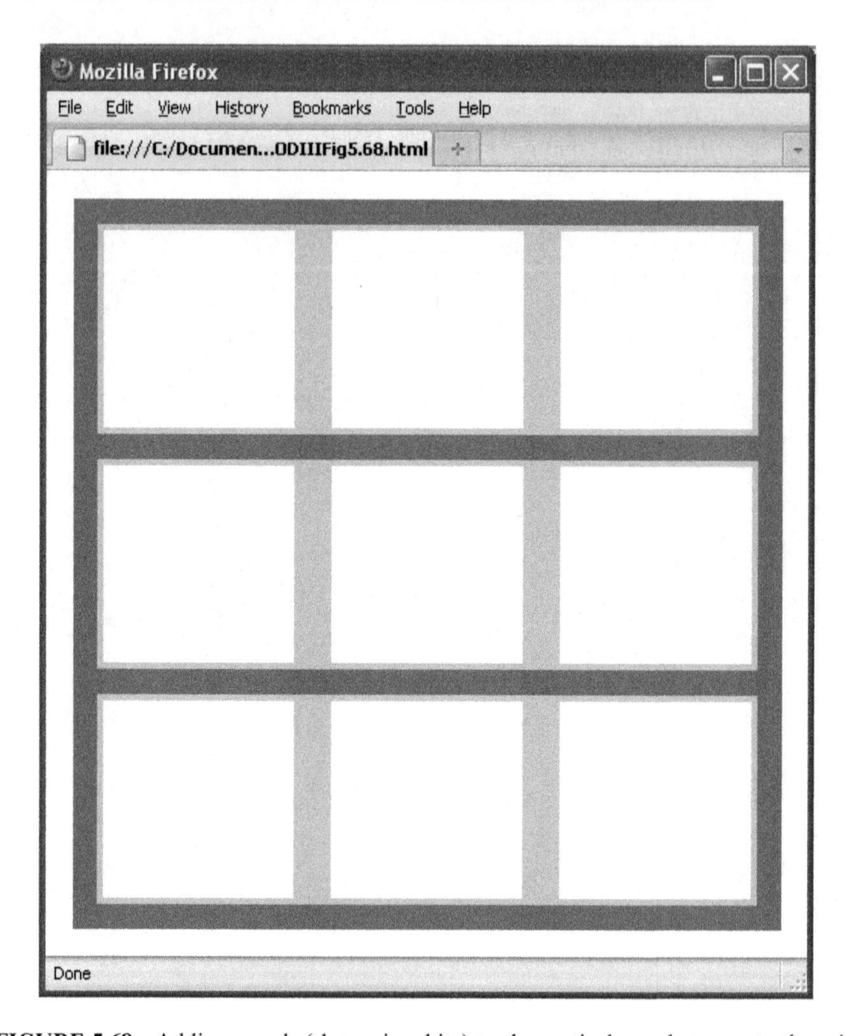


**FIGURE 5.68** Adding panels (shown in white) to the vertical panels to create the grid

```
var cellMatrix = cellRows.add(pv.Panel)
    .data(function(r) compoundProperties.map(function(c)
      ({parentColumnProperty:c, parentRowProperty:r})))
   .left(function() this.index * (scatterplotCellSize +
     spaceBetweenCells + (2*cellMargin)) + cellMargin)
   .width(scatterplotCellSize)
   .height(scatterplotCellSize)
   .top(cellMargin)
   .fillStyle("white");
```

Only the off-diagonal cells will be used to display the individual scatterplots (the diagonals will be used to name the axes). Panels are added to display the

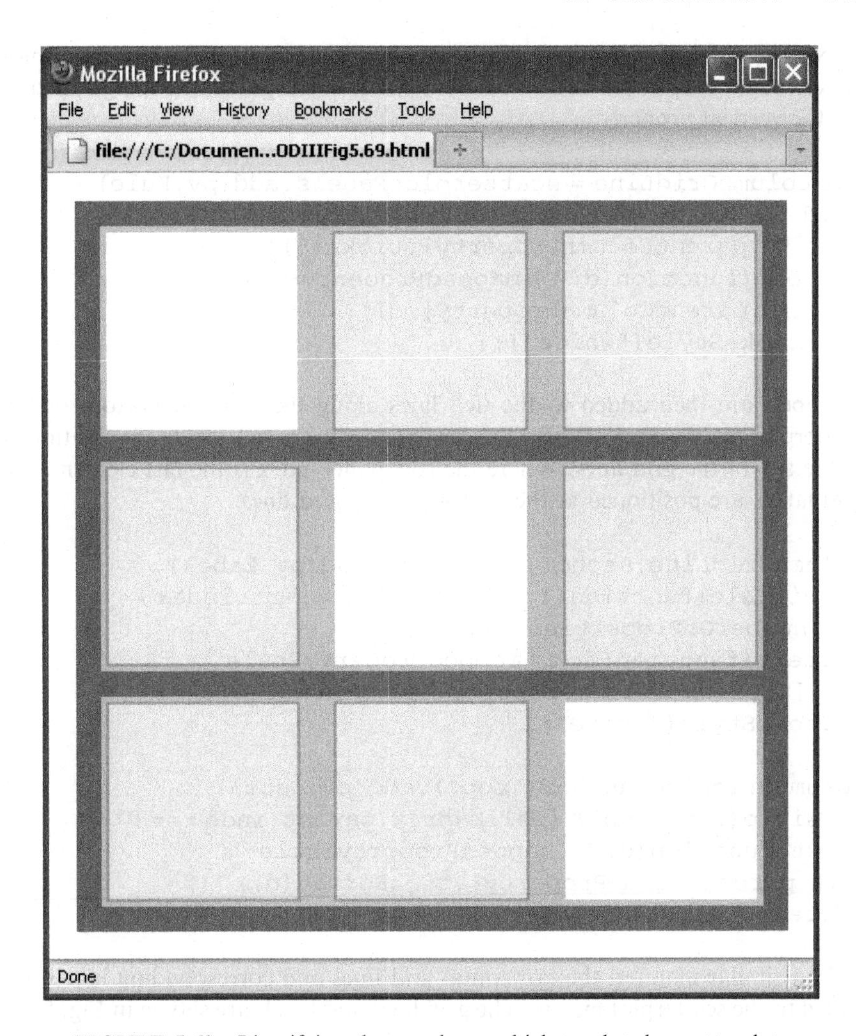


**FIGURE 5.69** Identifying the panels on which to plot the scatterplots

scatterplots. A function is used to determine whether any of the cells should be visible or not. Any cell where the corresponding column and row properties are not identical will be visible, as shown in the light gray shading with a dark gray border in Fig. 5.69.

```
var scatterplotPanels = cellMatrix.add(pv.Panel)
   .visible(function(d) d.parentColumnProperty!=
     d.parentRowProperty)
   .fillStyle("#eee")
   .strokeStyle("darkgray");
```

Next, a series of vertical grid lines are added to each of the visible scatterplot panels. A look-up is performed to identify the corresponding scaling function for the vertical property.

```
var columnGridLine = scatterplotPanels.add(pv.Rule)
   .data(function(t) mappedPropertyScale
      [t.parentColumnProperty].ticks(5))
   .left(function(d, t) mappedPropertyScale
      [t.parentColumnProperty](d))
   .strokeStyle("white");
```

Labels are then added to the tick lines along the top and bottom of the scatterplot matrix panel. Where the row index is 0, the tick labels are positioned at the top of the grid lines. Where the row panel index is the last element, the grid labels are positioned at the bottom of the grid lines.

```
columnGridLine.anchor("bottom").add(pv.Label)
   .visible(function() (cellMatrix.parent.index ==
     numberOfProperties - 1))
   .text(function(d, t) mappedPropertyScale
     [t.parentColumnProperty].tickFormat(d))
   .textStyle("white");

columnGridLine.anchor("top").add(pv.Label)
.visible(function() (cellMatrix.parent.index == 0))
.text(function(d, t) mappedPropertyScale
  [t.parentColumnProperty].tickFormat(d))
   .textStyle("white");
```

In a similar manner, the horizontal grid lines and corresponding labels are added to the scatterplot matrix. The grid lines and labels are shown in Fig. 5.70, drawn in white.

```
var rowGridLines = scatterplotPanels.add(pv.Rule)
   .data(function(t) mappedPropertyScale
     [t.parentRowProperty].ticks(5))
   .bottom(function(d, t) mappedPropertyScale
     [t.parentRowProperty](d))
   .strokeStyle("white");

rowGridLines.anchor("left").add(pv.Label)
   .visible(function() (cellMatrix.index == 0))
   .text(function(d, t) mappedPropertyScale
      [t.parentRowProperty].tickFormat(d))
   .textStyle("white");
```

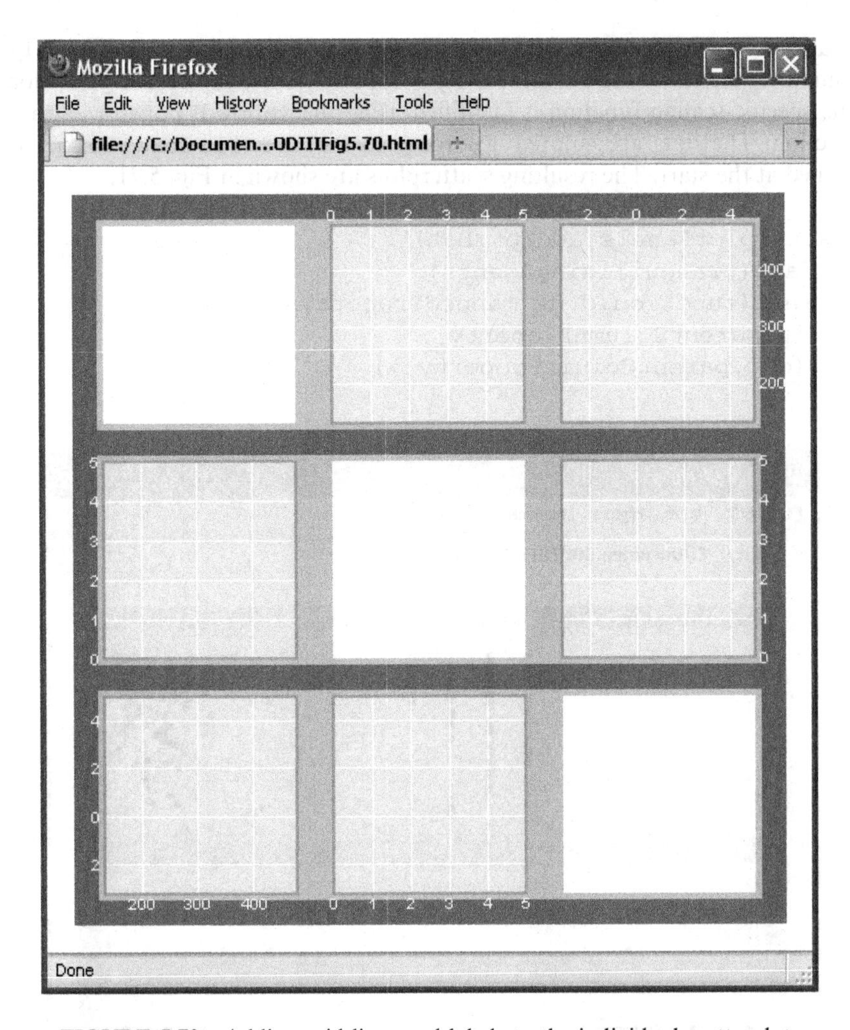


**FIGURE 5.70** Adding grid lines and labels to the individual scatterplots

```
rowGridLines.anchor("right").add(pv.Label)
   .visible(function() (cellMatrix.index ==
      numberOfProperties - 1))
   .text(function(d, t) mappedPropertyScale
      [t.parentRowProperty].tickFormat(d))
   .textStyle("white");
```

Each of the scatterplot panels to display is associated with two properties (`parentColumnProperty` and `parentRowProperty`), which are the x-axis and y-axis, respectively, of the scatterplot. Dots are placed onto each of the scatterplots corresponding to each observation in the `tranquilizingAgents` array. The horizontal and vertical mapping of the domain data onto an individual scatterplot is performed using a scaling function. The specific scaling function is identified using the `mappedPropertyScale` function. The dots are colored according to the `tranquilizerColor` list defined at the start. The resulting scatterplots are shown in Fig. 5.71.

```
scatterplotPanels.add(pv.Dot)
  .data(tranquilizingAgents)
  .left(function(d, t) mappedPropertyScale
    [t.parentColumnProperty]
    (d[t.parentColumnProperty]))
```

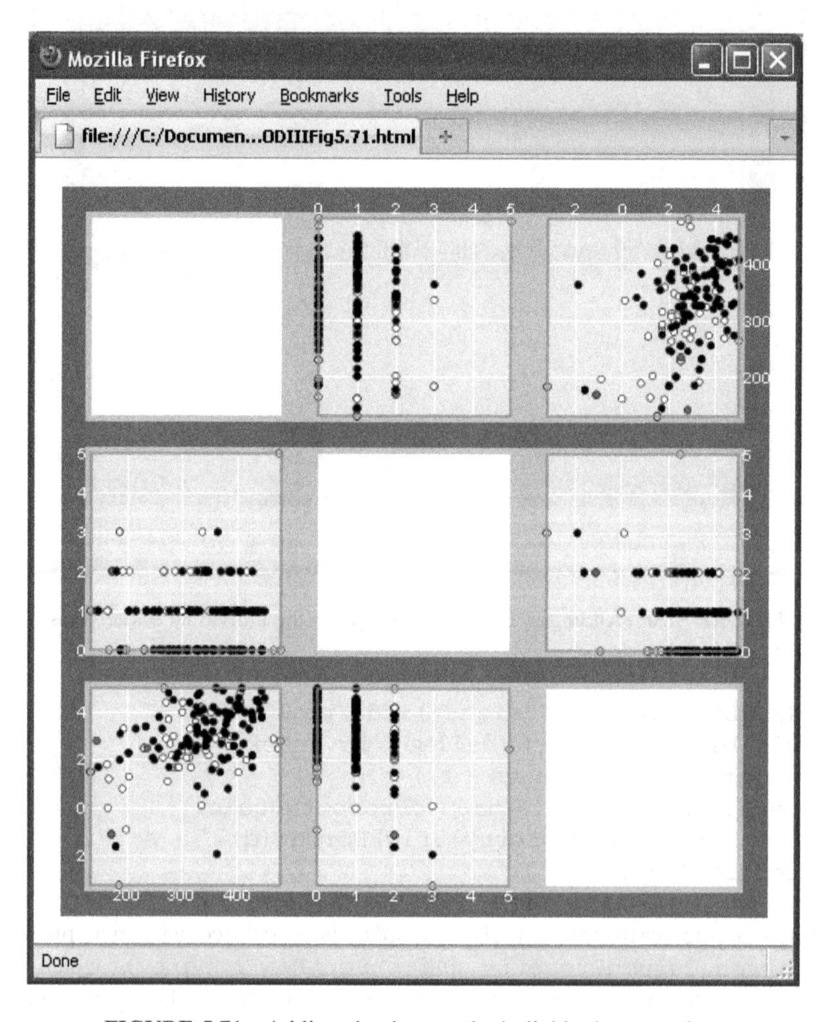


FIGURE 5.71 Adding the dots to the individual scatterplots

```
.bottom(function(d, t) mappedPropertyScale
  [t.parentRowProperty](d[t.parentRowProperty]))
.size(10)
.strokeStyle("gray")
.fillStyle(function(d) tranquilizerColor(d.type));
```

Property names are added to the diagonal panels for the axes, as shown in Fig. 5.72.

```
cellMatrix.anchor("center").add(pv.Label)
  .visible(function(t) t.parentColumnProperty==
    t.parentRowProperty)
```

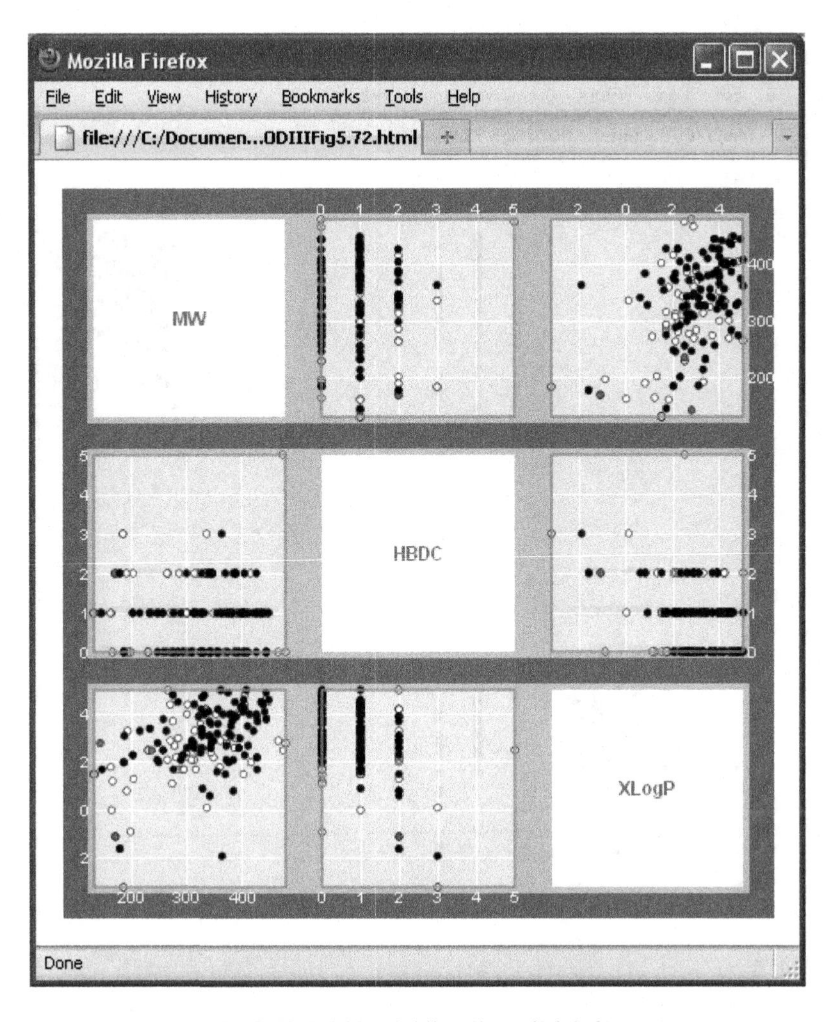


**FIGURE 5.72** Adding the axis labels

```
  .font("bold 12px sans-serif")
  .textStyle("gray")
  .text(function(t) t.parentColumnProperty);
```

Finally, a legend is added to the bottom of the scatterplot matrix to convey the meaning of the color-coding, as shown in Fig. 5.73.

```
scatterplotMatrixPanel.add(pv.Dot)
  .data(agents)
  .bottom(7)
  .left(function() (rowCellHeight) + (2 *
    spaceBetweenCells) + this.index * 80)
```

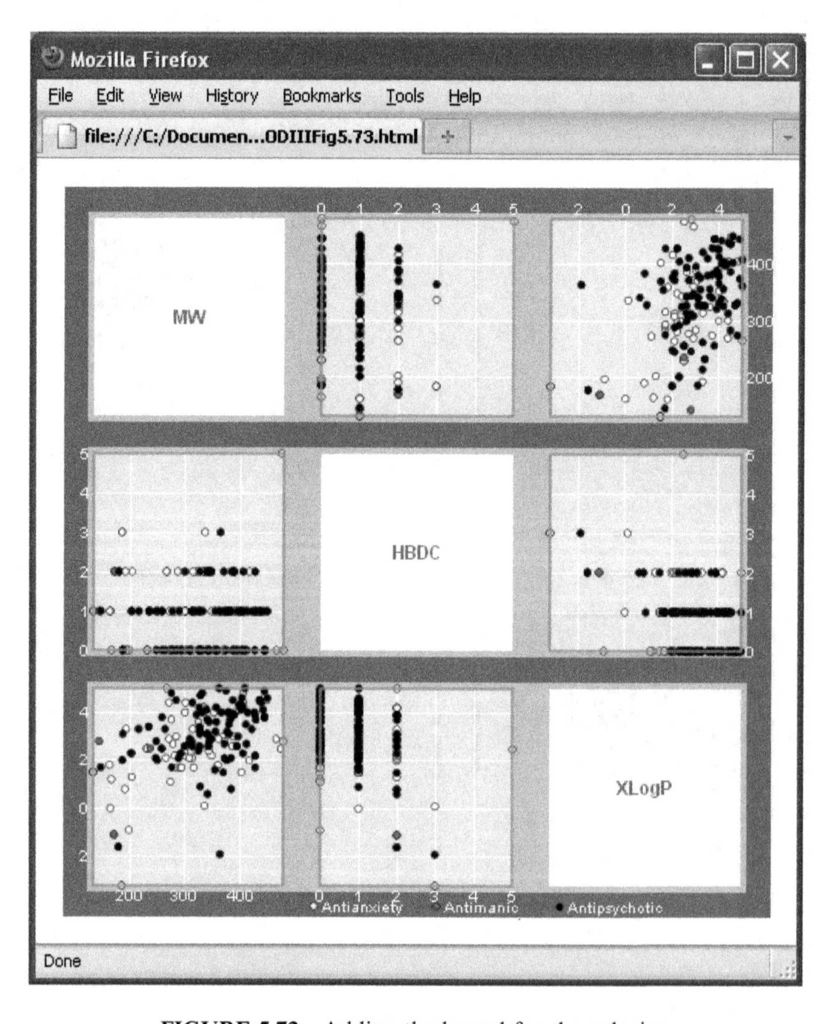


**FIGURE 5.73** Adding the legend for the coloring

```
    .size(5)
    .lineWidth(0.5)
    .strokeStyle("black")
    .fillStyle(tranquilizerColor)
  .anchor("right").add(pv.Label).textStyle("white");

scatterplotMatrixPanel.render();
```

### 5.7.5 Layout Management

In Protovis, you can create a large variety of rich visualizations by combining visual elements in different ways. Although this approach provides a great deal of flexibility, many types of visualizations are used often. To support these commonly used composite views, Protovis has a series of standardized techniques for these *layouts*, such as grids, hierarchies, and networks (hierarchies and networks will be reviewed in the next sections). This makes the development of commonly used visual organizations quicker and easier. These layouts share many similarities with the panel class used throughout.

One type of layout is the grid layout (`pv.Layout.Grid`), which creates a matrix of cells with each cell evenly spaced. This view corresponds to a 2-D array, for example, the grid

```
11 12 13 14
21 22 23 24
31 32 33 34
```

would be represented as a 2-D array:

```
[[11,12,13,14],[21,22,23,24],[31,32,33,34]].
```

In the following example, a 2-D array is specified (`heatmap`), and a panel is created (based on the size of the matrix defined). A grid layout is added to the panel, and the number of rows is set by assigning the heatmap. With these assignments, we are customizing the layout of the panel. The cells will be visualized using the bar mark, which is added to the cell directly (not the panel). A cell is referred to as a *mark prototype*, which contains default property settings for any marks that are assigned to it. In this example, the cells are colored based on their data values, with green colors for the lower numbers, red shades for the larger numbers, and gray for those in the middle, as shown in Fig. 5.74.

```
var heatmap = [[2,3,8,9,7,6,4,3],[4,6,8,9,5,6,4,3],[5,7,
  8,9,11,12,9,6,4,3],[3,4,6,8,9,8,7,5,4]];

var nosCols = heatmap[0].length,
    nosRows = heatmap.length;
```

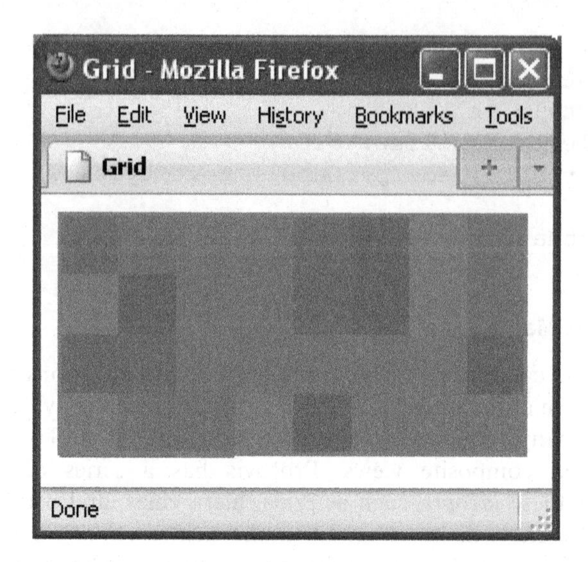


**FIGURE 5.74** Heatmap drawn using the grid layout

```
var vis = new pv.Panel()
   .width(nosCols * 30)
   .height(nosRows * 30)
   .margin(2)
   .strokeStyle("gray")
   .lineWidth(2);

vis.add(pv.Layout.Grid)
   .rows(heatmap)
  .cell.add(pv.Bar)
   .fillStyle(pv.Scale.linear()
     .domain(0, 4, 8, 12)
     .range("green", "gray", "red"));

vis.render();
```

### 5.7.6 Networks

A network is an organization of different nodes that are connected in some manner. To use such an organization, the nodes and connections between nodes need to be defined. In the following example, a series of financial transactions between companies is listed. First the companies are described (nodes in the network), along with a `typeOfBusiness` assignment (1 are financial companies, 2 are communications companies, 3 are pharmaceuticals, and 4 are energy). The format for the nodes must conform to the pv.Layout.Network.Node interface. Next, the relationships between the nodes are enumerated (links). Each link has a source and a target representing the link between the nodes. The first node ("Nations Bank Corp") is at index 0, and the second node ("BankAmerica Corp") is at index 1, and so on. The first link in the list has the source 0 and target 1, hence it describes a link between "Nations Bank Corp" and "BankAmercia Corp". The value associated with this link is 6.16 (meaning "Nations Bank Corp" purchased "BankAmerica Corp" for 6.16 × \$10 billions). The format for these links must conform with the pv.Layout.Network.Link interface.

```
var financialTransactions = {
 nodes:[
  {nodeName:"Nations Bank Corp", typeOfBusiness:1},
  {nodeName:"BankAmerica Corp", typeOfBusiness:1},
  {nodeName:"Travelers Group", typeOfBusiness:1},
  {nodeName:"Citycorp", typeOfBusiness:1},
  {nodeName:"ATT", typeOfBusiness:2},
  {nodeName:"Telecom Inc", typeOfBusiness:2},
  {nodeName:"Vodafone", typeOfBusiness:2},
  {nodeName:"Airtouch Comm", typeOfBusiness:2},
  {nodeName:"Aventis", typeOfBusiness:3},
  {nodeName:"SanofiSynthelabo", typeOfBusiness:3},
  {nodeName:"SBC Communications", typeOfBusiness:1},
  {nodeName:"Ameritech Corp", typeOfBusiness:1},
  {nodeName:"Exxon Corp", typeOfBusiness:4},
  {nodeName:"Mobile Corp", typeOfBusiness:4},
  {nodeName:"Mannesmann", typeOfBusiness:2},
  {nodeName:"Pfizer Inc", typeOfBusiness:3},
  {nodeName:"WarnerLambert Co", typeOfBusiness:3},
  {nodeName:"GlaxoWellcome", typeOfBusiness:3},
  {nodeName:"SmithKline Beecham", typeOfBusiness:3},
  {nodeName:"America Online", typeOfBusiness:2},
  {nodeName:"Time Warner", typeOfBusiness:2},
  {nodeName:"Bell Atlantic Corp", typeOfBusiness:2},
  {nodeName:"GTE Corp", typeOfBusiness:2},
  {nodeName:"Comcast Crop", typeOfBusiness:2},
  {nodeName:"ATT Broadband", typeOfBusiness:2},
  {nodeName:"Royal Dutch Petrol", typeOfBusiness:4},
  {nodeName:"Shell Trans Trade", typeOfBusiness:4},
  {nodeName:"BellSouth Corp", typeOfBusiness:2}
  ],
 links:[
  {source:0, target:1, value:6.16},
  {source:2, target:3, value:7.25},
  {source:4, target:5, value:6.99},
```

```
    {source:6, target:7, value:6.55},
    {source:8, target:9, value:6.56},
    {source:10, target:11, value:7.04},
    {source:12, target:13, value:8.51},
    {source:6, target:14, value:20.28},
    {source:15, target:16, value:8.88},
    {source:17, target:18, value:7.87},
    {source:19, target:20, value:18.16},
    {source:21, target:22, value:7.13},
    {source:23, target:24, value:7.29},
    {source:25, target:26, value:8.03},
    {source:4, target:27, value:8.94}
  ]
};
```

The following code, after creating the chart panel, sets up a panel according to the arc layout (`pv.Layout.Arc`). The nodes and links to be displayed are added to their respective mark prototypes. The arc is visualized using a line mark, and the nodes are color-coded according to the type of business, as shown in Fig. 5.75.

```
var chartPanel = new pv.Panel()
    .width(880)
    .height(410)
    .bottom(120);

var networkArc = chartPanel.add(pv.Layout.Arc)
.nodes(financialTransactions.nodes)
    .links(financialTransactions.links);

networkArc.link.add(pv.Line);

networkArc.node.add(pv.Dot)
    .radius(5)
    .fillStyle(pv.Colors.category19().by(function(d)
      d.typeOfBusiness))
    .strokeStyle("black");

networkArc.label.add(pv.Label);

chartPanel.render();
```

The `pv.Layout.Matrix` is an example of another layout approach for networks.

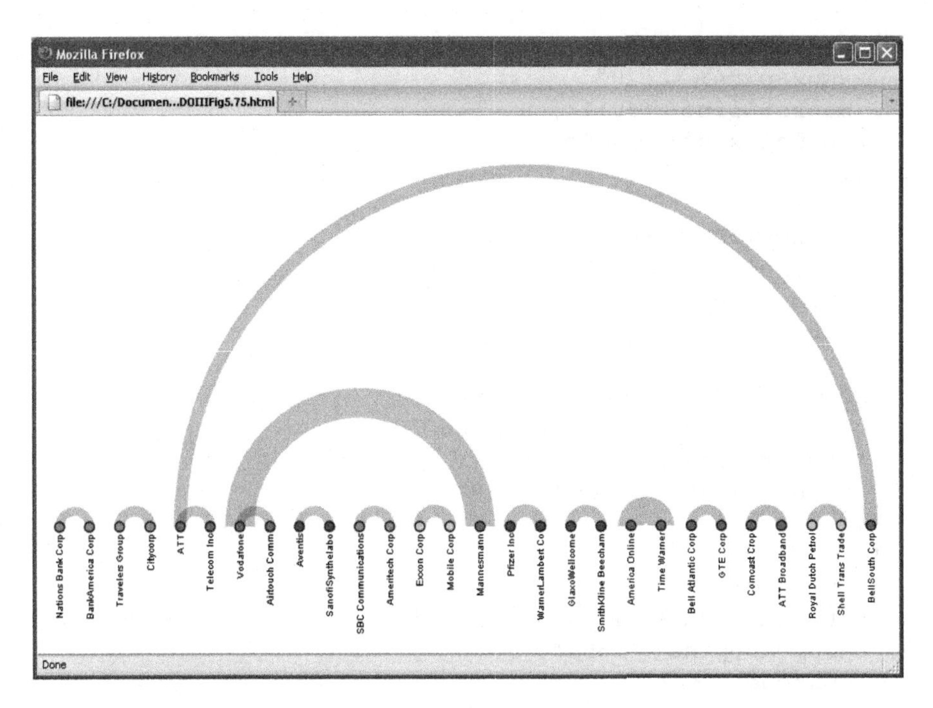


**FIGURE 5.75** Arc layout representing a network of transactions

### 5.7.7 Hierarchies

To create a hierarchical view in Protovis, the relationships between the nodes need to be defined. In a similar manner to reading an external data table, Protovis can also read hierarchies of information, such as a JavaScript hierarchy. In the following example, a hierarchy of central nervous system (CNS) medications is presented. The hierarchical relationships are depicted by embedding the subclasses within their parent classes. The Protovis class `pv.Dom` provides an internal representation for this hierarchical object to be used with different hierarchical layout managers. By default, the items that have values (such as `imipramine`) are the hierarchy leaves.

```
var CNSMedication = {
CNSMedication: {
 Antidepressants: {
  Tricyclics: {
   imipramine: "3696",
   amitriptyline: "2160",
   clomipramine: "2801"
  },
```

```
   Heterocyclics: {
    amoxapine: "2170",
    maprotiline: "4011",
    venlafaxine: "5656"
   },
   SSRI: {
    fluoxetine: "3386",
    sertraline: "68617",
    paroxetine: "43815",
    fluvoxamine: "5324346"
   },
   MAOs: {
    tranylcyprominesulfate: "26069",
    isocarboxazid: "3759",
    phenelzinesulfate: "61100"
   }
  },
 Anitmanic:{
   lithium: "3028194",
   carbamazepine: "2554",
   depakene: "3121"
  }
}
};
```

In the following example, a hierarchy of CNS medications is included in a JavaScript file. The `pv.dom` method will create the hierarchy object, and the top-level node "`CNSMedication`" is assigned as the root node. The hierarchy object is then used to determine the size of the panel, by calling the `nodes()` method to retrieve the list of all nodes, which is used to determine the required height of the panel. A layout for the panel (`indentPanelLayout`) is added to the chart's root panel. The nodes of the hierarchy are assigned to the `nodes` property. The line mark is added to the link object. If this is not added, there would be no explicit mark to link the nodes, although they would still be organized as a hierarchy through indentation. The individual nodes of the hierarchy are created by first adding a new panel, a dot, and then the node's label, anchored to the left and right, respectively. The resulting hierarchy is shown in Fig. 5.76.

```
var hierarchy = pv.dom(CNSMedication)
   .root("CNSMedication");

var chartPanel = new pv.Panel()
   .width(200)
   .height(function() (hierarchy.nodes().length + 1) * 15)
   .margin(10);
```

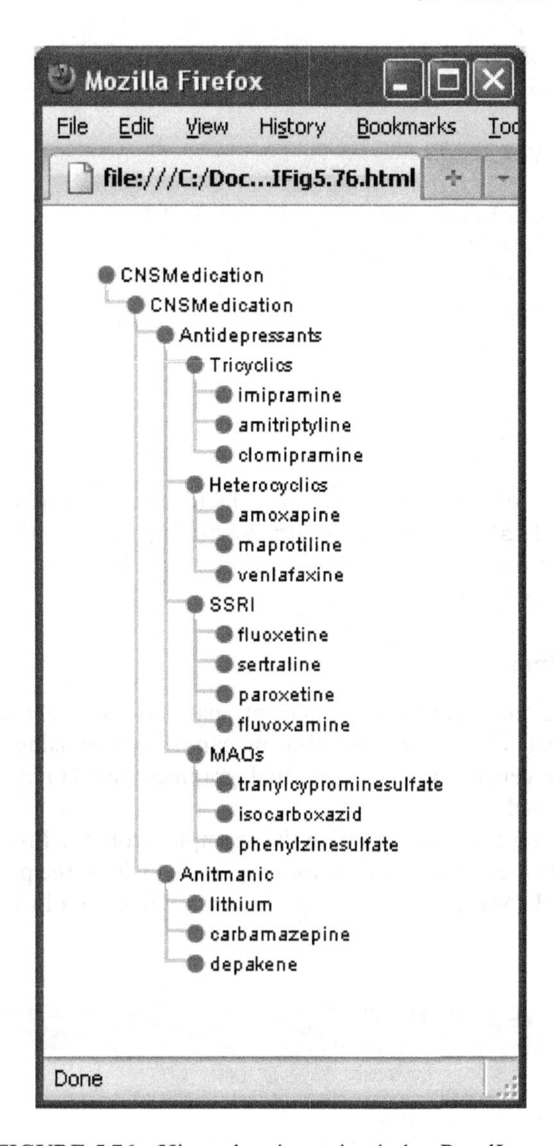


**FIGURE 5.76** Hierarchy view using indentPanelLayout

```
var indentPanelLayout = chartPanel.add(pv.Layout.Indent)
.nodes(function() hierarchy.nodes())
   .depth(15)
   .breadth(15);

indentPanelLayout.link.add(pv.Line);

var indentNode = indentPanelLayout.node.add(pv.Panel)
```

```
.top(function(n) n.y - 6)
  .height(15)
  .right(6)
  .strokeStyle(null);

indentNode.anchor("left").add(pv.Dot)
  .strokeStyle("#darkgray")
  .fillStyle("gray")
 .anchor("right").add(pv.Label)
  .text(function(n) n.nodeName);


chartPanel.render();
```

A number of layouts support hierarchical displays, including `pv.Layout.Cluster`, `pv.Layout.Pack`, `pv.Layout.Partition`, `pv.Layout.Tree`, and `pv.Layout.Treemap`. They are defined in the Protovis API documentation.

### 5.7.8 Sparklines

*Sparklines* refer to small data visualizations that are contained within a paragraph of text. This type of visualization can be created using a combination of Protovis for generating the individual graphics and HTML where paragraphs are detailed.

In the following example, based on the examples from the Protovis Web site, three small plots are defined as functions and called within the paragraph at the bottom of the HTML page. The resulting display is shown in Fig. 5.77.

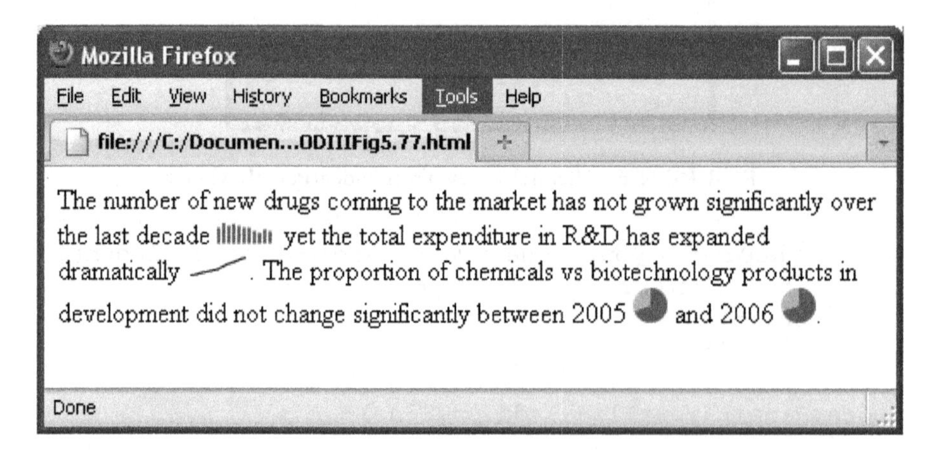


**FIGURE 5.77** Illustrations of sparkline, embedding graphics within the paragraphs

```
<html>
 <head>
  <script type="text/javascript" src=
    "../protovis.js"></script>
 </head>

<body>
 <script type="text/javascript+protovis">

var newDrugs = [36,46,40,42,39,35,34,29,24,33,29];

var expenditure = [100,105,120,130,140,150,175,195,210,
230,245];

var chemBio2005 = [70,30];
var chemBio2006 = [72,28];


function sparkline(data) {
 var n = data.length, width = n*3, height = 10,
   min = pv.min.index(data), max = pv.max.index(data);

 var vis = new pv.Panel().width(width).height(height)
   .margin(2);

vis.add(pv.Line)
    .data(data)
    .left(pv.Scale.linear(0, n - 1).range(0, width)
      .by(pv.index))
    .bottom(pv.Scale.linear(data).range(0, height))
    .strokeStyle("#000")
    .lineWidth(1);

 vis.render();
}

function sparkwedge(data) {
 var n = data.length, width = 20, height = 20;


 var vis = new pv.Panel().width(width).height(height);

 vis.add(pv.Wedge)
    .data(pv.normalize(data))
    .left(10)
```

```
    .bottom(10)
    .outerRadius(10)
    .angle(function(d) Math.PI * 2 * d);

 vis.render();
}

function sparkbar(data) {
 var width = 35, height = 12;

 var vis = new pv.Panel().width(width).height(height);

 vis.add(pv.Bar)
    .data(data)
    .width(2)
    .left(function() 3 * this.index)
    .height(function(d) Math.round(0.25 * d))
    .bottom(0);

 vis.render();
}
</script>

The number of new drugs coming to the market has not grown
significantly over the last decade
<script type="text/javascript+protovis">
  sparkbar(newDrugs); </script>
yet the total expenditure in R&D has expanded dramatically
  <script type="text/javascript+protovis">
  sparkline(expenditure); </script>.
The proportion of chemicals vs biotechnology products
in development did not change significantly between 2005
<script type="text/javascript+protovis">
  sparkwedge(chemBio2005); </script>
and 2006
<script type="text/javascript+protovis">
  sparkwedge(chemBio2005); </script>.
</body>
</html>
```

### 5.7.9 Exercises

5.7.9.1 Trellis plots are used to describe multiple dimensions concerning groups of data.The JavaScript file PI-PII-PIII-Clinical-Trials.js contains the number of drugs in phase I, phase II, and phase III for three fictitious pharmaceutical companies (Alphapharma, Betapharma, and Gammapharma). Two entries are shown here:

```
var clinicalTrials = [
{ numberOfDrugs: 10, clinicalTrial: "Phase I", disease:
  "cancer", company: "Alphapharm" },
{ numberOfDrugs: 8, clinicalTrial: "Phase II", disease:
  "cancer", company: "Alphapharm" },
```

Create the trellis plot as shown in Fig. 5.78.

## 5.8 INTERACTIVE PLOTS

### 5.8.1 Overview

In many situations, you may want to interact with a plot. You might want to understand more details concerning an element of the plot, such as the underlying data behind a bar within a bar chart. The JavaScript Protovis toolkit handles events such as mouse clicks and mouse moves. These events include the following: `"click"`, `"mousedown"`, `"mouseup"`, `"mouseover"`, `"mousemove"`, and `"mouseout"`. A full list is provided at www.w3.org/TR/SVGTiny12/interact.html#SVGEvents. The following section outlines the use of these events to provide interaction with different graphics.

### 5.8.2 Tooltips

A simple approach to annotating marks within a graphic is to assign a value to the mark's `title` property. In the following example, an image mark is added to the panel. The `title` property is set to "John Wiley & Sons, Inc.". This results in a tooltip appearing when the mouse cursor hovers over the mark, as shown in Fig. 5.79. Note that this approach only handles plain text tooltips.

```
var chartPanel = new pv.Panel()
   .width(200)
   .height(100);

chartPanel.add(pv.Image)
   .url("wiley-logo.bmp")
   .left(25)
   .bottom(25)
   .height(50)
   .width(151)
   .title("John Wiley & Sons, Inc.");

chartPanel.render();
```

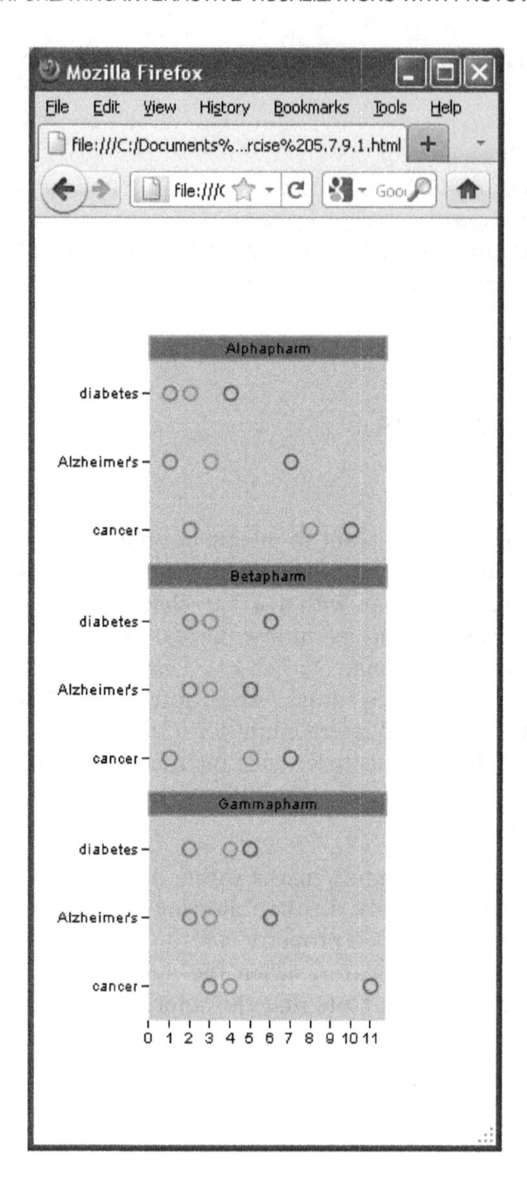


**FIGURE 5.78** Trellis plot for three fictitious pharmaceutical companies

### 5.8.3 Hyperlinks

Another simple approach to drilling down into details concerning one or more elements of a plot is to use hyperlinks to go to another Web page that contains more information. In the following example, an image is added to a panel. The mark's cursor property is set to `pointer`. Other cursor options are provided at www.w3.org/TR/CSS2/ui.html#propdef-cursor. They include

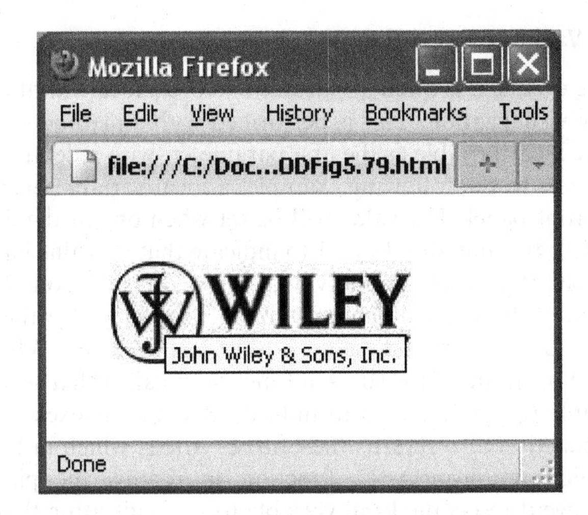


**FIGURE 5.79** Tooltip display when mouse cursor hovers over a mark

"auto", "crosshair", "move", "text", "wait", and "progress". The title is also set to explain the hyperlink action and to be presented as a tooltip when the mouse cursor hovers over the mark. In this example, three events are considered: "mouseover", "mouseout", and "click". When a "mouseover" event is captured, the value of the hyperlink status is set; when the "mouseout" event is captured, it is reset to present no information. A "click" event triggers the hyperlink action, performed by setting the `self.location` property to the URL where the Web page is to be redirected.

```
var chartPanel = new pv.Panel()
    .width(200)
    .height(100);

chartPanel.add(pv.Image)
    .url("wiley-logo.bmp")
    .left(25)
    .bottom(25)
    .height(50)
    .width(151)
    .cursor("pointer")
    .title("Hyperlink to John Wiley & Sons, Inc. website")
    .event("mouseover", function() self.status =
      "Go to \"http://www.wiley.com\"")
    .event("mouseout", function() self.status = "")
    .event("click", function() self.location =
      "http://www.wiley.com");

chartPanel.render();
```

### 5.8.4 Local Variables and Events

When handling events, it is often important to store a local variable concerning the state of the visualization. This is performed with a method `def` that takes the name of the local variable as the first parameter, with a default value as the second parameter. In the following example, a local variable `currOffset` is defined in the root panel. The value will be set when one of the data elements has been selected. It is initialized to −1 to indicate that no value has been set. A pie chart is created in which the individual slices are color-coded light gray if they correspond to the selected item and color-coded black for the other slices. Two events, `"mouseover"` and `"mouseout"`, are captured. When the mouse cursor is moved over one of the slices, a function is called that sets the value of the local variable (`currOffset`) to indicate that the indexed slice has been selected. The function also returns the `chartPanel`, which in turn results in the panel calling the `render()` function to redraw the pie chart. The `"mouseout"` event resets the local variable to −1, indicating that no value is currently selected. Because the function also returns the panel, the pie chart is redrawn, and all the chart slices are drawn in black because `currOffset` is now set to −1. Fig. 5.80 shows an example view with the mouse cursor moved into the gray slice, whereas Fig. 5.81 illustrates the results of a `"mouseout"` event.

```
var chartPanel = new pv.Panel()
  .def("currOffset", -1)
  .width(200)
  .height(200);

var data = [26,32,75,26,34], sum = pv.sum(data);

var wedge = chartPanel.add(pv.Wedge)
  .data(data)
  .left(75)
  .bottom(75)
  .outerRadius(70)
  .angle(function(d) d / sum * 2 * Math.PI)
  .strokeStyle("white")
  .fillStyle(function() chartPanel.currOffset() ==
    this.index ?
         "lightgray" : "black")
  .event("mouseover", function() chartPanel.currOffset
    (this.index))
 .event("mouseout", function() chartPanel.currOffset(-1))
  .anchor("center").add(pv.Label)
  .textStyle(function() chartPanel.currOffset() ==
    this.index ? "black" : "white");

chartPanel.render();
```

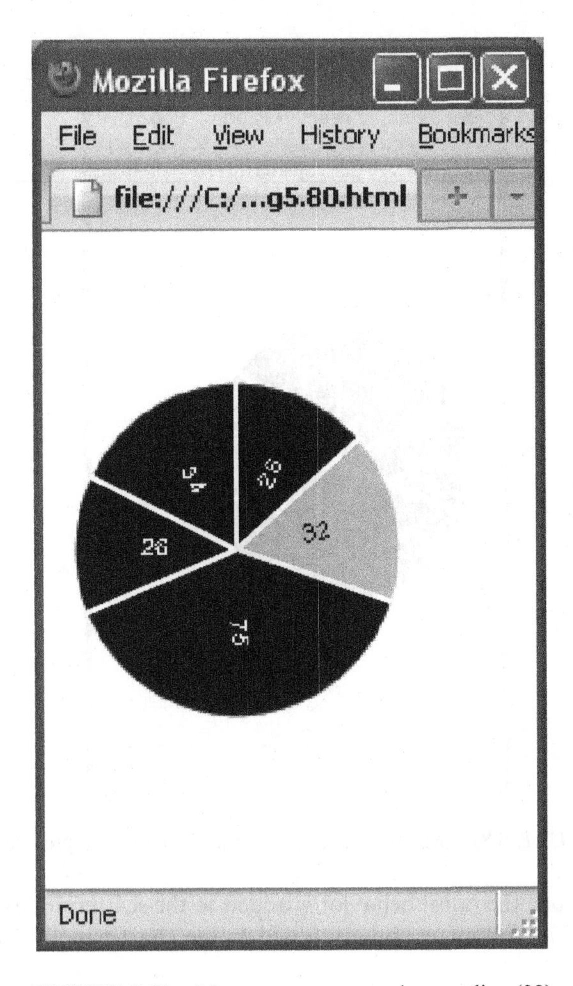


**FIGURE 5.80** Mouse cursor moved over slice (32)

### 5.8.5 Behavior

Protovis includes a number of built-in classes to handle events for a series of common types of interactions, including drag, pan, point, resize, select, and zoom. Depending on how an event is initiated, a behavior is registered based on this initiating event, such as a `"mouseover"` event. The `pv.Behavior.point` is an example of one of these built-in Protovis behavior classes. With this class, the behavior identifies whether the mouse cursor is close to a mark, which will trigger an event. Where marks are close together, this class identifies the closest mark to the mouse cursor (such as in a dense region of dots in a scatterplot). Where marks are part of a continuous trend (such as a line mark), the class also identifies the closest mark.

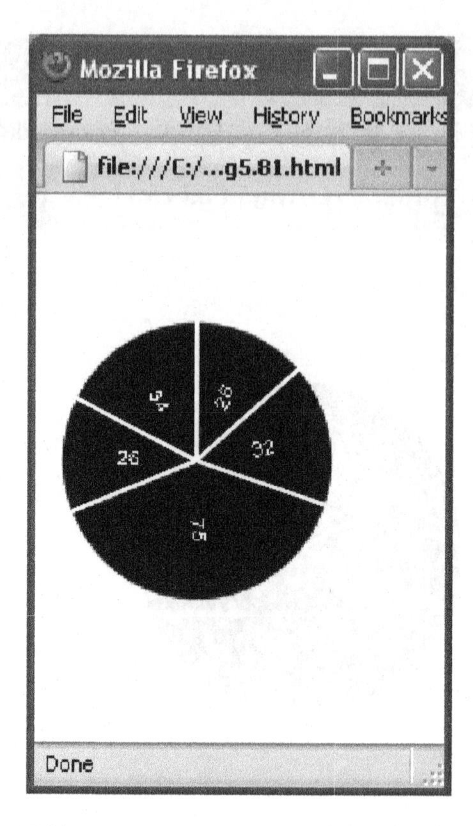


**FIGURE 5.81** Mouse cursor moved away from the pie chart

In this example, the point behavior is added to the scatterplot examples from Section 5.6.3. The following code is added to the chart panel's declaration.

```
.event("all")
.event("mousemove", pv.Behavior.point());
```

This entry ensures that any "mousemove" events that are performed within the panel of the scatterplot are identified. A specified behavior (`pv.Behavior.point()`) is bound to the `mousemove` event and is handled within the marks that are added to the panel. An optional property can be assigned to this point to change the radius of the circle within which the event should be considered. The default is 30 pixels; however, an "Infinity" value can be assigned where the entire panel should be considered. As an aside, the `.event("all")` is used here to catch the first event when the panel is translucent.

A local variable is defined to identify the selected scatterplot dot.

```
.def("active", -1)
```

Two events are added to the dot marks: "point" and "unpoint". In this example, the "point" event identifies the mark closest to the mouse cursor (by default, within a 30-pixel radius). This event is handled by setting the this.active property to the current index position, which is the selected scatterplot dot. This code also returns the parent panel, which is redrawn with the render method. Similarly, an "unpoint" event is handled to recognize when the mouse cursor is moved away. Here, the active property is reset to −1.

```
.event("point", function() this.active
  (this.index).parent)
.event("unpoint", function() this.active(-1).parent)
```

A label is drawn next to the scatterplot dot; however, it is only visible when the active value equals the current index, that is, the selected scatterplot dot.

```
.anchor("right").add(pv.Label)
  .visible(function() this.anchorTarget().active() ==
    this.index)
  .text(function(d) d.id);
```

The following code produces the scatterplot, as shown in Fig. 5.82, where a label (e.g., "47979") is placed next to the dot when a cursor is moved close to it.

```
var panelWidth    = 400,
    panelHeight   = 400,
    margin        = 30
    maxMolWtValue = pv.max(tranquilizingAgents,
      function(d)d.MW)
    xScale        = pv.Scale.linear(0, maxMolWtValue )
      .range(0, panelWidth).nice(),
    minALogPValue = pv.min(tranquilizingAgents,
      function(d)d.XLogP)
    maxALogPValue = pv.max(tranquilizingAgents,
      function(d)d.XLogP)
    yScale        = pv.Scale.linear(minALogPValue,
      maxALogPValue).range(0, panelHeight).nice();

var chartPanel = new pv.Panel()
  .width(panelWidth)
  .height(panelHeight)
  .margin(margin)
  .event("all")
  .event("mousemove", pv.Behavior.point());
```

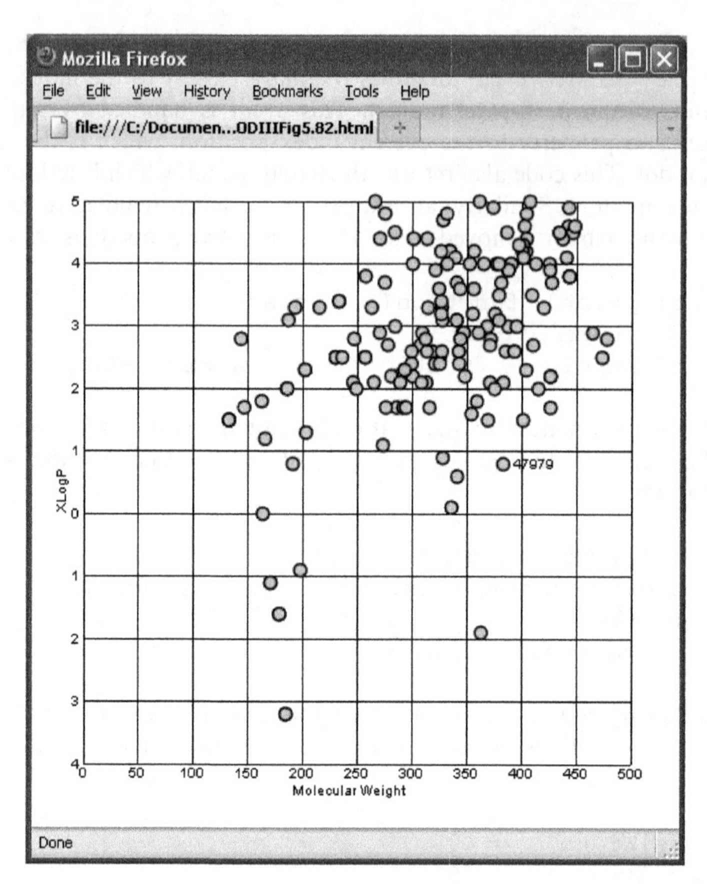


**FIGURE 5.82** Scatterplot dot annotated with the observation id (47979) using the point behavior

```
chartPanel.add(pv.Rule)
   .data(yScale.ticks())
   .bottom(yScale)
  .anchor("left").add(pv.Label)
   .text(yScale.tickFormat);

chartPanel.add(pv.Label)
  .text("XLogP")
  .left(-10)
  .bottom(175)
  .textAngle(-Math.PI/2);

chartPanel.add(pv.Rule)
   .data(xScale.ticks())
```

```
    .left(xScale)
  .anchor("bottom").add(pv.Label)
    .text(xScale.tickFormat);

chartPanel.add(pv.Label)
     .text("Molecular Weight")
     .bottom(-25)
     .left(150);

chartPanel.add(pv.Dot)
    .def("active", -1)
    .data(tranquilizingAgents)
    .left(function(d) xScale(d.MW))
    .bottom(function(d) yScale(d.XLogP))
    .fillStyle("lightgray")
    .strokeStyle("black")
   .event("point", function() this.active(this.index).parent)
    .event("unpoint", function() this.active(-1).parent)
  .anchor("right").add(pv.Label)
    .visible(function() this.anchorTarget().active() ==
     this.index)
    .text(function(d) d.CID);

chartPanel.render();
```

In situations where only one dimension needs to be considered, the x-dimension or y-dimension can be `"collapse"`. `pv.Behavior.point` is one of six different types of behaviors defined in Protovis, which include drag, pan, resize, select, and zoom (see the API description for more details). Events can also be generated using JavaScript controls, such as buttons, external to the Protovis code (see examples on the Protovis Web site for illustrations).

### 5.8.6 Exercises

5.8.6.1 Add a tooltip to the box-and-whisker plot created in Section 5.6.2 to generate the text shown in Fig. 5.83.

5.8.6.2 Modify the scatterplot visualization described in Section 5.8.5 where, by clicking on the scatterplot point, the user is redirected to a Web page with the details on the specific tranquilizing agent. The CID property is a unique identification for each of the agents. The PubChem database contains the underlying information on each CID record. For example, the URL http://pubchem.ncbi.nlm.nih.gov/summary/summary.cgi?cid=441233 will display the record for the agent whose CID is 441233 (the first observation in the list).

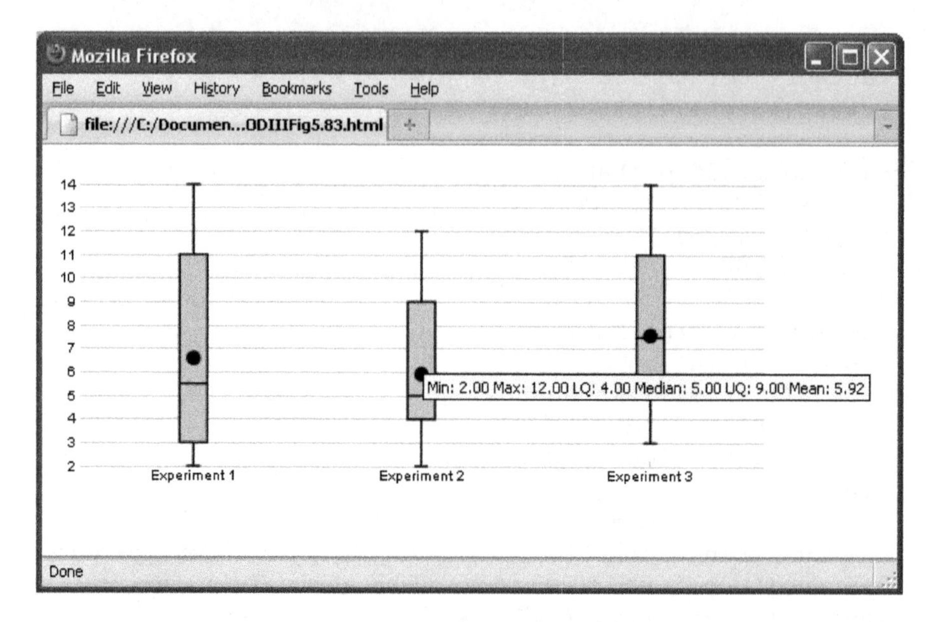


**FIGURE 5.83** Tooltips added to the box-and-whisker plot

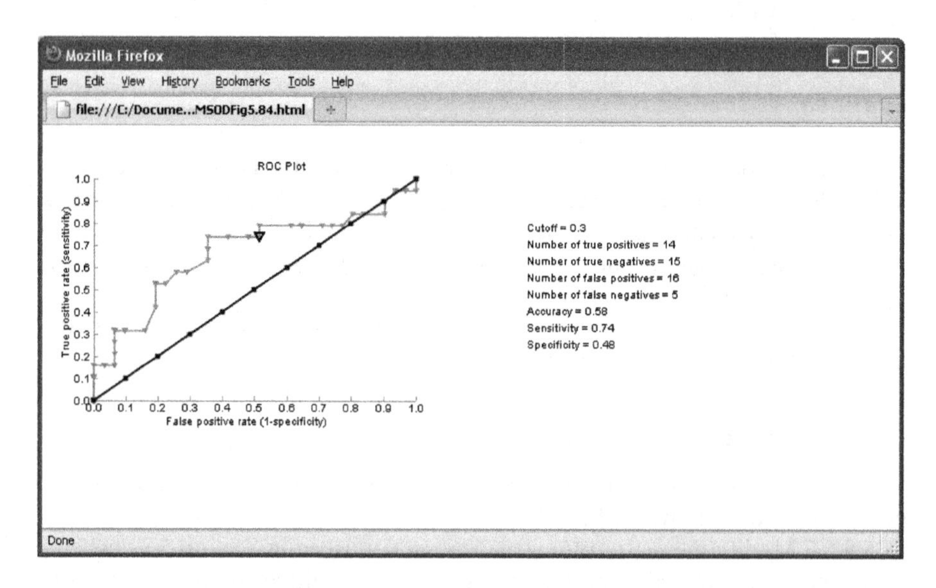


**FIGURE 5.84** Interactive ROC plot

5.8.6.3 Using the ROC plot created in Exercise 5.6.4.4, generate an interactive ROC plot so that moving a cursor close to a point in the plot provides more information on the cutoff and the quality of the underlying logistic regression model at that point, as shown in Fig. 5.84.

**TABLE 5.1 Summary of the Different Mark's Properties**

| | Data | Positioning | | | | Size | | Style | | | Tooltips | Visible |
|---|---|---|---|---|---|---|---|---|---|---|---|---|
| | `data` | `top` | `bottom` | `left` | `right` | `Width` | `height` | `fillStyle` | `strokeStyle` | `lineWidth` | `title` | `visible` |
| Bar | • | • | • | • | • | • | • | • | • | • | • | • |
| Label* | • | • | • | • | • | | | | | | • | • |
| Dot** | • | • | • | • | • | | | • | • | • | • | • |
| Line | • | • | • | • | • | | | | • | • | • | • |
| Area | • | • | • | • | • | • | • | • | • | • | • | • |
| Wedge*** | • | • | • | • | • | | | • | • | • | • | • |
| Image | • | • | • | • | • | • | • | | | | • | • |

* Other label properties: textAlign, textBaseline, textMargin, textAngle, font, textStyle
** Other dot properties: size, shape
*** Other wedge properties: startAngle, endAngle, angle, innerRadius, outerRadius

## 5.9 PROTOVIS SUMMARY

The Protovis graphical framework is composed of *panels* (canvases within which visualizations are drawn) and *marks* (specific objects that are drawn). Properties that can be set to customize these marks are summarized in Table 5.1.

The use of these properties to position a mark on a panel is summarized in Fig. 5.85.

Colors can be set using

- Named colors, for example, `"black"`, `"maroon"`, `"lightgray"`
- RGB format, for example, `rgb(24,243,117)`, `rgb(40%,60%,80%)`, or `rgba (30%,35%,40%,0.6)`
- HSL format, for example, `hsl(130,80%,40%)` or `hsla (140,70%,55%,0.3)`
- Protovis classes, for example, `pv.Color.Rgb`, `pv.Color.Hsl`

Text can be formatted with

- `pv.Format.date` (formats dates using the `strftime` function from C)
- `pv.Format.time` (formats time)
- `pv.Format.number` (adds appropriate commas and truncates digits)

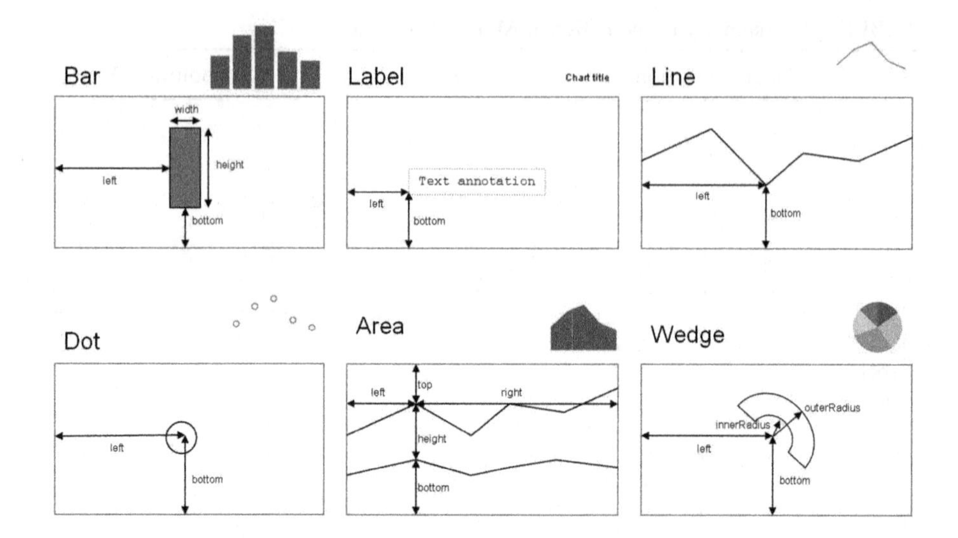


**FIGURE 5.85** Summary of marks and how to position marks on a panel

**TABLE 5.2 Summary of Common Scaling Functions in Protovis**

| Protovis Class | Description |
|---|---|
| `pv.Scale.linear` | Maps continuous data onto dimensions of the chart such as location or color |
| `pv.Scale.ordinal` | Maps categorical values onto discrete colors or position on the chart |
| `pv.Scale.quantile` | Maps continuous data onto a discrete range |

Plots can be annotated with

- Axes and grid lines (`pv.Rule`)
- A mark in a specific position close to another mark (`pv.Anchor`)

Table 5.2 summarizes a number of commonly used scaling functions in Protovis.

Table 5.3 describes a number of the main functions to handle data in Protovis, and Table 5.4 describes some of the commonly used mathematical operations.

Protovis has a number of standardized techniques for *layouts*, such as grids, hierarchies, and networks.

Protovis includes a number of built-in classes to handle events for a series of common types of interactions, including drag, pan, point, resize, select, and zoom. The `pv.Behavior.point` is an example of one of these built-in Protovis behavior classes.

**TABLE 5.3 Selected Functions to Handle Data in Protovis**

| Protovis Class or Method | Description |
|---|---|
| `pv.uniq` | Returns a list of unique values |
| `pv.normalize` | Generates an array that is normalized where all elements add up to 1 |
| `pv.blend` | Creates a single array by concatenating each nested array |
| `pv.transpose` | Converts an m × n matrix to an n × m matrix |
| `pv.dict` | Mapping function indexed using a key |

**TABLE 5.4 Selected Mathematical Operation in Protovis**

| Protovis Class or Method | Description |
|---|---|
| `pv.range` | The difference between the minimum and maximum value |
| `pv.min` | The minimum value |
| `pv.max` | The maximum value |
| `pv.sum` | Summation of the values in the list |
| `pv.mean` | The average value calculated from a list |
| `pv.median` | The value in the center, when the list is ordered. |

## 5.10 FURTHER READING

As discussed earlier, the most important accompanying documentation to this chapter is the detailed API provided on the Protovis Web site (http://mbostock.github.com/protovis/jsdoc/). The Website also contains useful summaries and examples. A discussion group on the Protovis language currently contains technical discussions at http://groups.google.com/group/protovis.

A number of additional online tutorials have been created that describe the Protovis language and potential applications, including the Knight Digital Media Center data visualization tutorials (http://multimedia.journalism.berkeley.edu/), a three-part tutorial at eagereyes.org (http://eagereyes.org/tutorials/protovis-primer-part-1, http://eagereyes.org/tutorials/protovis-primer-part-2, http://eagereyes.org/tutorials/protovis-primer-part-3), and a five-part tutorial at www.jeromecukier.net/?p=429.

An overview on Protovis has been documented in "Declarative Language Design for Interactive Visualization" (Heer & Bostock, 2010) with background to the language provided in "Prefuse: A toolkit for interactive information visualization" (Heer et al., 2005) and "Software design patterns for information visualization" (Heer & Agrawala, 2006). Background on toolkit designs for building graphical applications is provided in "Toolkit design for interactive structured graphics" (Bederson et al., 2004).

A lot of useful accompanying information is also on wikipedia around the JavaScript language (http://en.wikipedia.org/wiki/JavaScript) and JSON file format (http://en.wikipedia.org/wiki/JSON).

Additional utilities for converting CSV files to JSON files are available at http://www.cparker15.com/utilities/csv-to-json/ and http://shancarter.com/data_converter/.

A new JavaScript library D3.js (http://mbostock.github.com/d3/) is being developed for building custom visualizations to be displayed in a Web browser, with a focus on interactive visualization.

# APPENDIX A

# EXERCISE CODE EXAMPLES

5.1.4.1. Follow the instructions in Section 5.1.2, and create the visualization as shown in Fig. 5.2.

The following is an example implementation with Fig. A.1 showing the resulting screenshot.

```
<html>
 <head>
  <script type="text/javascript" src=
    "../protovis.js"></script>
 </head>
 <body>
  <script type="text/javascript+protovis">
   new pv.Panel()
      .width(150)
      .height(100)
     .add(pv.Bar)
      .data([1.4,2.3,2.7,1.6,0.8])
      .bottom(0)
      .width(20)
      .height(function(d) d * 25)
      .left(function() this.index * 30)
```

*Making Sense of Data III: A Practical Guide to Designing Interactive Data Visualizations*, First Edition. Glenn J. Myatt and Wayne P. Johnson.
© 2011 John Wiley & Sons, Inc. Published 2011 by John Wiley & Sons, Inc.

**FIGURE A.1** Exercise 5.1.4.1 screenshot

```
    .root.render();
  </script>
 </body>
</html>
```

5.1.4.2. Change the array values from `[1.4,2.3,2.7,1.6,0.8]` to `[1.4,2.3,2.7,1.6,0.8,1.4]`, change the width from 150 to 180, and save the file. Refresh the Web browser.

The following is an example implementation with Fig. A.2 showing the resulting screenshot.

```
<html>
 <head>
  <script type="text/javascript" src=
    "../protovis.js"></script>
 </head>
 <body>
  <script type="text/javascript+protovis">
   new pv.Panel()
     .width(180)
     .height(100)
    .add(pv.Bar)
     .data([1.4,2.3,2.7,1.6,0.8,1.4])
     .bottom(0)
     .width(20)
     .height(function(d) d * 25)
     .left(function() this.index * 30)
```

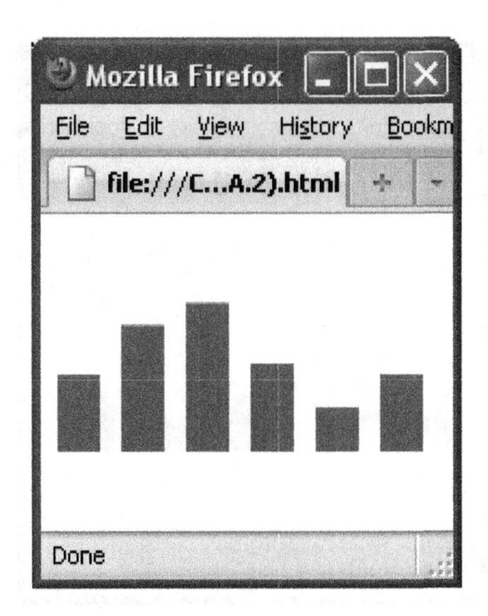


**FIGURE A.2** Exercise 5.1.4.2 screenshot

```
      .root.render();
    </script>
  </body>
</html>
```

5.1.4.3. Change the function for the `height` field to `function(d) d * 20`, save the file, and refresh the Web browser.

The following is an example implementation with Fig. A.3 showing the resulting screenshot.

```
<html>
  <head>
    <script type="text/javascript" src=
      "../protovis.js"></script>
  </head>
  <body>
    <script type="text/javascript+protovis">
     new pv.Panel()
        .width(200)
        .height(100)
      .add(pv.Bar)
        .data([1.4,2.3,2.7,1.6,0.8,1.4])
        .bottom(0)
        .width(20)
        .height(function(d) d * 20)
        .left(function() this.index * 30)
```

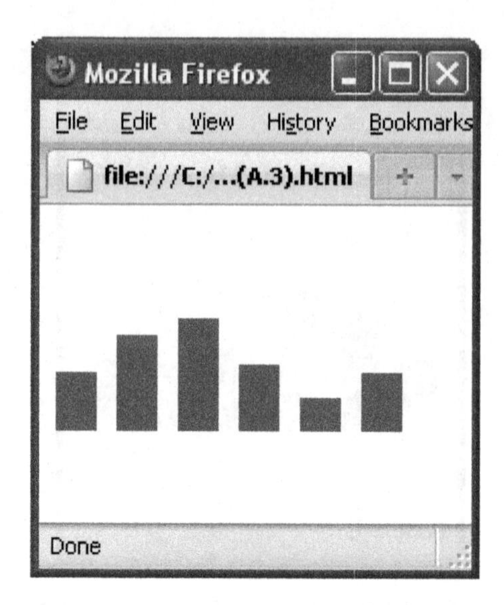


**FIGURE A.3** Exercise 5.1.4.3 screenshot

```
      .root.render();
    </script>
  </body>
</html>
```

5.1.4.4. Replace the word `bottom` with `top`, save the file, and refresh the Web browser.

The following is an example implementation with Fig. A.4 showing the resulting screenshot.

```
<html>
  <head>
    <script type="text/javascript" src=
      "../protovis.js"></script>
  </head>
  <body>
    <script type="text/javascript+protovis">
     new pv.Panel()
        .width(200)
        .height(100)
      .add(pv.Bar)
        .data([1.4,2.3,2.7,1.6,0.8,1.4])
        .top(0)
        .width(20)
```

**FIGURE A.4** Exercise 5.1.4.4 screenshot

```
      .height(function(d) d * 20)
      .left(function() this.index * 30)
    .root.render();
  </script>
 </body>
</html>
```

5.1.4.5. Create a new label visualization by removing the block of code starting with `.add(pv.Bar)` and ending with `.left(function() this.index * 30)`, change the panel width to 100, change the panel height to 50, and add the four lines:

```
.add(pv.Label)
 .top(20)
 .left(10)
 .text("Exercise example")
```

Save the file, and refresh the Web browser.

The following is an example implementation with Fig. A.5 showing the resulting screenshot.

```
<html>
 <head>
  <script type="text/javascript" src=
    "../protovis.js"></script>
 </head>
```

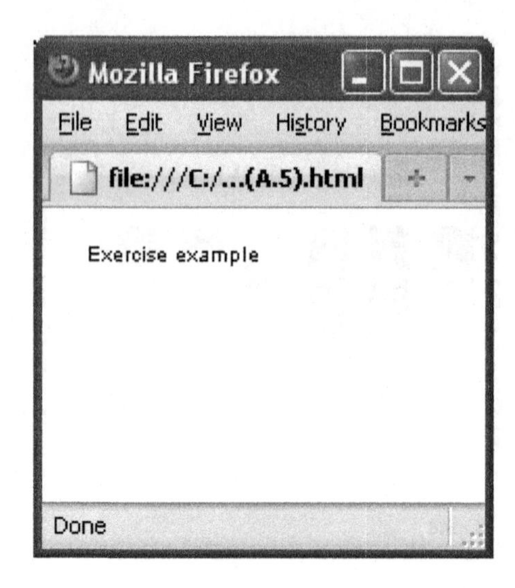


**FIGURE A.5** Exercise 5.1.4.5 screenshot

```
  <body>
   <script type="text/javascript+protovis">
    new pv.Panel()
       .width(200)
       .height(100)
     .add(pv.Label)
       .top(20)
       .left(10)
       .text("Exercise example")
     .root.render();
   </script>
  </body>
</html>
```

5.1.4.6. Change the `left` property value to 140, save the file, and refresh the Web browser.

The following is an example implementation with Fig. A.6 showing the resulting screenshot.

```
<html>
  <head>
   <script type="text/javascript" src=
     "../protovis.js"></script>
  </head>
  <body>
   <script type="text/javascript+protovis">
    new pv.Panel()
```

**FIGURE A.6** Exercise 5.1.4.6 screenshot

```
      .width(200)
      .height(100)
    .add(pv.Label)
      .top(20)
      .left(140)
      .text("Exercise example")
    .root.render();
  </script>
 </body>
</html>
```

5.1.4.7. Change the width of the panel to 250, save the file, and refresh the Web browser.

The following is an example implementation with Fig. A.7 showing the resulting screenshot.

```
<html>
 <head>
  <script type="text/javascript" src=
    "../protovis.js"></script>
 </head>
 <body>
  <script type="text/javascript+protovis">
   new pv.Panel()
      .width(250)
      .height(100)
    .add(pv.Label)
```

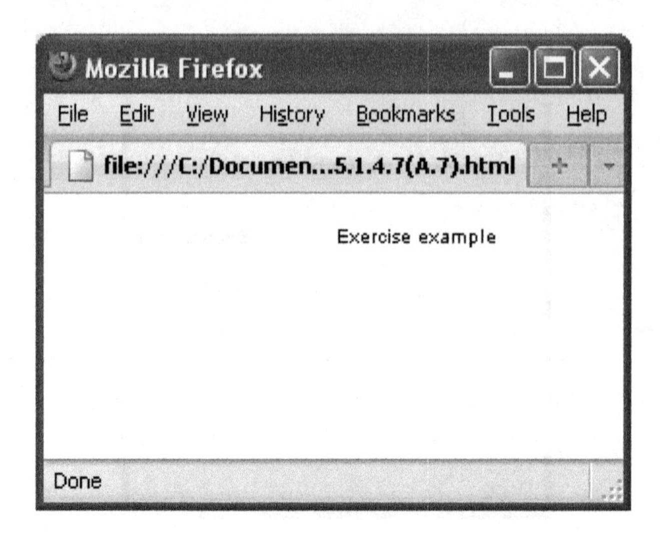


**FIGURE A.7** Exercise 5.1.4.7 screenshot

```
      .top(20)
      .left(140)
      .text("Exercise example")
     .root.render();
   </script>
  </body>
</html>
```

5.2.6.1 Create the chart as shown in Fig. 5.10.

The following is an example implementation with Fig. A.8 showing the resulting screenshot.

```
<html>
  <head>
   <script type="text/javascript" src=
     "../protovis.js"></script>
  </head>

  <body>

   <script type="text/javascript+protovis">

     var panelWidth = 200, panelHeight = 150, barWidth = 20;
     var barChartData = [1.2,4.3,2.3,0.9,5.2];
    var barColor = function(d) (d < 2) || (d > 5) ? "black" :
      "gray";
```

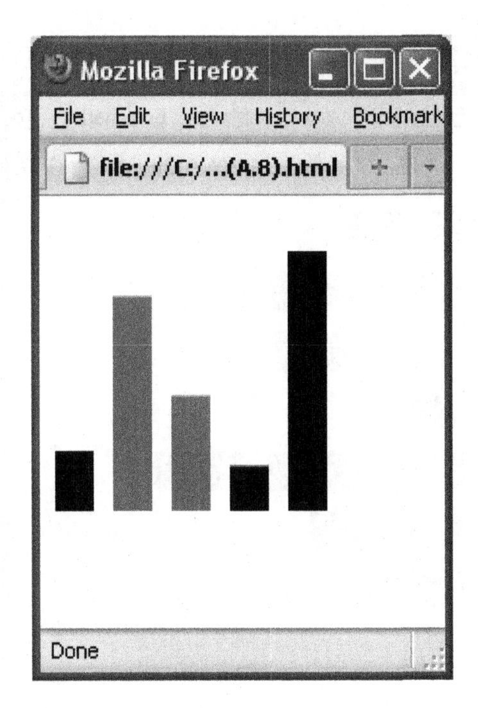


**FIGURE A.8** Exercise 5.2.6.1 screenshot

```
    var chartPanel = new pv.Panel()
      .width(panelWidth )
      .height(panelHeight );

    chartPanel.add(pv.Bar)
      .data(barChartData)
      .bottom(0)
      .width(barWidth)
      .height(function(d) d * 25)
      .fillStyle(function(d) barColor(d))
      .left(function() this.index * 30)
    .root.render();

  </script>
 </body>
</html>
```

5.2.6.2 Create a chart for the data `[3,6,5,2,8,4,3]` where the bars are presented as shown in Fig. 5.11.

The following is an example implementation with Fig. A.9 showing the resulting screenshot.

**FIGURE A.9** Exercise 5.2.6.2 screenshot

```
<html>
 <head>
  <script type="text/javascript" src=
    "../protovis.js"></script>
 </head>
 <body>
  <script type="text/javascript+protovis">
   new pv.Panel()
      .width(200)
      .height(300)
     .add(pv.Bar)
      .data([3,6,5,2,8,4,3] )
      .left(0)
      .height(20)
      .width(function(d) d * 20)
```

```
      .top(function() this.index * 30)
    .root.render();
  </script>
 </body>
</html>
```

5.2.6.3 For the chart generated in Exercise 5.2.6.2, color the bars "orange" for data values greater than 5; otherwise, color the bars "lightblue".

The following is an example implementation with Fig. A.10 showing the resulting screenshot.

```
<html>
 <head>
  <script type="text/javascript" src=
    "../protovis.js"></script>
 </head>
```

**FIGURE A.10** Exercise 5.2.6.3 screenshot

```
  <body>
   <script type="text/javascript+protovis">
    new pv.Panel()
       .width(200)
       .height(300)
     .add(pv.Bar)
       .data([3,6,5,2,8,4,3] )
       .left(0)
       .height(20)
       .width(function(d) d * 20)
       .top(function() this.index * 30)
       .fillStyle(function(d) d > 5 ? "orange" : "lightblue")
     .root.render();
   </script>
  </body>
</html>
```

5.2.6.4 For the chart created in Exercise 5.2.6.2, color even-valued bars "`lightgray`" and odd valued bars "`darkgray`". Note that the mod function (`d%2`) will return 0 if the value is even and 1 if the value is odd.

The following is an example implementation with Fig. A.11 showing the resulting screenshot.

```
<html>
  <head>
   <script type="text/javascript" src=
     "../protovis.js"></script>
  </head>
  <body>
   <script type="text/javascript+protovis">
    new pv.Panel()
       .width(200)
       .height(300)
     .add(pv.Bar)
       .data([3,6,5,2,8,4,3] )
       .left(0)
       .height(20)
       .width(function(d) d * 20)
       .top(function() this.index * 30)
         .fillStyle(function(d) (d % 2) == 0 ? "lightgray" :
           "darkgray")
     .root.render();
   </script>
  </body>
</html>
```

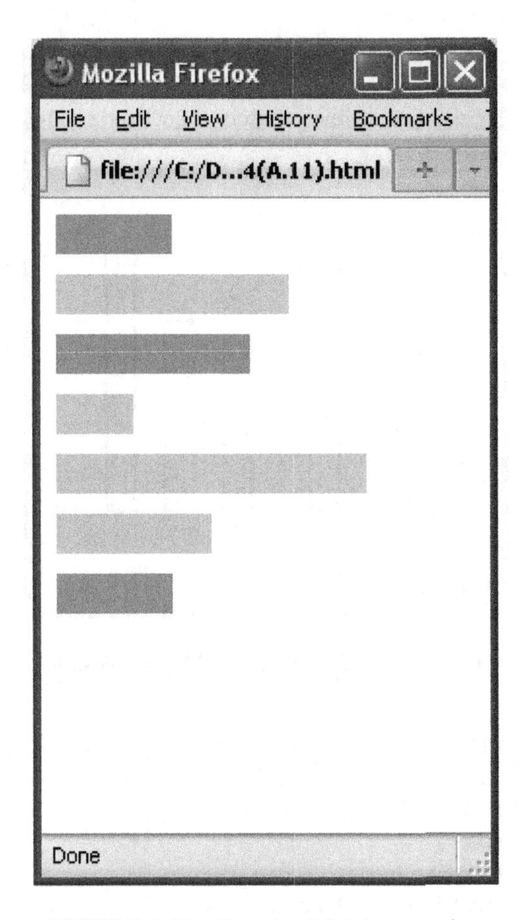


**FIGURE A.11** Exercise 5.2.6.4 screenshot

5.3.8.1 Create a bar chart aligned with the x-axis from the array `[3,5,6,8,9,8,11]` with a "`lightblue`" bar, a border drawn in "`blue`", and data value labels.

The following is an example implementation with Fig. A.12 showing the resulting screenshot.

```
<html>
 <head>
  <script type="text/javascript" src=
    "../protovis.js"></script>
 </head>
 <body>
  <script type="text/javascript+protovis">
   new pv.Panel()
     .width(200)
```

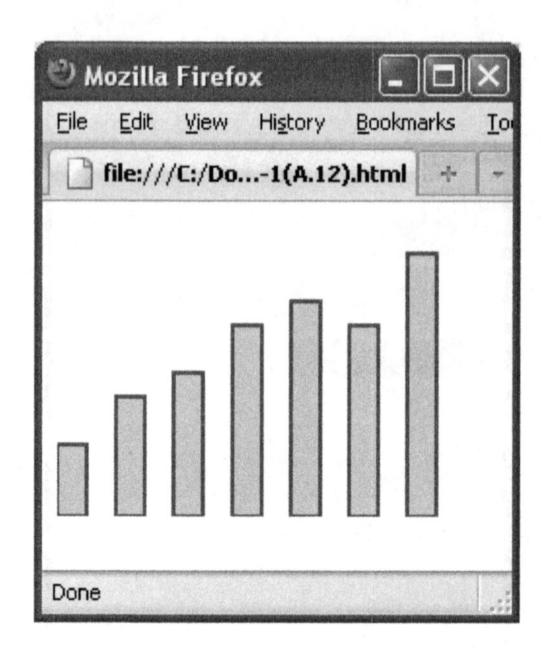


**FIGURE A.12** Exercise 5.3.8.1 screenshot

```
      .height(150)
    .add(pv.Bar)
      .data([3,5,6,8,9,8,11])
      .bottom(0)
      .width(15)
      .left(function() this.index * 30)
      .height(function(d) d * 12)
      .fillStyle("lightblue")
      .strokeStyle("blue")
    .root.render();
  </script>
 </body>
</html>
```

5.3.8.2 Create a bar chart aligned with the y-axis with the same attributes as Exercise 5.3.8.1.

The following is an example implementation with Fig. A.13 showing the resulting screenshot.

```
<html>
 <head>
  <script type="text/javascript" src=
    "../protovis.js"></script>
 </head>
```

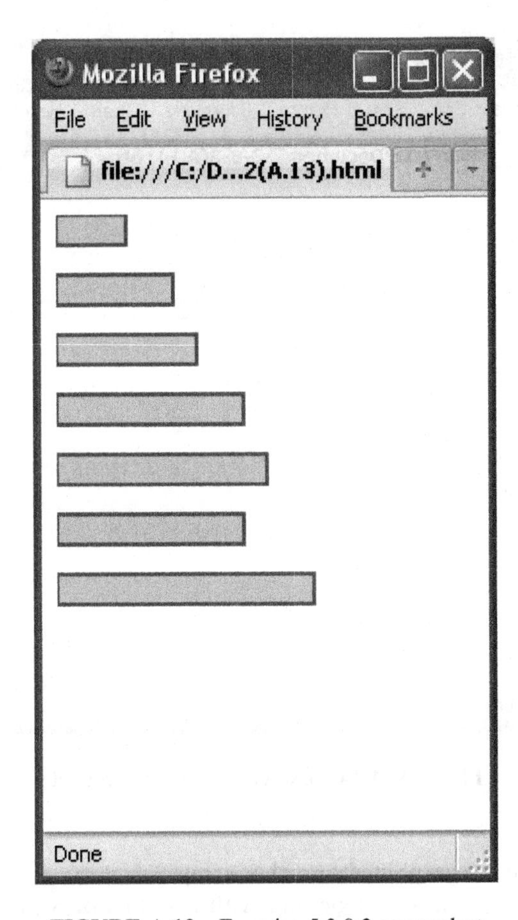


**FIGURE A.13** Exercise 5.3.8.2 screenshot

```
<body>
  <script type="text/javascript+protovis">
   new pv.Panel()
      .width(200)
      .height(300)
    .add(pv.Bar)
      .data([3,5,6,8,9,8,11])
      .left(0)
      .height(15)
      .top(function() this.index * 30)
      .width(function(d) d * 12)
      .fillStyle("lightblue")
      .strokeStyle("blue")
    .root.render();
</script>
</body>
</html>
```

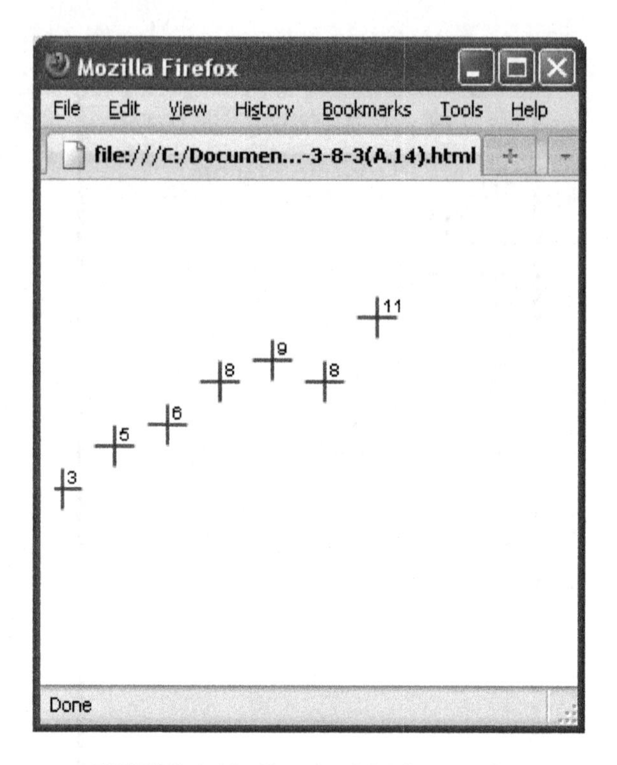


**FIGURE A.14** Exercise 5.3.8.3 screenshot

5.3.8.3 Create a graphic from the array `[3,5,6,8,9,8,11]` using `"maroon"` crosses at a 45° angle and with a radius of 8 pixels to represent the data values. Add a label to each of the data values.

The following is an example implementation with Fig. A.14 showing the resulting screenshot.

```
<html>
 <head>
  <script type="text/javascript" src=
    "../protovis.js"></script>
 </head>
 <body>
  <script type="text/javascript+protovis">
   new pv.Panel()
      .width(300)
      .height(200)
     .add(pv.Dot)
      .data([3,5,6,8,9,8,11])
      .left(function() this.index * 30 + 5)
```

```
      .bottom(function(d) d * 12)
      .strokeStyle("maroon")
      .shape("cross")
      .radius(8)
      .angle(Math.PI / 4)
    .add(pv.Label)
    .root.render();
  </script>
 </body>
</html>
```

5.3.8.4 Create a line plot from the array `[4.3,5.4,7.3,6.9,10.3,11.5]` using the cardinal interpolate option, and color the line "`darkgreen`" with a 2-pixel width.

The following is an example implementation with Fig. A.15 showing the resulting screenshot.

```
<html>
 <head>
  <script type="text/javascript" src=
    "../protovis.js"></script>
 </head>
```

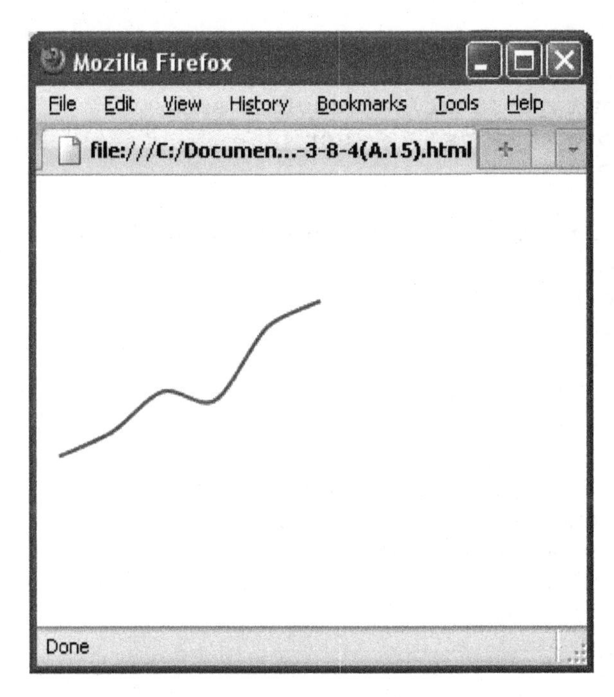


**FIGURE A.15** Exercise 5.3.8.4 screenshot

```
 <body>
  <script type="text/javascript+protovis">
   new pv.Panel()
      .width(300)
      .height(200)
     .add(pv.Line)
      .data([4.3,5.4,7.3,6.9,10.3,11.5])
      .left(function() this.index * 30 + 5)
      .bottom(function(d) d * 12)
      .strokeStyle("darkgreen")
      .lineWidth(2)
      .interpolate("cardinal")
     .root.render();
  </script>
 </body>
</html>
```

5.3.8.5 Create five different line plots based on Exercise 5.3.8.4 with different tension values: 0, 0.25, 0.5, 0.75, 1.

The following is an example implementation (where tension is 0) with Fig. A.16 showing the resulting screenshot.

```
<html>
 <head>
  <script type="text/javascript" src=
    "../protovis.js"></script>
 </head>
 <body>
  <script type="text/javascript+protovis">
   new pv.Panel()
      .width(300)
      .height(200)
     .add(pv.Line)
      .data([4.3,5.4,7.3,6.9,10.3,11.5])
      .left(function() this.index * 30 + 5)
      .bottom(function(d) d * 12)
      .strokeStyle("darkgreen")
      .lineWidth(2)
      .interpolate("cardinal")
      .tension(0)
     .root.render();
  </script>
 </body>
</html>
```

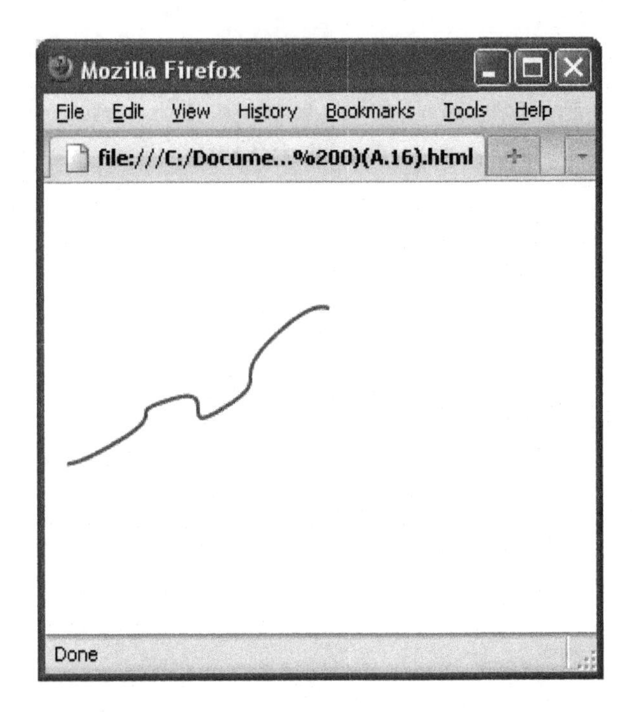


**FIGURE A.16** Exercise 5.3.8.5 screenshot (tension is 0)

The following is an example implementation (where tension is 0.25) with Fig. A.17 showing the resulting screenshot.

```
<html>
 <head>
  <script type="text/javascript" src=
    "../protovis.js"></script>
 </head>
 <body>
  <script type="text/javascript+protovis">
   new pv.Panel()
      .width(300)
      .height(200)
    .add(pv.Line)
      .data([4.3,5.4,7.3,6.9,10.3,11.5])
      .left(function() this.index * 30 + 5)
      .bottom(function(d) d * 12)
      .strokeStyle("darkgreen")
      .lineWidth(2)
      .interpolate("cardinal")
```

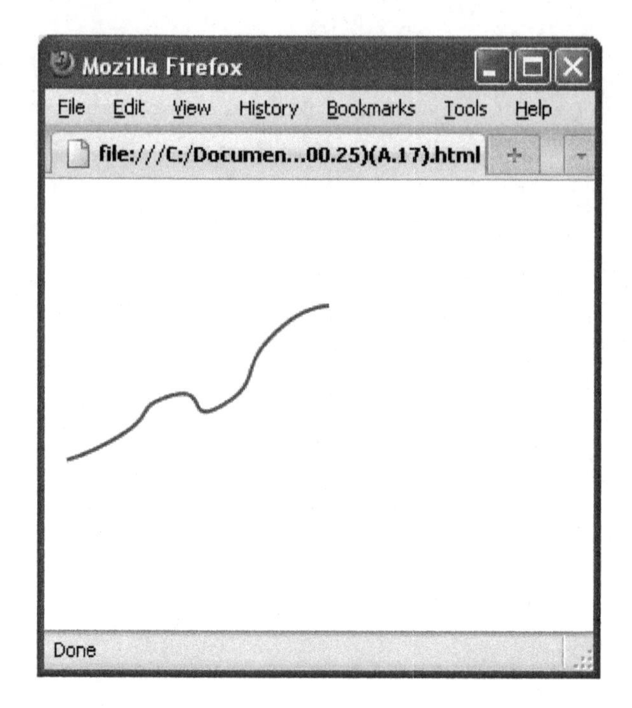


**FIGURE A.17** Exercise 5.3.8.5 screenshot (tension is 0.25)

```
      .tension(0.25)
    .root.render();
  </script>
 </body>
</html>
```

The following is an example implementation (where tension is 0.5) with Fig. A.18 showing the resulting screenshot.

```
<HTML>
 <head>
  <script type="text/javascript" src=
    "../protovis.js"></script>
 </head>
 <body>
  <script type="text/javascript+protovis">
   new pv.Panel()
      .width(300)
      .height(200)
    .add(pv.Line)
      .data([4.3,5.4,7.3,6.9,10.3,11.5])
      .left(function() this.index * 30 + 5)
```

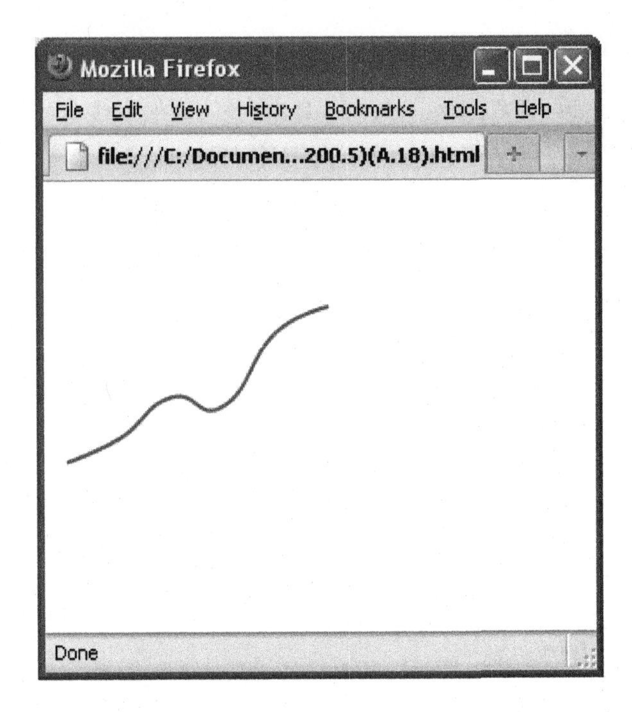


**FIGURE A.18** Exercise 5.3.8.5 screenshot (tension is 0.5)

```
      .bottom(function(d) d * 12)
      .strokeStyle("darkgreen")
      .lineWidth(2)
      .interpolate("cardinal")
      .tension(0.5)
    .root.render();
  </script>
 </body>
</html>
```

The following is an example implementation (where tension is 0.75) with Fig. A.19 showing the resulting screenshot.

```
<HTML>
 <head>
  <script type="text/javascript" src=
    "../protovis.js"></script>
 </head>
 <body>
  <script type="text/javascript+protovis">
   new pv.Panel()
      .width(300)
```

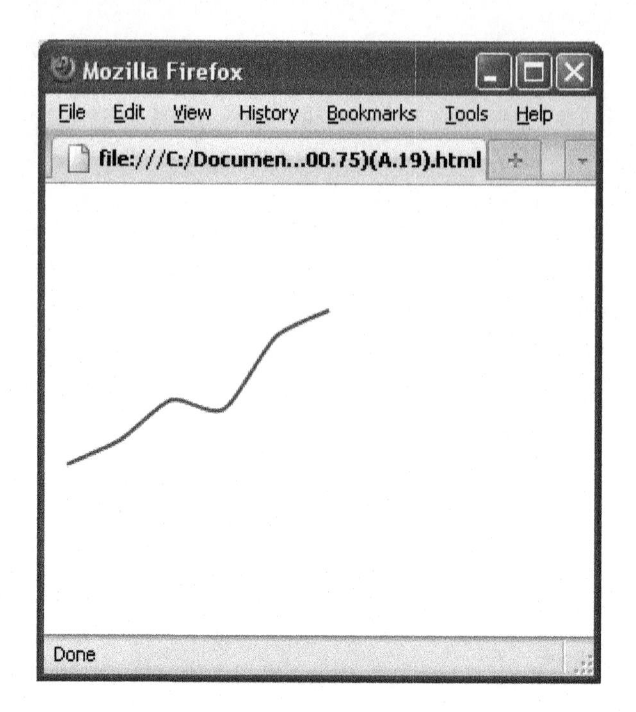


**FIGURE A.19** Exercise 5.3.8.5 screenshot (tension is 0.75)

```
      .height(200)
     .add(pv.Line)
      .data([4.3,5.4,7.3,6.9,10.3,11.5])
      .left(function() this.index * 30 + 5)
      .bottom(function(d) d * 12)
      .strokeStyle("darkgreen")
      .lineWidth(2)
      .interpolate("cardinal")
      .tension(0.75)
     .root.render();
   </script>
  </body>
</html>
```

The following is an example implementation (where tension is 1) with Fig. A.20 showing the resulting screenshot.

```
<HTML>
  <head>
   <script type="text/javascript" src=
     "../protovis.js"></script>
  </head>
```

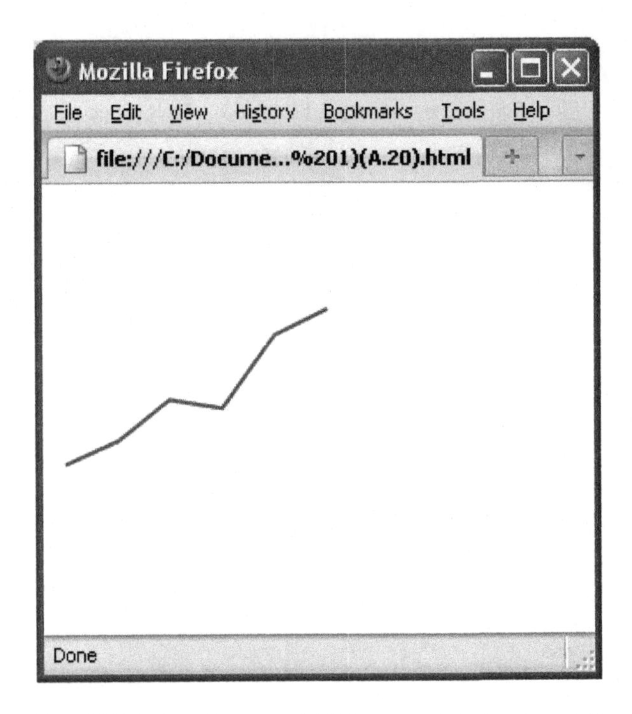


**FIGURE A.20** Exercise 5.3.8.5 screenshot (tension is 1)

```
  <body>
   <script type="text/javascript+protovis">
    new pv.Panel()
       .width(300)
       .height(200)
      .add(pv.Line)
       .data([4.3,5.4,7.3,6.9,10.3,11.5])
       .left(function() this.index * 30 + 5)
       .bottom(function(d) d * 12)
       .strokeStyle("darkgreen")
       .lineWidth(2)
       .interpolate("cardinal")
       .tension(1)
      .root.render();
   </script>
  </body>
</html>
```

5.3.8.6 Create six different line plots based on Exercise 5.3.8.4, where the `interpolate` property is set to "`linear`", "`step-before`", "`step-after`", "`polar`", "`polar-reverse`", and "`basis`", respectively.

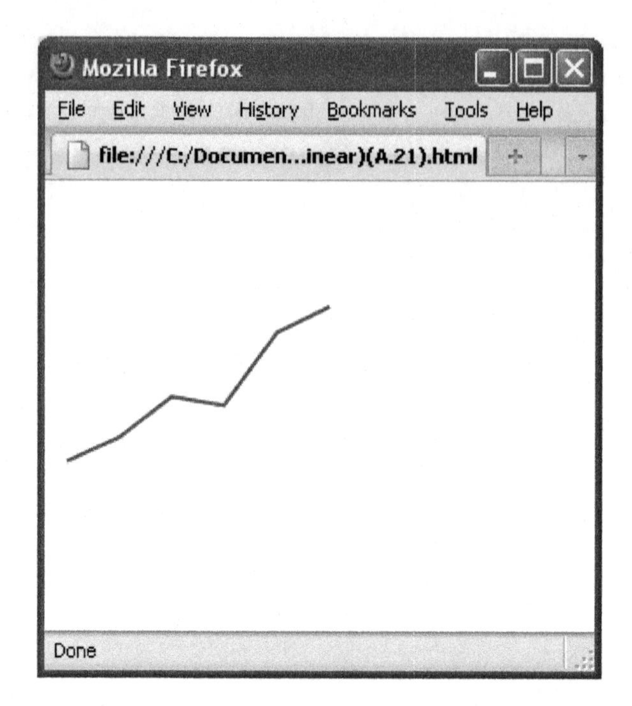


**FIGURE A.21** Exercise 5.3.8.6 screenshot (interpolate property is "`linear`")

The following is an example implementation (interpolate property is "`linear`") with Fig. A.21 showing the resulting screenshot.

```
<HTML>
 <head>
  <script type="text/javascript" src=
    "../protovis.js"></script>
 </head>
 <body>
  <script type="text/javascript+protovis">
   new pv.Panel()
      .width(300)
      .height(200)
     .add(pv.Line)
      .data([4.3,5.4,7.3,6.9,10.3,11.5])
      .left(function() this.index * 30 + 5)
      .bottom(function(d) d * 12)
      .strokeStyle("darkgreen")
      .lineWidth(2)
```

```
      .interpolate("linear")
     .root.render();
   </script>
  </body>
</html>
```

The following is an example implementation (interpolate property is "step-before") with Fig. A.22 showing the resulting screenshot.

```
<HTML>
 <head>
  <script type="text/javascript" src=
    "../protovis.js"></script>
 </head>
 <body>
  <script type="text/javascript+protovis">
   new pv.Panel()
      .width(300)
      .height(200)
     .add(pv.Line)
      .data([4.3,5.4,7.3,6.9,10.3,11.5])
```

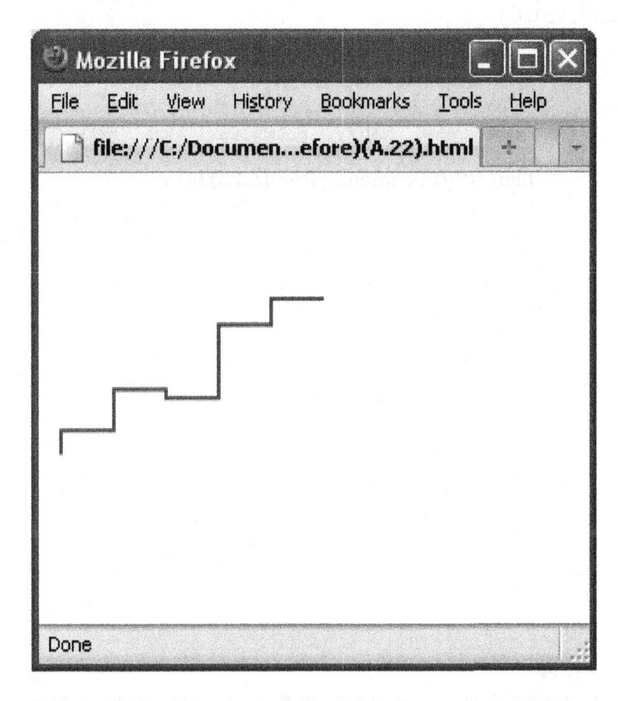


**FIGURE A.22** Exercise 5.3.8.6 screenshot (interpolate property is "step-before")

```
      .left(function() this.index * 30 + 5)
      .bottom(function(d) d * 12)
      .strokeStyle("darkgreen")
      .lineWidth(2)
      .interpolate("step-before")
    .root.render();
  </script>
 </body>
</html>
```

The following is an example implementation (interpolate property is "`step-after`") with Fig. A.23 showing the resulting screenshot.

```
<HTML>
 <head>
  <script type="text/javascript" src=
    "../protovis.js"></script>
 </head>
 <body>
  <script type="text/javascript+protovis">
   new pv.Panel()
      .width(300)
```

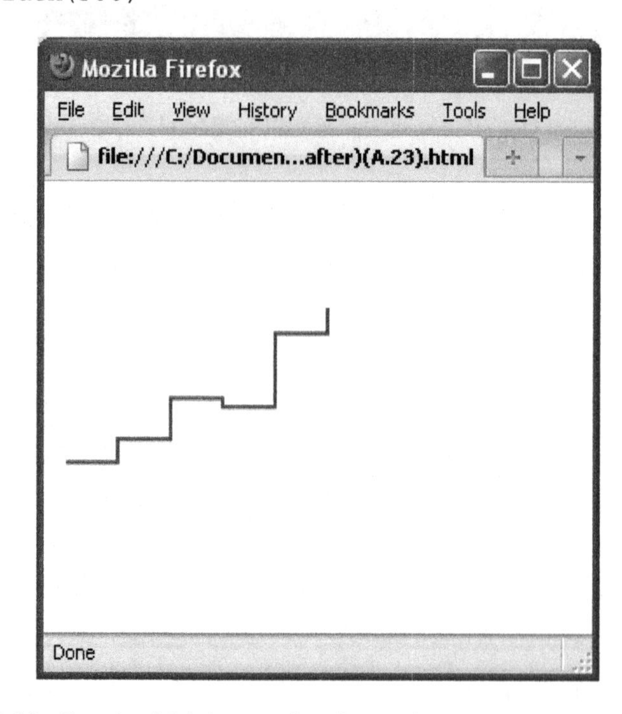


**FIGURE A.23** Exercise 5.3.8.6 screenshot (interpolate property is "`step-after`")

```
      .height(200)
    .add(pv.Line)
      .data([4.3,5.4,7.3,6.9,10.3,11.5])
      .left(function() this.index * 30 + 5)
      .bottom(function(d) d * 12)
      .strokeStyle("darkgreen")
      .lineWidth(2)
      .interpolate("step-after")
    .root.render();
  </script>
 </body>
</html>
```

The following is an example implementation (interpolate property is "polar") with Fig. A.24 showing the resulting screenshot.

```
<HTML>
 <head>
  <script type="text/javascript" src=
    "../protovis.js"></script>
 </head>
```

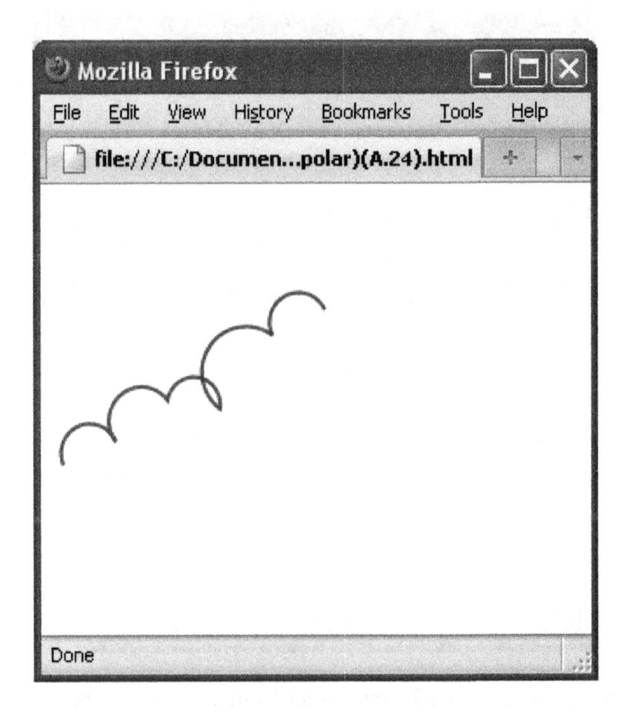


**FIGURE A.24** Exercise 5.3.8.6 screenshot (interpolate property is "polar")

```
  <body>
   <script type="text/javascript+protovis">
    new pv.Panel()
       .width(300)
       .height(200)
     .add(pv.Line)
       .data([4.3,5.4,7.3,6.9,10.3,11.5])
       .left(function() this.index * 30 + 5)
       .bottom(function(d) d * 12)
       .strokeStyle("darkgreen")
       .lineWidth(2)
       .interpolate("polar")
     .root.render();
   </script>
  </body>
</html>
```

The following is an example implementation (interpolate property is "`polar-reverse`") with Fig. A.25 showing the resulting screenshot.

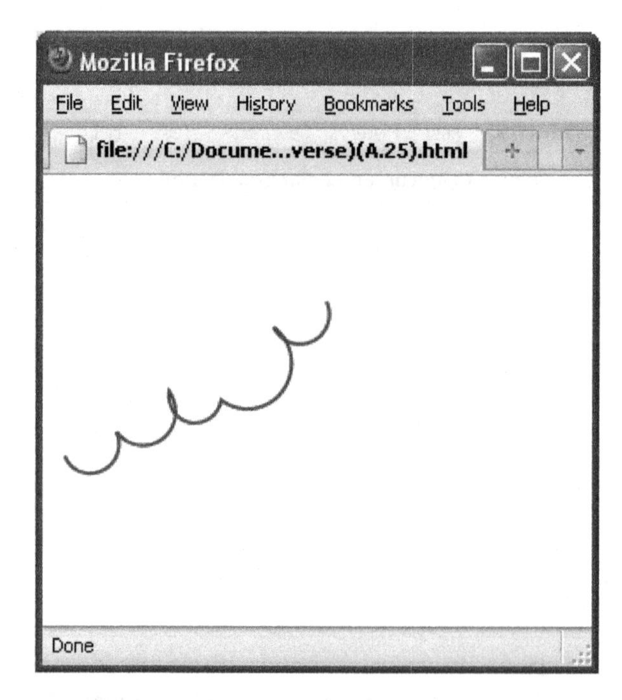


**FIGURE A.25** Exercise 5.3.8.6 screenshot (interpolate property is "`polar-reverse`")

```
<HTML>
 <head>
  <script type="text/javascript" src=
    "../protovis.js"></script>
 </head>
 <body>
  <script type="text/javascript+protovis">
   new pv.Panel()
      .width(300)
      .height(200)
     .add(pv.Line)
      .data([4.3,5.4,7.3,6.9,10.3,11.5])
      .left(function() this.index * 30 + 5)
      .bottom(function(d) d * 12)
      .strokeStyle("darkgreen")
      .lineWidth(2)
      .interpolate("polar-reverse")
     .root.render();
  </script>
 </body>
</html>
```

The following is an example implementation (interpolate property is "`basis`") with Fig. A.26 showing the resulting screenshot.

```
<HTML>
 <head>
  <script type="text/javascript" src=
    "../protovis.js"></script>
 </head>
 <body>
  <script type="text/javascript+protovis">
   new pv.Panel()
      .width(300)
      .height(200)
     .add(pv.Line)
      .data([4.3,5.4,7.3,6.9,10.3,11.5])
      .left(function() this.index * 30 + 5)
      .bottom(function(d) d * 12)
      .strokeStyle("darkgreen")
      .lineWidth(2)
      .interpolate("basis")
     .root.render();
  </script>
 </body>
</html>
```

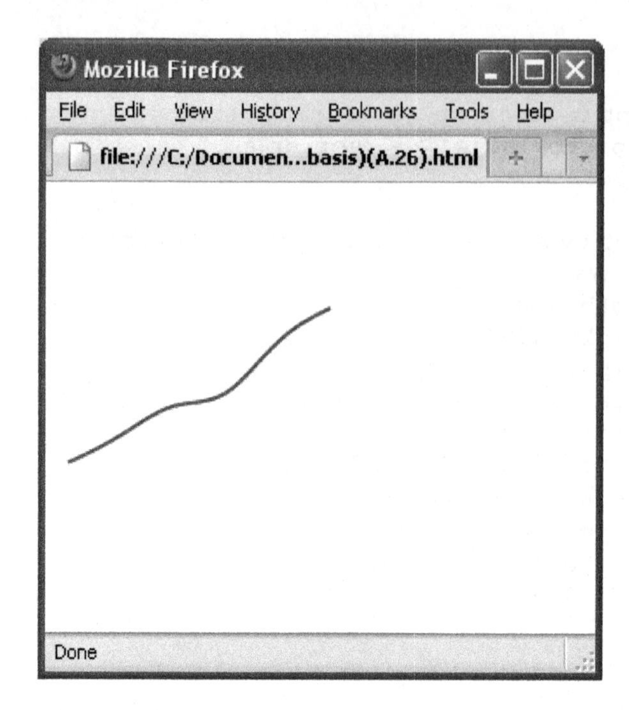


**FIGURE A.26** Exercise 5.3.8.6 screenshot (interpolate property is "basis")

5.3.8.7 Create a vertically aligned area plot from the array [43.6,54.8, 47.2,34,7,58.6,34.1] where the left polyline is along the y-axis, and the variation in the data is shown in the right polyline. The inner region of the chart should be colored "steelblue" with "darkblue" border.

The following is an example implementation with Fig. A.27 showing the resulting screenshot.

```
<HTML>
 <head>
  <script type="text/javascript" src=
    "../protovis.js"></script>
 </head>
 <body>
  <script type="text/javascript+protovis">
   new pv.Panel()
      .width(200)
      .height(300)
     .add(pv.Area)
      .data([43.6,54.8,47.2,34,7,58.6,34.1])
      .left(0)
      .width(function(d) d * 3)
      .top(function() this.index * 50)
```

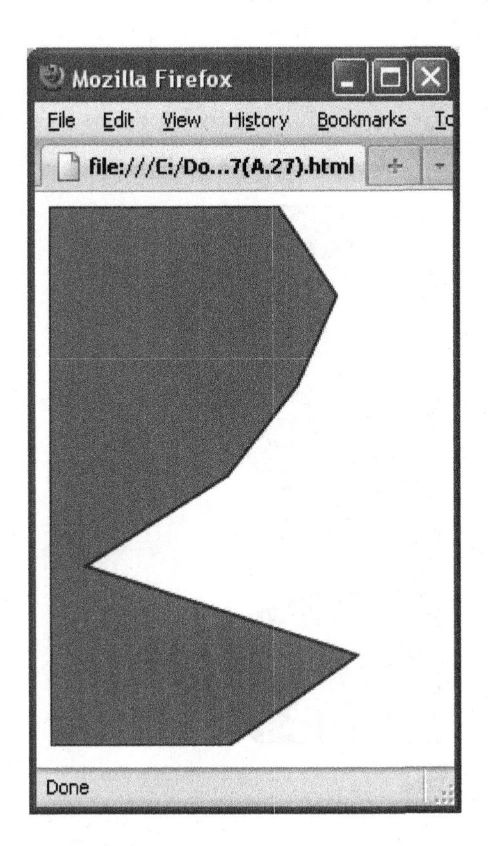


**FIGURE A.27** Exercise 5.3.8.7 screenshot

```
      .fillStyle("steelblue")
      .strokeStyle("darkblue")
    .root.render();
  </script>
 </body>
</html>
```

5.3.8.8 Create seven different plots by modifying the plot created in Exercise 5.3.8.7 so that each chart uses a different interpolate option (“`linear`”, “`step-before`”, “`step-after`”, “`polar`”, “`polar-reverse`”, “`basis`”, and “`cardinal`”).

The following is an example implementation (interpolate property is “`linear`”) with Fig. A.28 showing the resulting screenshot.

```
<HTML>
 <head>
  <script type="text/javascript" src=
    "../protovis.js"></script>
 </head>
```

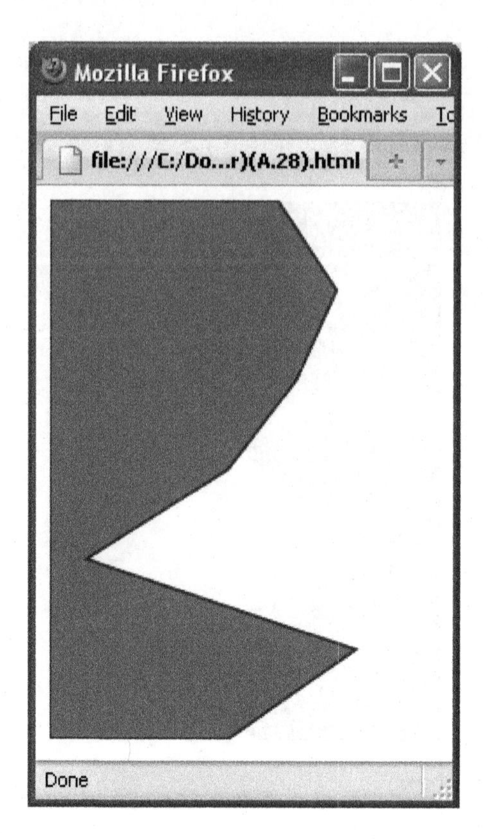


**FIGURE A.28** Exercise 5.3.8.8 screenshot (interpolate property is "`linear`")

```
 <body>
  <script type="text/javascript+protovis">
   new pv.Panel()
      .width(200)
      .height(300)
    .add(pv.Area)
      .data([43.6,54.8,47.2,34,7,58.6,34.1])
      .left(0)
      .width(function(d) d * 3)
      .top(function() this.index * 50)
      .fillStyle("steelblue")
      .strokeStyle("darkblue")
      .interpolate("linear")
    .root.render();
  </script>
 </body>
</html>
```

The following is an example implementation (interpolate property is "step-before") with Fig. A.29 showing the resulting screenshot.

```
<HTML>
 <head>
  <script type="text/javascript" src=
    "../protovis.js"></script>
 </head>
 <body>
  <script type="text/javascript+protovis">
   new pv.Panel()
      .width(200)
      .height(300)
     .add(pv.Area)
      .data([43.6,54.8,47.2,34,7,58.6,34.1])
      .left(0)
      .width(function(d) d * 3)
```

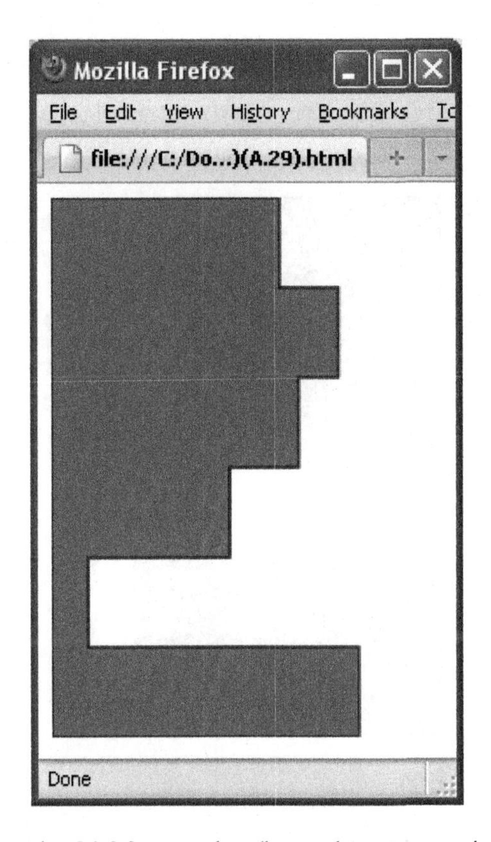


**FIGURE A.29** Exercise 5.3.8.8 screenshot (interpolate property is "step-before")

```
      .top(function() this.index * 50)
      .fillStyle("steelblue")
      .strokeStyle("darkblue")
      .interpolate("step-before")
    .root.render();
  </script>
 </body>
</html>
```

The following is an example implementation (interpolate property is "step-after") with Fig. A.30 showing the resulting screenshot.

```
<HTML>
 <head>
  <script type="text/javascript" src=
    "../protovis.js"></script>
 </head>
```

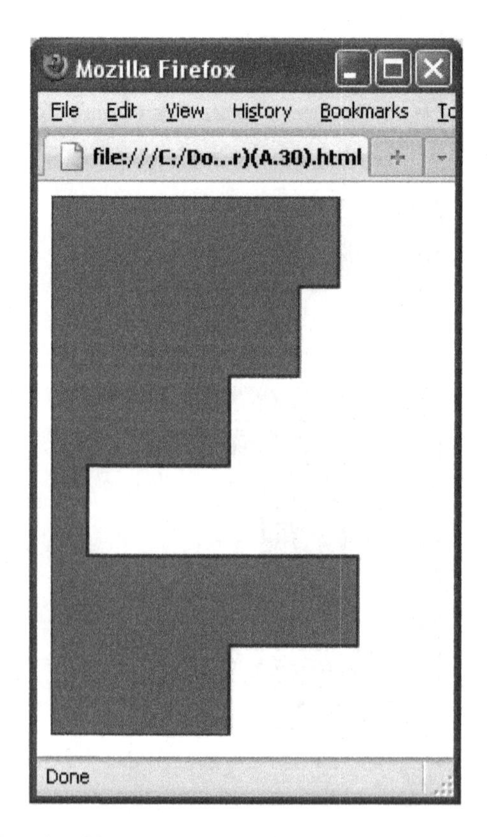


**FIGURE A.30** Exercise 5.3.8.8 screenshot (interpolate property is "step-after")

```
 <body>
  <script type="text/javascript+protovis">
   new pv.Panel()
      .width(200)
      .height(300)
     .add(pv.Area)
      .data([43.6,54.8,47.2,34,7,58.6,34.1])
      .left(0)
      .width(function(d) d * 3)
      .top(function() this.index * 50)
      .fillStyle("steelblue")
      .strokeStyle("darkblue")
      .interpolate("step-after")
     .root.render();
  </script>
 </body>
</html>
```

The following is an example implementation (interpolate property is "polar") with Fig. A.31 showing the resulting screenshot. It should be noted that "polar" is not a supported option for the area mark and hence Fig. A.31 shows the linear default view.

```
<HTML>
 <head>
  <script type="text/javascript" src=
    "../protovis.js"></script>
 </head>
 <body>
  <script type="text/javascript+protovis">
   new pv.Panel()
      .width(200)
      .height(300)
     .add(pv.Area)
      .data([43.6,54.8,47.2,34,7,58.6,34.1])
      .left(0)
      .width(function(d) d * 3)
      .top(function() this.index * 50)
      .fillStyle("steelblue")
      .strokeStyle("darkblue")
      .interpolate("polar")
     .root.render();
  </script>
 </body>
</html>
```

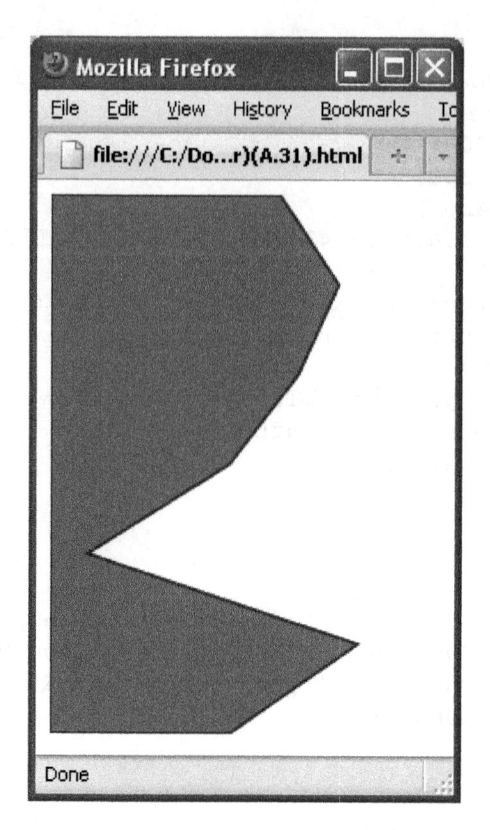


**FIGURE A.31** Exercise 5.3.8.8 screenshot (interpolate property is "`polar`")

The following is an example implementation (interpolate property is "`polar-reverse`") with Fig. A.32 showing the resulting screenshot. It should be noted that "`polar-reverse`" is not a supported option for the area mark and hence Fig. A.32 shows the linear default view.

```
<HTML>
 <head>
  <script type="text/javascript" src=
    "../protovis.js"></script>
 </head>
 <body>
  <script type="text/javascript+protovis">
   new pv.Panel()
      .width(200)
      .height(300)
     .add(pv.Area)
      .data([43.6,54.8,47.2,34,7,58.6,34.1])
      .left(0)
```

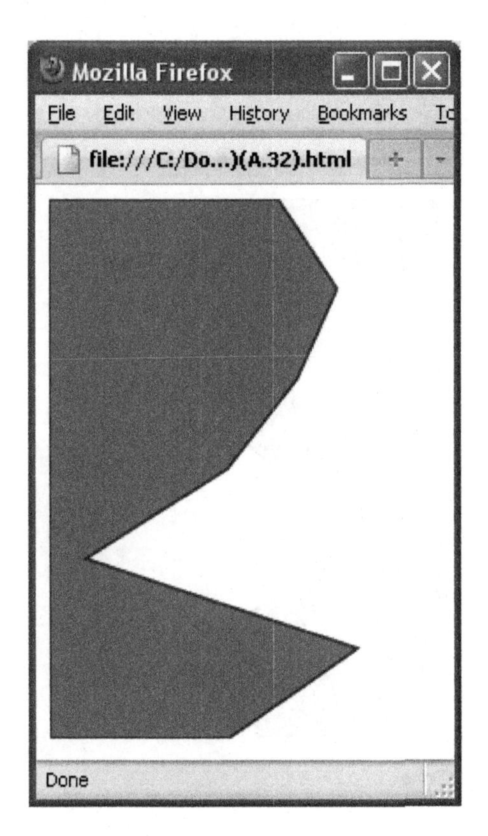


**FIGURE A.32** Exercise 5.3.8.8 screenshot (interpolate property is "`polar-reverse`")

```
      .width(function(d) d * 3)
      .top(function() this.index * 50)
      .fillStyle("steelblue")
      .strokeStyle("darkblue")
      .interpolate("polar-reverse")
    .root.render();
  </script>
 </body>
</html>
```

The following is an example implementation (interpolate property is "`basis`") with Fig. A.33 showing the resulting screenshot.

```
<HTML>
 <head>
  <script type="text/javascript" src=
    "../protovis.js"></script>
 </head>
```

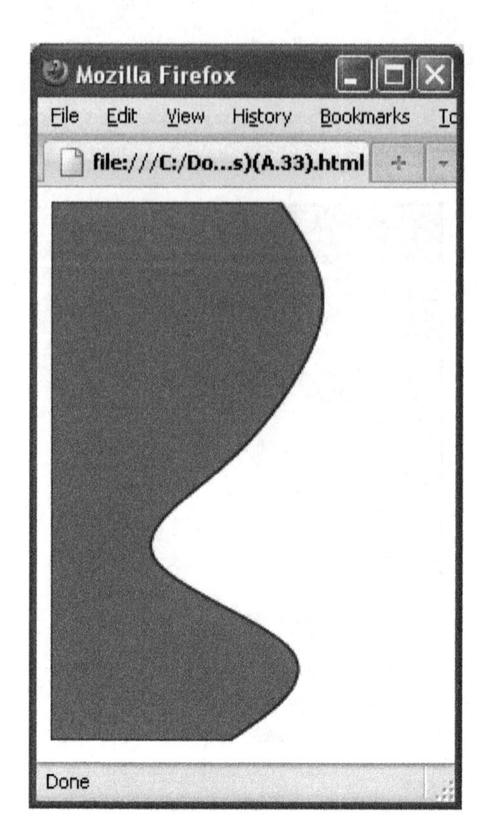


**FIGURE A.33** Exercise 5.3.8.8 screenshot (interpolate property is "basis")

```
 <body>
  <script type="text/javascript+protovis">
   new pv.Panel()
      .width(200)
      .height(300)
    .add(pv.Area)
      .data([43.6,54.8,47.2,34,7,58.6,34.1])
      .left(0)
      .width(function(d) d * 3)
      .top(function() this.index * 50)
      .fillStyle("steelblue")
      .strokeStyle("darkblue")
      .interpolate("basis")
    .root.render();
  </script>
 </body>
</html>
```

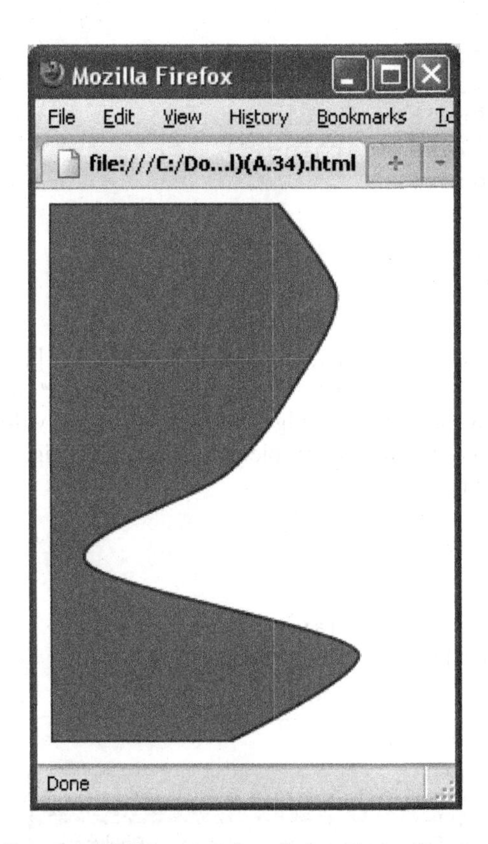


**FIGURE A.34** Exercise 5.3.8.8 screenshot (interpolate property is "`cardinal`")

The following is an example implementation (interpolate property is "`cardinal`") with Fig. A.34 showing the resulting screenshot.

```
<HTML>
 <head>
  <script type="text/javascript" src=
    "../protovis.js"></script>
 </head>
 <body>
  <script type="text/javascript+protovis">
   new pv.Panel()
      .width(200)
      .height(300)
     .add(pv.Area)
      .data([43.6,54.8,47.2,34,7,58.6,34.1])
      .left(0)
      .width(function(d) d * 3)
```

```
      .top(function() this.index * 50)
      .fillStyle("steelblue")
      .strokeStyle("darkblue")
      .interpolate("cardinal")
    .root.render();
  </script>
 </body>
</html>
```

5.3.8.9 Create a donut plot corresponding to the array [4, 5, 7, 8, 2] with the colors `"white"`, `"lightgray"`, `"gray"`, `"darkgray"`, and `"black"`, and also add a label to each of the wedges showing the data value.

The following is an example implementation with Fig. A.35 showing the resulting screenshot.

```
<HTML>
 <head>
  <script type="text/javascript" src=
    "../protovis.js"></script>
 </head>
```

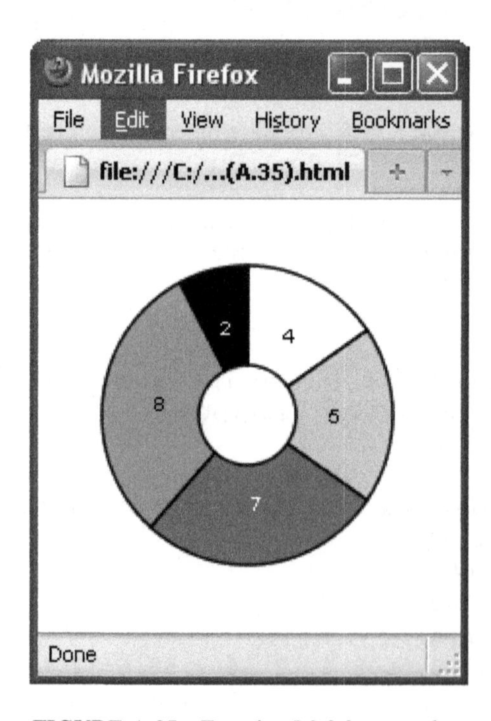


**FIGURE A.35** Exercise 5.3.8.9 screenshot

```
<body>
 <script type="text/javascript+protovis">

  var dataArray = [4,5,7,8,2];

   var chartPanel = new pv.Panel()
      .width(200)
      .height(200);

   var wedge = chartPanel.add(pv.Wedge)
      .data(dataArray)
      .left(100)
      .bottom(100)
      .outerRadius(75)
      .innerRadius(25)
     .angle(function(d) (d/pv.sum(dataArray))*Math.PI*2)
      .strokeStyle("black")
        .fillStyle(pv.colors("white", "lightgray", "gray",
          "darkgray", "black"));

   wedge.add(pv.Label)
    .left(function() 45 * Math.cos(wedge.midAngle()) + 100)
     .bottom(function() -45 * Math.sin(wedge.midAngle()) +
       100)
         .textStyle(function() this.index == 4 ? "white" :
           "black")
      .textAlign("center")
      .textBaseline("middle");

   chartPanel.render();

  </script>
 </body>
</html>
```

5.3.8.10 Create an image using the wiley-logo.bmp file that is 302 pixels wide and100 pixels high with a black border of 10 pixels.

The following is an example implementation with Fig. A.36 showing the resulting screenshot.

```
<HTML>
 <head>
  <script type="text/javascript" src=
    "../protovis.js"></script>
 </head>
```

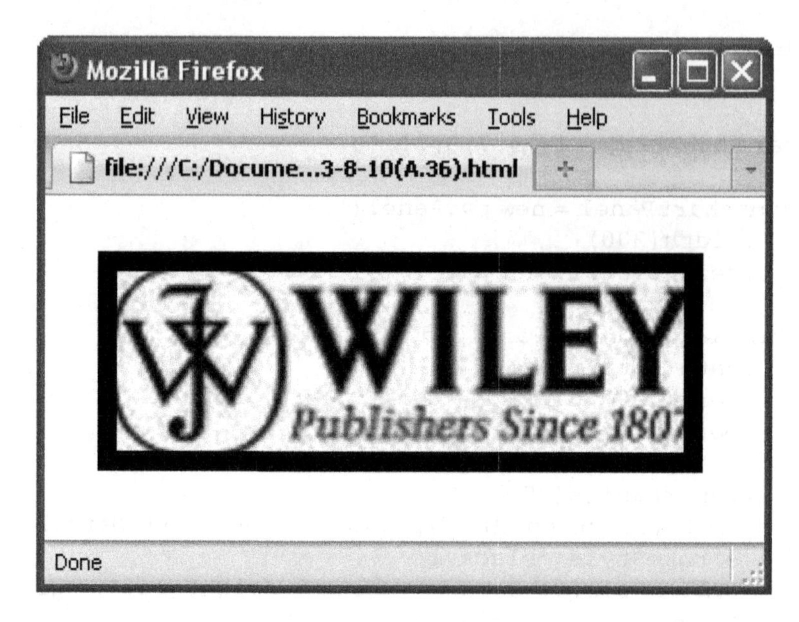


**FIGURE A.36** Exercise 5.3.8.10 screenshot

```
 <body>
  <script type="text/javascript+protovis">

   new pv.Panel()
      .width(350)
      .height(150)

     .add(pv.Image)
      .url("wiley-logo.bmp")
      .left(25)
      .bottom(25)
      .height(100)
      .width(302)
      .lineWidth(10)
      .strokeStyle("black")

     .root.render();

  </script>
 </body>
</html>
```

5.4.6.1 Create a bar chart for the array `[2,5,6,8,4,9]` where values above 8 are set to a blue color defined using the RGB color scale, values below 3 are set to a red color defined using the RGB scale, and all other values are defined as a light gray (again defined using the RGB scale).

The following is an example implementation with Fig. A.37 showing the resulting screenshot.

```
<HTML>
 <head>
  <script type="text/javascript" src=
    "../protovis.js"></script>
 </head>
 <body>
  <script type="text/javascript+protovis">
   new pv.Panel()
      .width(200)
      .height(150)
     .add(pv.Bar)
      .data([2,5,6,8,4,9])
      .bottom(0)
      .width(15)
      .left(function() this.index * 30)
      .height(function(d) d * 12)
```

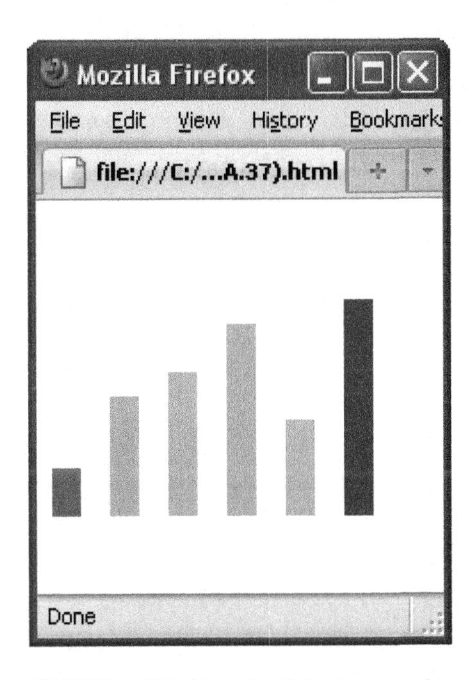


**FIGURE A.37** Exercise 5.4.6.1 screenshot

```
        .fillStyle(function(d) d>8 ? "rgb(0,0,255)" : (d<3 ?
          "rgb(255,0,0)" : "rgb(200,200,200)"))
      .root.render();
    </script>
  </body>
</html>
```

5.4.6.2 Create the same bar chart as in Exercise 5.4.6.1, but use the HSL color format to display the colors.

The following is an example implementation with Fig. A.38 showing the resulting screenshot.

```
<HTML>
  <head>
    <script type="text/javascript" src=
      "../protovis.js"></script>
  </head>
  <body>
    <script type="text/javascript+protovis">
     new pv.Panel()
        .width(200)
        .height(150)
      .add(pv.Bar)
```

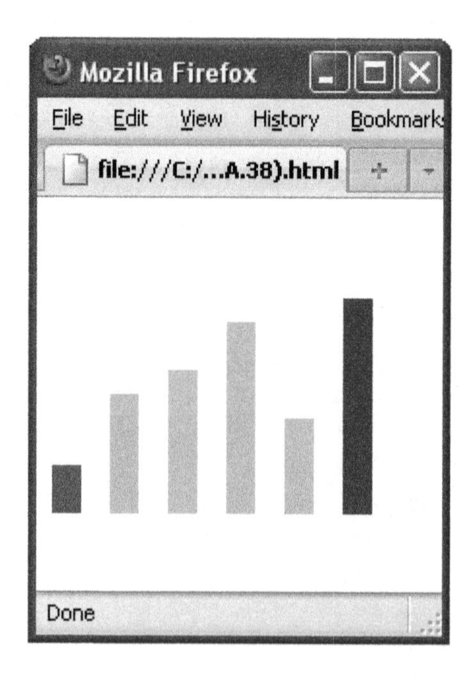


FIGURE A.38 Exercise 5.4.6.2 screenshot

```
      .data([2,5,6,8,4,9])
      .bottom(0)
      .width(15)
      .left(function() this.index * 30)
      .height(function(d) d * 12)
    .fillStyle(function(d) d>8 ? "hsl(240,50%,50%)" : (d<3
      ? "hsl(360,50%,50%)" : "hsl(0,0%,80%)"))
    .root.render();
  </script>
 </body>
</html>
```

5.4.6.3 For the following array of dates in January 2011 `["2011 01 01", "2011 01 02", "2011 01 03", "2011 01 04", 2011 01 05"]`, display on the screen the full date (e.g., Monday 02 January 2011).

The following is an example implementation with Fig. A.39 showing the resulting screenshot.

```
<HTML>
 <head>
  <script type="text/javascript" src=
    "../protovis.js"></script>
 </head>
```

**FIGURE A.39** Exercise 5.4.6.3 screenshot

```
 <body>
  <script type="text/javascript+protovis">

  var panelHeight = 100, panelWidth  = 200, margin = 10;
  var dateList = ["2011 01 01", "2011 01 02", "2011 01 03",
    "2011 01 04", "2011 01 05"];
  var dateInputFormat  = pv.Format.date("%y %m %d");
  var dateOutputFormat = pv.Format.date("%A %d %B %Y");

  var chartPanel = new pv.Panel()
   .width(panelWidth)
   .height(panelHeight)
   .margin(margin);

  chartPanel.add(pv.Label)
    .data(dateList.map(function(d) dateInputFormat.parse
      (d)))
   .text(function(d) dateOutputFormat(d))
   .left(10)
   .top(function() this.index*20);

   chartPanel.render();

  </script>
 </body>
</html>
```

5.4.6.4 Create a horizontally aligned area plot (colored "`lightblue`") and anchor a 3-pixel-wide green line to the top of the chart for the following array: `[3,5,7,8,7,9,14]`.

The following is an example implementation with Fig. A.40 showing the resulting screenshot.

```
<HTML>
 <head>
  <script type="text/javascript" src=
    "../protovis.js"></script>
 </head>
 <body>
  <script type="text/javascript+protovis">

    var areaPanel = new pv.Panel()
      .width(220)
      .height(150);
```

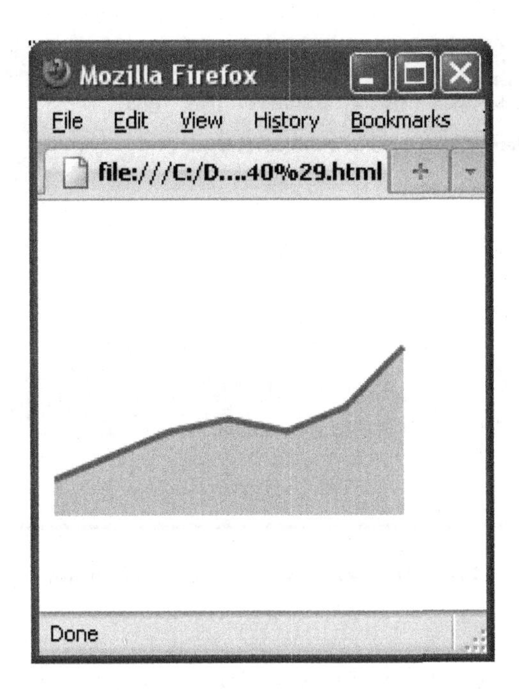


**FIGURE A.40** Exercise 5.4.6.4 screenshot

```
    var area = areaPanel.add(pv.Area)
      .data([3,5,7,8,7,9,14])
      .fillStyle("lightblue")
      .left(function() this.index * 30)
      .height(function(d) d*6)
      .bottom(0);

    area.anchor("top").add(pv.Line)
      .strokeStyle("green")
      .lineWidth(3);

    areaPanel.render();

  </script>
 </body>
</html>
```

5.4.6.5 Create a bar chart for the array `[5,7,6,2,8,9,6,4,3,6,7,3,8,6,5,4,10,2,6,8,6,5,4,3,6]`, anchoring a label to the top of the chart where the value is greater than 8.

The following is an example implementation with Fig. A.41 showing the resulting screenshot.

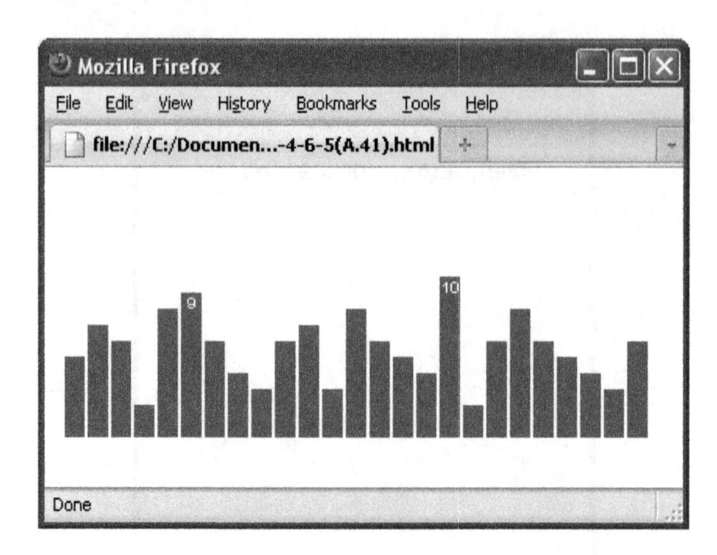


**FIGURE A.41** Exercise 5.4.6.5 screenshot

```
<HTML>
 <head>
  <script type="text/javascript" src=
    "../protovis.js"></script>
 </head>
 <body>
  <script type="text/javascript+protovis">

    var barPanel = new pv.Panel()
      .width(400)
      .height(160);

    var bars = barPanel.add(pv.Bar)
     .data([5,7,6,2,8,9,6,4,3,6,7,3,8,6,5,4,10,2,6,8,6,
       5,4,3,6])
      .width(13)
      .left(function() this.index * 15 + 5)
      .height(function(d) d * 10)
      .bottom(0);

    bars.anchor("top").add(pv.Label)
      .textStyle("white")
      .visible(function(d) d > 8);

    barPanel.render();

  </script>
 </body>
</html>
```

5.4.6.6 Create a plot with the dot mark for the same data array as in Exercise 5.4.6.5, positioning the label above the dot for those values greater than 8.

The following is an example implementation with Fig. A.42 showing the resulting screenshot.

```
<HTML>
 <head>
  <script type="text/javascript" src=
    "../protovis.js"></script>
 </head>
 <body>
  <script type="text/javascript+protovis">

    var dotPanel = new pv.Panel()
       .width(400)
       .height(160);

    var dots = dotPanel.add(pv.Dot)
       .data([5,7,6,2,8,9,6,4,3,6,7,3,8,6,5,4,10,2,6,8,
            6,5,4,3,6])
       .size(8)
       .left(function() this.index * 15 + 5)
       .bottom(function(d) d * 10)
       .add(pv.Line);

    dots.anchor("top").add(pv.Label)
```

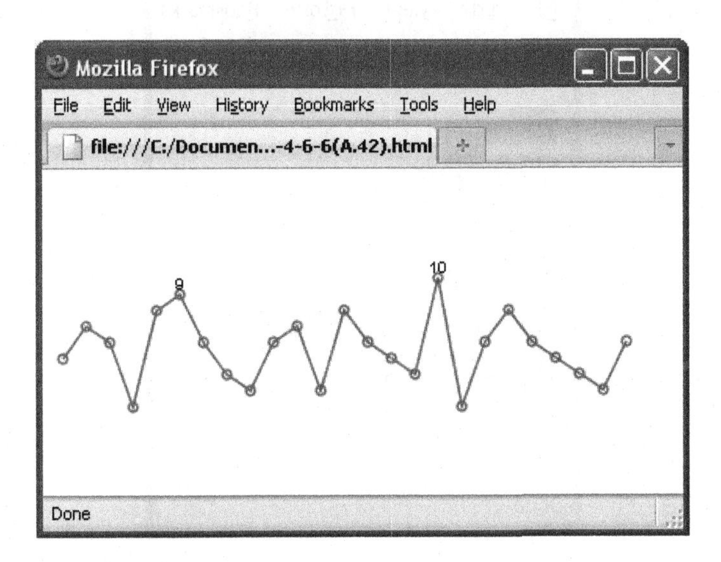


**FIGURE A.42** Exercise 5.4.6.6 screenshot

```
        .visible(function(d) d > 8);

      dotPanel.render();

   </script>
  </body>
</html>
```

5.4.6.7 Create a bar chart aligned with the y-axis using the linear scaling functions for both axes for the array `[124.5, 286.43, 134.76, 255.39, 461.38, 336.26]`.

The following is an example implementation with Fig. A.43 showing the resulting screenshot.

```
<HTML>
<head>
<script type="text/javascript" src=
  "../protovis.js"></script>
</head>
<body>
<script type="text/javascript+protovis">

    var panelWidth = 200, panelHeight = 150;
```

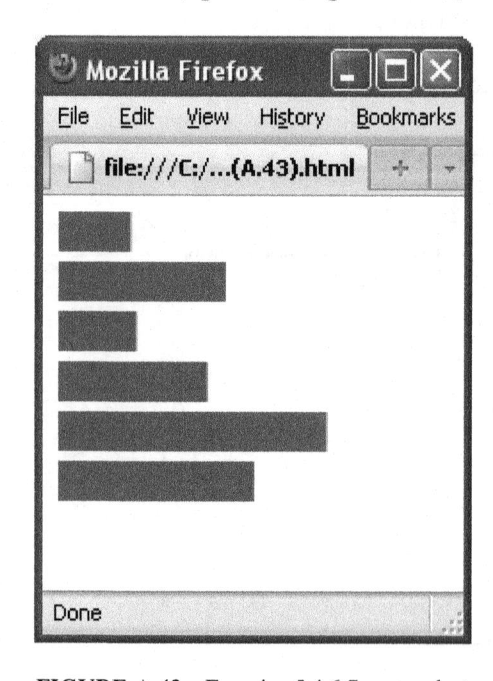


FIGURE A.43 Exercise 5.4.6.7 screenshot

```
     var xScaleMapping = pv.Scale.linear(0, 500).range(0,
panelHeight);

     var barPanel = new pv.Panel()
       .width(panelWidth)
       .height(panelHeight);

     var bars = barPanel.add(pv.Bar)
       .data([124.5, 286.43, 134.76, 255.39, 461.38, 336.26])
       .top(function() this.index * 25)
       .height(20)
       .left(0)
       .width(function(d) xScaleMapping(d));

     barPanel.render();

   </script>
  </body>
</html>
```

5.4.6.8 Create a bar chart aligned with the y-axis chart using the log scaling functions for the array `[0.95, 28.63, 1.34, 245.69, 461.38, 336.26]`.

The following is an example implementation with Fig. A.44 showing the resulting screenshot.

```
<HTML>
  <head>
   <script type="text/javascript" src=
     "../protovis.js"></script>
  </head>
  <body>
   <script type="text/javascript+protovis">

     var panelWidth = 200, panelHeight = 150;

       var xScaleMapping = pv.Scale.log(0.9, 500).range(0,
panelHeight);

     var barPanel = new pv.Panel()
       .width(panelWidth)
       .height(panelHeight);

     var bars = barPanel.add(pv.Bar)
       .data([0.95, 28.63, 1.34, 245.69, 461.38, 336.26])
       .top(function() this.index * 25)
```

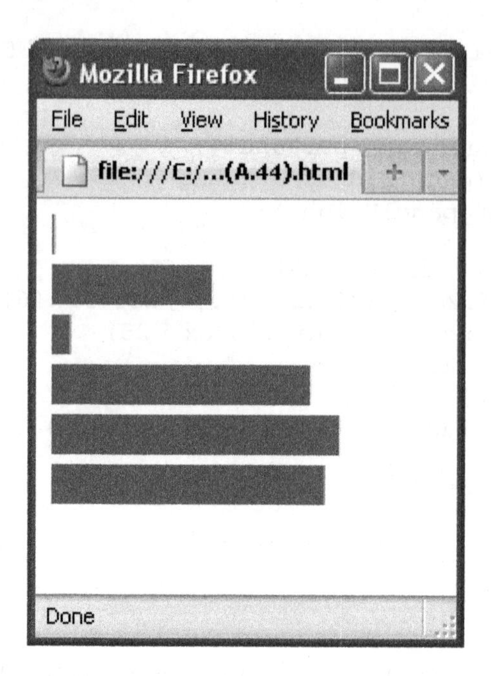


**FIGURE A.44** Exercise 5.4.6.8 screenshot

```
        .height(20)
        .left(0)
        .width(function(d) xScaleMapping(d));

      barPanel.render();

    </script>
  </body>
</html>
```

5.5.5.1 A candlestick plot is often used to represent financial data, such as the prices of stocks over a period of time. The plot presents information on the high and low values for the day (using a line) and the opening and closing values (using a bar). If the stock or fund closes higher than the opening value, the color of the bar is white, whereas if the stock closes lower, the bar is filled in (using dark gray in this example).

The following data captures information on the price for the Standard and Poors 500 stock index for the month of January 2011. This data is located in the file CandlestickData.js

```
var standardAndPoorsData = [
{day: "2011 01 03", open: 1257.62, close: 1271.87, high:
1276.17, low: 1257.62},
{day: "2011 01 04", open: 1272.62, close: 1270.20, high:
1274.12, low: 1262.66},
```

```
{day: "2011 01 05", open: 1268.78, close: 1276.56, high:
1277.63, low: 1265.36},
{day: "2011 01 06", open: 1276.29, close: 1273.85, high:
1278.17, low: 1270.43},
{day: "2011 01 07", open: 1274.41, close: 1271.50, high:
1276.83, low: 1261.70},
{day: "2011 01 10", open: 1270.84, close: 1269.75, high:
1271.52, low: 1262.18},
{day: "2011 01 11", open: 1272.58, close: 1274.48, high:
1277.25, low: 1269.62},
{day: "2011 01 12", open: 1275.65, close: 1283.96, high:
1286.87, low: 1275.65},
{day: "2011 01 13", open: 1285.78, close: 1283.76, high:
1286.70, low: 1280.47},
{day: "2011 01 14", open: 1283.90, close: 1293.24, high:
1293.24, low: 1281.24},
{day: "2011 01 18", open: 1293.22, close: 1295.02, high:
1296.06, low: 1290.16},
{day: "2011 01 19", open: 1294.52, close: 1281.92, high:
1294.60, low: 1278.92},
{day: "2011 01 20", open: 1280.85, close: 1280.26, high:
1283.35, low: 1271.26},
{day: "2011 01 21", open: 1283.63, close: 1283.35, high:
1291.21, low: 1282.07},
{day: "2011 01 24", open: 1283.29, close: 1290.84, high:
1291.93, low: 1282.47},
{day: "2011 01 25", open: 1288.17, close: 1291.18, high:
1291.26, low: 1281.07},
{day: "2011 01 26", open: 1291.97, close: 1292.63, high:
1299.74, low: 1291.97},
{day: "2011 01 27", open: 1297.51, close: 1299.54, high:
1301.29, low: 1294.41},
{day: "2011 01 28", open: 1299.63, close: 1276.34, high:
1302.67, low: 1275.10}
];
```

Using this data, create the candlestick plot as shown in Fig. 5.57.

The following is an example implementation with Fig. A.45 showing the resulting screenshot.

```
<HTML>
 <head>
  <script type="text/javascript" src=
    "../protovis.js"></script>
 <script type="text/javascript" src=
    "candlestickData.js"></script>
 </head>
```

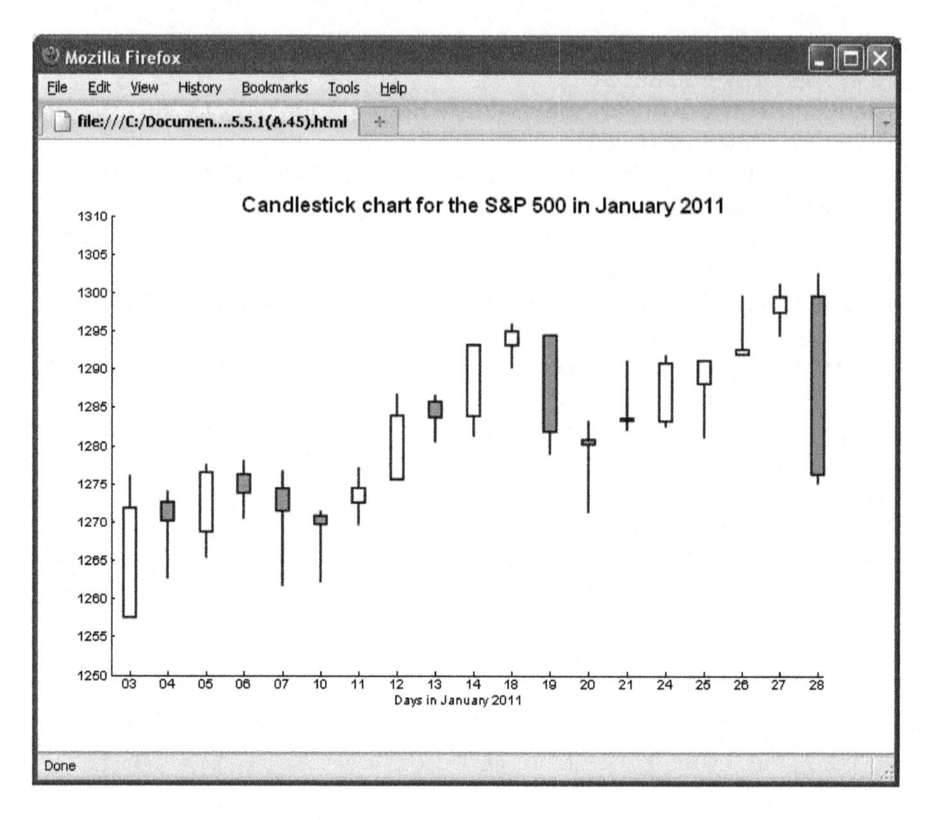


FIGURE A.45 Exercise 5.5.5.1 screenshot

```
<body>

 <script type="text/javascript+protovis">

var panelHeight   = 350
    panelWidth    = 560
    margin        = 50
    barWidth      = 10
    legendLeft    = 670
    legendTop     = 90;

var standardAndPoorsScale = pv.Scale.linear(1250,1310)
  .range(0,panelHeight);

var dateInputFormat  = pv.Format.date("%y %m %d");
var dateOutputFormat = pv.Format.date("%d");


var chartPanel = new pv.Panel()
```

```
    .width(panelWidth)
    .height(panelHeight)
    .margin(margin);

chartPanel.add(pv.Label)
 .text("Candlestick chart for the S&P 500 in January 2011")
  .top(0)
  .left(100)
  .font("bold 12pt sans-serif");

chartPanel.add(pv.Dot)
  .data(standardAndPoorsData)
  .shape("tick")
  .bottom(function(d) standardAndPoorsScale (d.low))
    .size(function(d) (standardAndPoorsScale(d.high) -
      standardAndPoorsScale(d.low)))
  .left(function() this.index * 30 + 10 + (barWidth/2))
  .strokeStyle("black");

chartPanel.add(pv.Bar)
  .data(standardAndPoorsData)
  .bottom(function(d) standardAndPoorsScale (d.open <
   d.close ? d.open : d.close))
  .height(function(d) (d.open < d.close ?
     standardAndPoorsScale (d.close) -
     standardAndPoorsScale (d.open) :
     standardAndPoorsScale (d.open) -
     standardAndPoorsScale (d.close)))
  .left(function() this.index * 30 + 10)
  .width(barWidth)
    .fillStyle(function(d) d.open < d.close ? "white" :
      "darkgray")
  .strokeStyle("black");

chartPanel.add(pv.Rule)
  .left(0);

chartPanel.add(pv.Rule)
   .data(standardAndPoorsScale.ticks())
   .bottom(standardAndPoorsScale)
   .left(0)
   .width(3)
    .anchor("left").add(pv.Label);

chartPanel.add(pv.Rule)
```

```
    .bottom(0)
    .left(0)
    .width(panelWidth);

  chartPanel.add(pv.Rule)
    .data(standardAndPoorsData)
    .bottom(0)
    .left(function() this.index * 30 + 10 + (barWidth/2))
    .height(3)
     .anchor("bottom").add(pv.Label)
     .text(function(d) dateOutputFormat(dateInputFormat.
       parse (d.day)));

  chartPanel.add(pv.Label)
   .text("Days in January 2011")
   .bottom(-25)
   .left(220);

   chartPanel.render();

   </script>
  </body>
</html>
```

5.6.4.1 Re-create the histogram visualization from Section 5.6.1, and change the number of bins to (i) 7 and (ii) approximately 40.

The following is an example implementation where the number of bins is 7 with Fig. A.46 showing the resulting screenshot.

```
<HTML>
 <head>
  <script type="text/javascript" src=
    "../protovis.js"></script>
 <script type="text/javascript" src=
    "Tranquilizers.js"></script>
 </head>

 <body>

  <script type="text/javascript+protovis">

  var panelHeight   = 300
      panelWidth    = 400
      margin        = 20
      maxMolWtValue = pv.max(tranquilizingAgents,
function(d) d.MW)
```

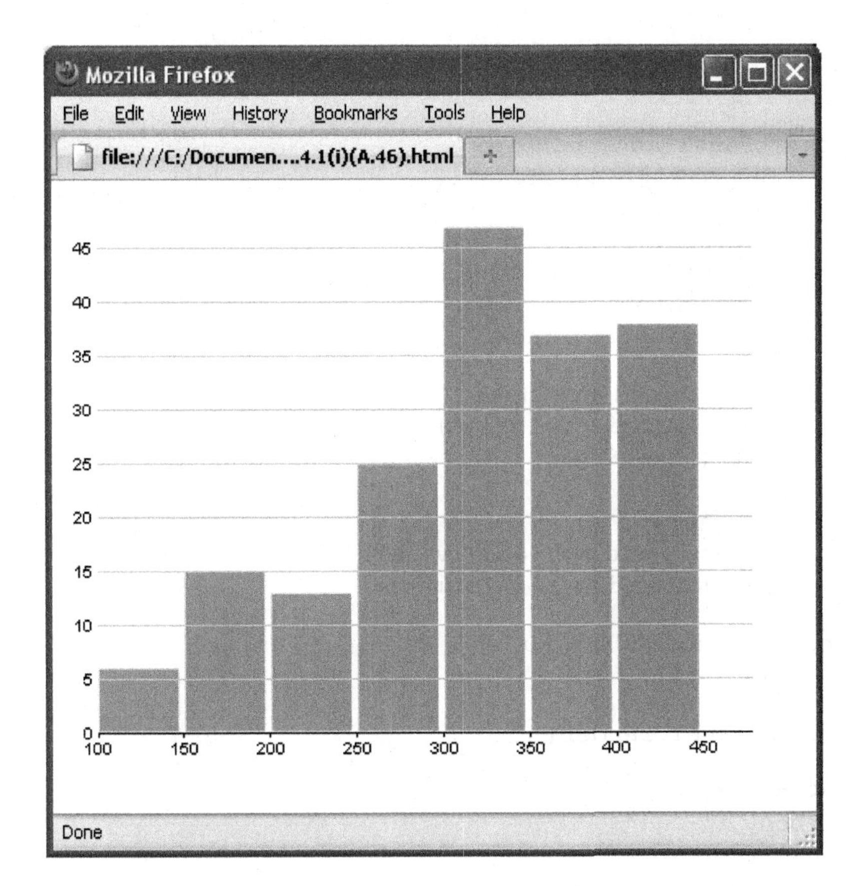


**FIGURE A.46** Exercise 5.6.4.1 screenshot (number of bins is 7)

```
      xMapping      = pv.Scale.linear(100, maxMolWtValue)
.range(0, panelWidth)
      histogramBins = pv.histogram(tranquilizingAgents,
function(d)d.MW).bins(xMapping.ticks(7))
      yMapping      = pv.Scale.linear(0, pv.max
(histogramBins, function(d) d.y)).range(0, panelHeight);

  var chartPanel = new pv.Panel()
      .width(panelWidth)
      .height(panelHeight)
      .margin(margin);

  chartPanel.add(pv.Bar)
     .data(histogramBins)
     .bottom(0)
     .left(function(d) xMapping(d.x))
     .width(50)
     .height(function(d) yMapping(d.y))
```

```
    .fillStyle("darkgray")
    .strokeStyle("white");

  chartPanel.add(pv.Rule)
    .data(yMapping.ticks(10))
    .bottom(yMapping)
    .strokeStyle("lightgray")
   .anchor("left").add(pv.Label)
    .text(yMapping.tickFormat)
    .strokeStyle("black");

  chartPanel.add(pv.Rule)
    .data(xMapping.ticks())
    .left(xMapping)
    .bottom(-3)
    .height(3)
   .anchor("bottom").add(pv.Label)
    .text(xMapping.tickFormat);

  chartPanel.add(pv.Rule)
   .bottom(0);

  chartPanel.render();

  </script>
 </body>
</html>
```

The following is an example implementation where the number of bins is approximately 40 with Fig. A.47 showing the resulting screenshot.

```
<HTML>
 <head>
  <script type="text/javascript" src=
    "../protovis.js"></script>
 <script type="text/javascript" src=
    "Tranquilizers.js"></script>
 </head>

 <body>

  <script type="text/javascript+protovis">

  var panelHeight  = 300
      panelWidth   = 400
      margin       = 20
```

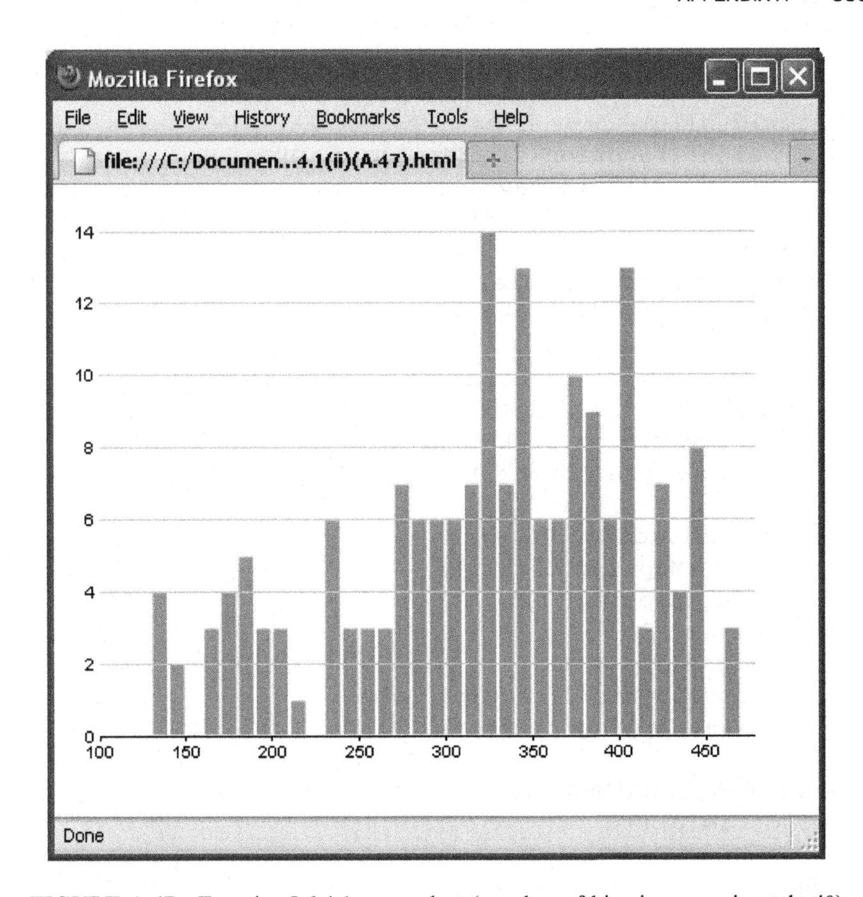


**FIGURE A.47** Exercise 5.6.4.1 screenshot (number of bins is approximately 40)

```
  maxMolWtValue = pv.max(tranquilizingAgents,
    function(d) d.MW)
  xMapping      = pv.Scale.linear(100, maxMolWtValue)
    .range(0, panelWidth)
  histogramBins = pv.histogram(tranquilizingAgents,
      function(d)d.MW).bins(xMapping.ticks(40))
  yMapping      = pv.Scale.linear(0, pv
    .max(histogramBins, function(d) d.y))
    .range(0, panelHeight);

var chartPanel = new pv.Panel()
    .width(panelWidth)
    .height(panelHeight)
    .margin(margin);

chartPanel.add(pv.Bar)
   .data(histogramBins)
```

```
      .bottom(0)
      .left(function(d) xMapping(d.x))
      .width(10)
      .height(function(d) yMapping(d.y))
      .fillStyle("darkgray")
      .strokeStyle("white");

  chartPanel.add(pv.Rule)
      .data(yMapping.ticks(10))
      .bottom(yMapping)
      .strokeStyle("lightgray")
    .anchor("left").add(pv.Label)
      .text(yMapping.tickFormat)
      .strokeStyle("black");

  chartPanel.add(pv.Rule)
      .data(xMapping.ticks())
      .left(xMapping)
      .bottom(-3)
      .height(3)
    .anchor("bottom").add(pv.Label)
      .text(xMapping.tickFormat);

  chartPanel.add(pv.Rule)
    .bottom(0);

   chartPanel.render();

   </script>
  </body>
</html>
```

5.6.4.2 Create a frequency histogram for the property`XLogP` (distribution coefficient).

The following is an example implementation where the data is `XLogP` with Fig. A.48 showing the resulting screenshot.

```
<HTML>
  <head>
   <script type="text/javascript" src=
     "../protovis.js"></script>
   <script type="text/javascript" src=
     "Tranquilizers.js"></script>
  </head>
```

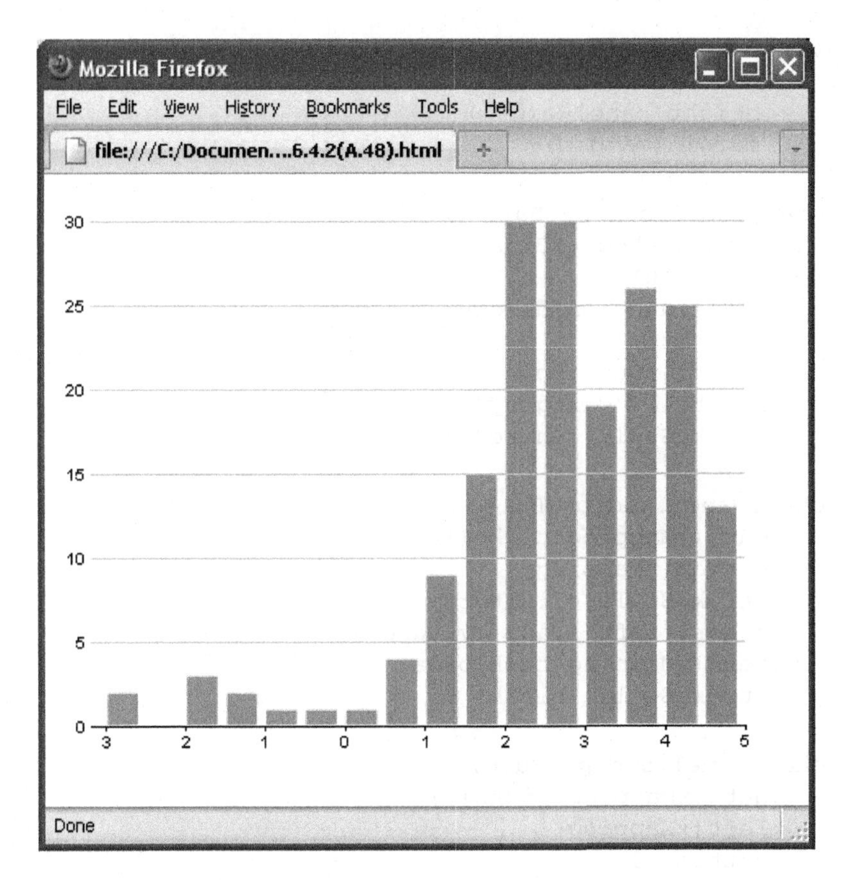


**FIGURE A.48** Exercise 5.6.4.2 screenshot

```
<body>

 <script type="text/javascript+protovis">

 var panelHeight  = 300
     panelWidth   = 400
     margin       = 20
     minXLogPValue = pv.min(tranquilizingAgents,
       function(d) d.XLogP)
     maxXLogPValue = pv.max(tranquilizingAgents,
       function(d) d.XLogP)
     xMapping      = pv.Scale.linear(minXLogPValue,
       maxXLogPValue ).range(0, panelWidth)
     histogramBins   = pv.histogram(tranquilizingAgents,
       function(d)d.XLogP).bins(xMapping.ticks(20))
     yMapping  = pv.Scale.linear(0, pv.max(histogramBins,
       function(d) d.y)).range(0, panelHeight);
```

```
  var chartPanel = new pv.Panel()
      .width(panelWidth)
      .height(panelHeight)
      .margin(margin);

  chartPanel.add(pv.Bar)
     .data(histogramBins)
     .bottom(0)
     .left(function(d) xMapping(d.x))
     .width(20)
     .height(function(d) yMapping(d.y))
     .fillStyle("darkgray")
     .strokeStyle("white");

  chartPanel.add(pv.Rule)
     .data(yMapping.ticks(10))
     .bottom(yMapping)
     .strokeStyle("lightgray")
    .anchor("left").add(pv.Label)
     .text(yMapping.tickFormat)
     .strokeStyle("black");

  chartPanel.add(pv.Rule)
     .data(xMapping.ticks())
     .left(xMapping)
     .bottom(-3)
     .height(3)
    .anchor("bottom").add(pv.Label)
     .text(xMapping.tickFormat);

  chartPanel.add(pv.Rule)
    .bottom(0);

   chartPanel.render();

   </script>
  </body>
</html>
```

5.6.4.3 Create a scatterplot using the tranquilizer data where each observations is color-coded according to the type of tranquilizer: red for antianxiety, green for antimanic, and blue for antipsychotic.

The following is an example implementation with Fig. A.49 showing the resulting screenshot.

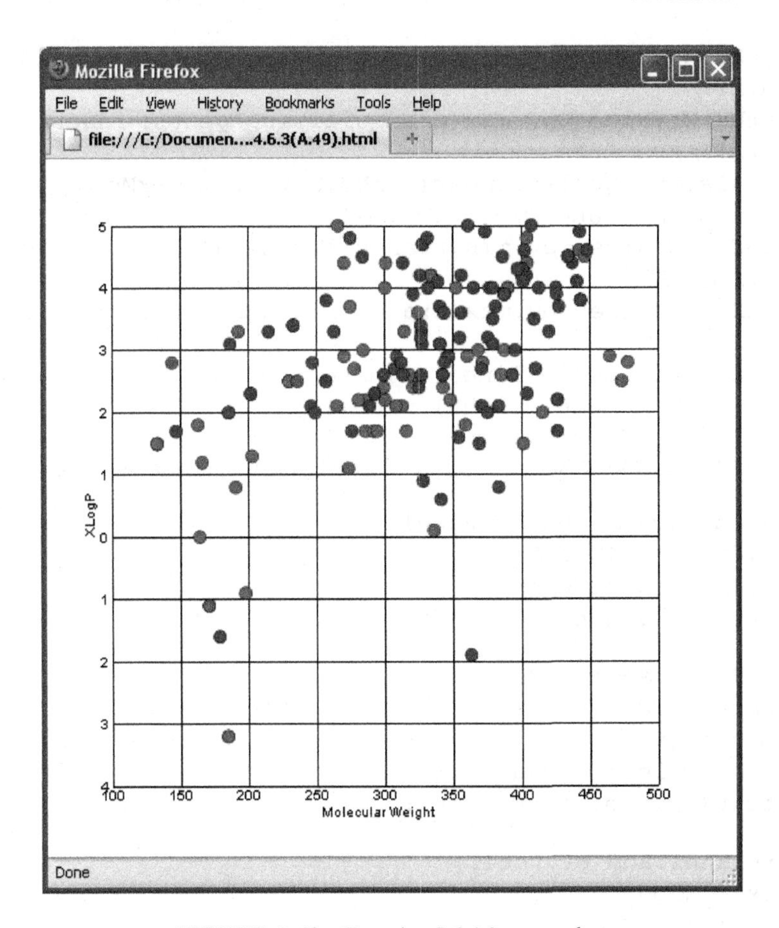


**FIGURE A.49** Exercise 5.6.4.3 screenshot

```
<HTML>
 <head>
  <script type="text/javascript" src=
    "../protovis.js"></script>
  <script type="text/javascript" src=
    "Tranquilizers.js"></script>
 </head>

 <body>

  <script type="text/javascript+protovis">

var panelWidth  = 400,
    panelHeight = 400,
    margin      = 40
```

```
  minMolWtValue = pv.min(tranquilizingAgents, function(d)
    d.MW)
  maxMolWtValue = pv.max(tranquilizingAgents, function(d)
    d.MW)
  xScale = pv.Scale.linear(minMolWtValue , maxMolWtValue )
    .range(0, panelWidth).nice(),
  minXLogPValue = pv.min(tranquilizingAgents, function(d)
    d.XLogP)
  maxXLogPValue = pv.max(tranquilizingAgents, function(d)
    d.XLogP)
  yScale = pv.Scale.linear(minXLogPValue, maxXLogPValue)
    .range(0, panelHeight).nice();

var tranquilizerColor = pv.colors("red", "green", "blue");

var chartPanel = new pv.Panel()
   .width(panelWidth)
   .height(panelHeight)
   .margin(margin);

chartPanel.add(pv.Rule)
   .data(yScale.ticks())
   .bottom(yScale)
  .anchor("left").add(pv.Label)
   .text(yScale.tickFormat);

chartPanel.add(pv.Label)
  .text("XLogP")
  .left(-10)
  .bottom(175)
  .textAngle(-Math.PI/2);

chartPanel.add(pv.Rule)
   .data(xScale.ticks())
   .left(xScale)
  .anchor("bottom").add(pv.Label)
   .text(xScale.tickFormat);

chartPanel.add(pv.Label)
    .text("Molecular Weight")
    .bottom(-25)
    .left(150);

chartPanel.add(pv.Dot)
   .data(tranquilizingAgents)
   .left(function(d) xScale(d.MW))
```

```
    .bottom(function(d) yScale(d.XLogP))
    .fillStyle("lightgray")
    .lineWidth(0.5)
    .strokeStyle("black")
    .fillStyle(function(d) tranquilizerColor(d.type));

chartPanel.render();

    </script>
  </body>
</html>
```

5.6.4.4 A logistic regression model (Myatt & Johnson, 2009) was generated to predict whether an observation is positive (1) or negative (0). The logistic regression model calculates a probability. A cutoff can be used to assign the calculated prediction probability as positive or negative. For example, using a 0.5 cutoff will assign values greater than or equal to 0.5 as positive and values less than 0.5 as negative. A ROC plot (Receiver Operating Characteristic) assesses the quality of a model. Different cut-off values can be used, the false positive rate (or sensitivity) is plotted along the x-axis, and the false positive rate (or 1-specificity) is plotted along the y-axis, as shown in Fig. 5.61. The plot should be to the left of the diagonal line, with better models having plots close to the top-left corner.

A Dataset of actual values, along with the predicted probabilities, is provided in the file actualAndPredicted.js. A few examples are shown here:

```
var actualVSPredicted = [
{ actual: 0 , predicted: 0.411},
{ actual: 0 , predicted: 0.305},
{ actual: 0 , predicted: 0.485},
{ actual: 0 , predicted: 0.191},
 . . .
];
```

The number of true positives (*TP*) is the number of observations that are correctly predicted as positive. The number of true negatives (*TN*) is the number of observations that are correctly predicted as negative. The number of false positives (*FP*) is the number of observations that are predicted positive but are in fact negative. The number of false negatives (*FN*) is the number of observations that are predicted as negative but are in fact positive. These numbers change with different cut-off values. Sensitivity is calculated as $TP / (TP+FN)$; specificity is calculated as $TN / (TN+FP)$.

Create the ROC plot as shown in Fig. 5.61.

The following is an example implementation with Fig. A.50 showing the resulting screenshot.

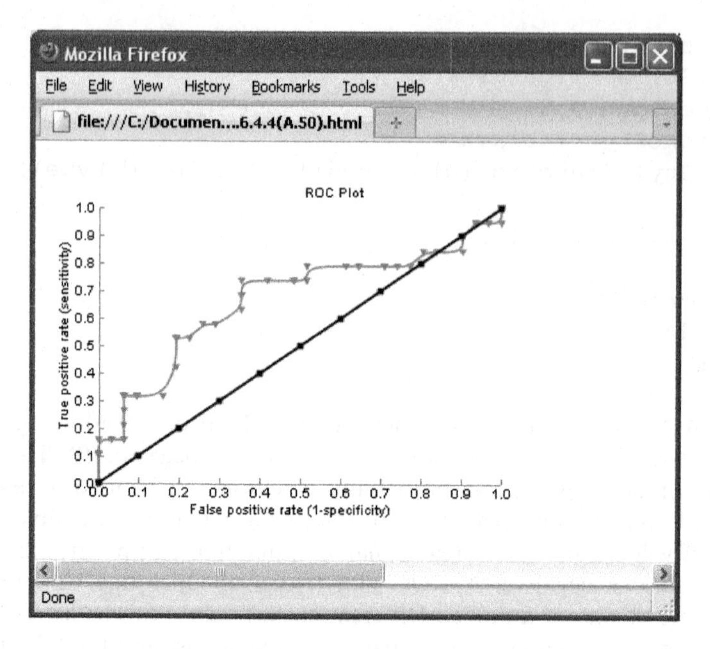


FIGURE A.50 Exercise 5.6.4.4 screenshot

```
<HTML>
 <head>
  <script type="text/javascript" src=
    "../protovis.js"></script>
  <script type="text/javascript" src=
    "actualAndProbability.js"></script>
 </head>
 <body>
  <script type="text/javascript+protovis">


var width = 300, height = 200, margin = 40;

var cutoffs = pv.range(0,1.01,0.01);

var truePositive = function(actual, predicted, cutoff)
   ((predicted >= cutoff) && (actual == 1)) ? 1 : 0;
var trueNegative = function(actual, predicted, cutoff)
   ((predicted < cutoff) && (actual == 0)) ? 1 : 0;
var falsePositive = function(actual, predicted, cutoff)
   ((predicted >= cutoff) && (actual == 0)) ? 1 : 0;
var falseNegative = function(actual, predicted, cutoff)
   ((predicted < cutoff) && (actual == 1)) ? 1 : 0;
```

```
var numberTruePositives = function(cutoff)
   pv.sum(actualVSPredicted.map(function(d) truePositive
(d.actual, d.predicted, cutoff)));
var numberTrueNegatives = function(cutoff)
   pv.sum(actualVSPredicted.map(function(d) trueNegative
(d.actual, d.predicted, cutoff)));
var numberFalsePositives = function(cutoff)
   pv.sum(actualVSPredicted.map(function(d)
      falsePositive(d.actual, d.predicted, cutoff)));
var numberFalseNegatives = function(cutoff)
   pv.sum(actualVSPredicted.map(function(d)
      falseNegative(d.actual, d.predicted, cutoff)));

var sensitivity = function(cutoff)
     (numberTruePositives(cutoff) / (numberTruePositives
      (cutoff) + numberFalseNegatives(cutoff)));
var specificity = function(cutoff)
     (numberTrueNegatives(cutoff) / (numberTrueNegatives
      (cutoff) + numberFalsePositives(cutoff)));

var xScaleMapping = pv.Scale.linear(0,1).range(0,width),
  yScaleMapping = pv.Scale.linear(0,1).range(0,height);

var chartPanel = new pv.Panel().width(width + 400).height
(height).margin(margin);

chartPanel.add(pv.Label)
   .left(150)
   .top(-5)
   .text("ROC Plot");

chartPanel.add(pv.Rule)
 .strokeStyle("lightgray")
 .left(0);

chartPanel.add(pv.Rule)
 .strokeStyle("lightgray")
 .width(xScaleMapping(1))
 .bottom(0);

chartPanel.add(pv.Rule)
  .data(yScaleMapping.ticks(10))
  .bottom(yScaleMapping)
  .left(0)
  .width(4)
  .strokeStyle("lightgray")
```

```
 .anchor("left").add(pv.Label)
  .text(yScaleMapping.tickFormat);

chartPanel.add(pv.Label)
 .text("True positive rate (sensitivity)")
 .left(-20)
 .bottom(35)
 .textAngle(-Math.PI/2);

chartPanel.add(pv.Rule)
  .data(xScaleMapping.ticks(10))
  .left(xScaleMapping)
  .bottom(0)
  .height(4)
  .strokeStyle("lightgray")
 .anchor("bottom").add(pv.Label)
  .text(yScaleMapping.tickFormat);

chartPanel.add(pv.Label)
  .text("False positive rate (1-specificity)")
  .bottom(-25)
  .left(65);

chartPanel.add(pv.Line)
  .data(cutoffs)
  .bottom(function(d) yScaleMapping(sensitivity(d)))
  .left(function(d) xScaleMapping(1-specificity(d)))
  .strokeStyle("lightblue")
      .interpolate("basis")
  .add(pv.Dot)
   .shape("triangle")
   .size(2)
   .strokeStyle("lightblue");

chartPanel.add(pv.Line)
   .data(pv.range(0,1.1,0.1))
   .bottom(function(d) yScaleMapping((d)))
   .left(function(d) xScaleMapping((d)))
   .strokeStyle("green")
   .add(pv.Dot)
    .shape("square")
    .size(2);

chartPanel.render();

   </script>
  </body>
</html>
```

5.7.9.1 Trellis plots are used to describe multiple dimensions concerning groups of data.The JavaScript file PI-PII-PIII-Clinical-Trials.js contains the number of drugs in phase I, phase II, and phase III for three fictitious pharmaceutical companies (Alphapharma, Betapharma, and Gammapharma). Two entries are shown here:

```
var clinicalTrials = [
{ numberOfDrugs: 10, clinicalTrial: "Phase I", disease:
"cancer", company: "Alphapharm" },
{ numberOfDrugs: 8, clinicalTrial: "Phase II", disease:
"cancer", company: "Alphapharm" },
```

Create the trellis plot as shown in Fig. 5.78.

The following is an example implementation with Fig. A.51 showing the resulting screenshot.

```
<HTML>
 <head>
  <script type="text/javascript" src=
    "../protovis.js"></script>
  <script type="text/javascript" src=
    "PI-PII-PIII-Clinical-Trials.js"></script>
 </head>

 <body>
  <script type="text/javascript+protovis">

var uniqueCompanies = pv.uniq(clinicalTrials, function(d)
  d.company);
var uniqueDiseases = pv.uniq(clinicalTrials, function(d)
  d.disease);
var uniqueClinicalTrials = pv.uniq(clinicalTrials,
  function(d) d.clinicalTrial);

var cellHeight = 150;
var chartHeight = cellHeight*uniqueCompanies.length,
  chartWidth = 150, margin = 70;

var maxDrugCount = pv.max(clinicalTrials, function(d)
  d.numberOfDrugs);

var diseaseScale = pv.Scale.ordinal(uniqueDiseases).split
  (0,(cellHeight-15));
var colorScale = pv.Scale.ordinal(uniqueClinicalTrials);
var drugCountScale = pv.Scale.linear(0,maxDrugCount)
  .range(0,chartWidth);
```

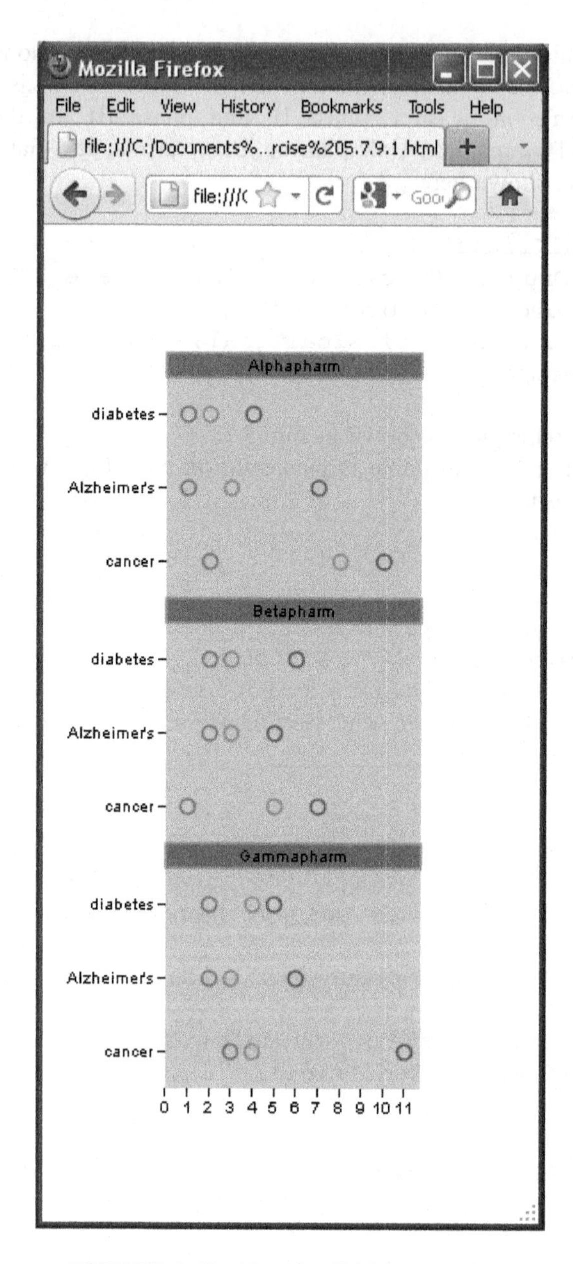


**FIGURE A.51** Exercise 5.7.9.1 screenshot

```
var colorScale = pv.Scale.ordinal(uniqueClinicalTrials)
.split("red", "green", "blue");

var chartPanel = new pv.Panel()
   .width(chartWidth+10)
```

```
  .height(chartHeight)
  .margin(margin)
  .add(pv.Bar)
  .fillStyle("lightgray");

var companyCells = chartPanel.add(pv.Panel)
  .data(uniqueCompanies)
  .left(0)
  .top(function() this.index * cellHeight)
  .height(cellHeight);

companyCells.add(pv.Bar)
  .top(0)
  .height(15)
  .strokeStyle("darkgray")
  .fillStyle("gray")
 .anchor("center").add(pv.Label)
   .strokeStyle("darkgray");

companyCells.add(pv.Rule)
 .data(uniqueDiseases)
 .left(-5)
 .width(5)
 .bottom(function(d) diseaseScale(d))
 .anchor("left").add(pv.Label).text(function (d) d);

var clinicalTrialCells = companyCells.add(pv.Panel)
  .data(uniqueClinicalTrials)
  .bottom(0)
  .left(0)
  .height(cellHeight-15);

clinicalTrialCells.add(pv.Dot)
  .def("cti", function() this.parent.index)
  .def("coi", function() this.parent.parent.index)
  .data(clinicalTrials)
  .visible(function(d) ( (d.clinicalTrial ==
    uniqueClinicalTrials[this.cti()]) && (d.company ==
    uniqueCompanies[this.coi()]) ) )
  .bottom(function(d) diseaseScale(d.disease))
  .left(function(d) drugCountScale(d.numberOfDrugs));

chartPanel.add(pv.Rule)
  .data(drugCountScale.ticks())
  .bottom(-5)
```

```
  .height(5)
  .left(function(d) drugCountScale(d))
  .anchor("bottom").add(pv.Label);

chartPanel.render();

   </script>
  </body>
</html>
```

5.8.6.1 Add a tooltip to the box-and-whisker plot created in Section 5.6.2 to generate the text shown in Fig. 5.83.

The following is an example implementation with Fig. A.52 showing the resulting screenshot.

```
<HTML>
  <head>
   <script type="text/javascript" src=
     "../protovis.js"></script>
  </head>
  <body>
   <script type="text/javascript+protovis">

var experiments = ["Experiment 1","Experiment 2",
  "Experiment 3"];
```

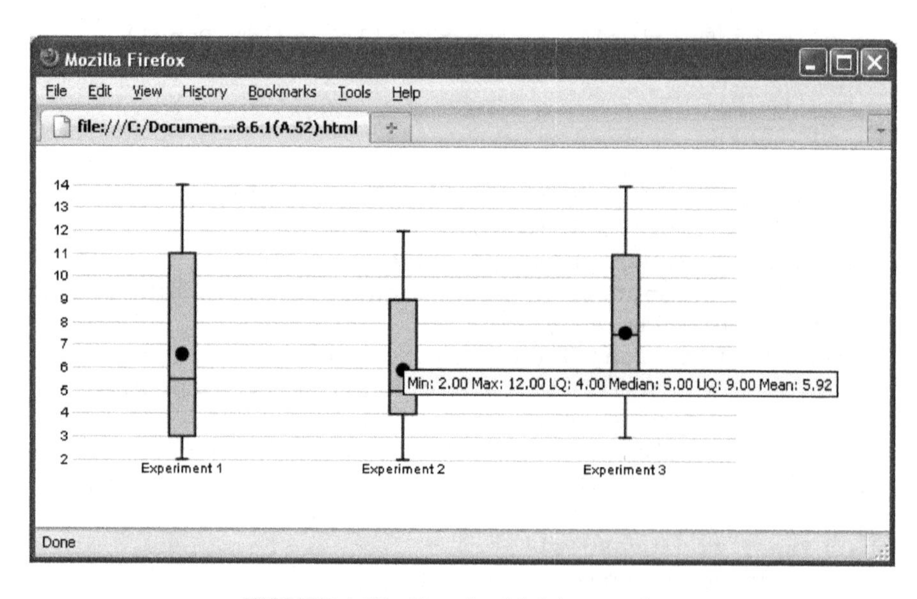


**FIGURE A.52** Exercise 5.8.6.1 screenshot

```
var dataValues = [[3,2,3,2,5,6,3,8,9,14,11,13],
                  [5,3,2,5,6,4,8,9,10,3,12,4],
                  [3,6,5,8,7,3,9,10,14,12,11,3]];

var width = 500, height = 200, tailWidth = 10, barWidth = 20,
  meanDotSize = 2;
var fillColor = "lightgray", lineColor = "black";

var numberFormat = pv.Format.number().fractionDigits(2);


var maxValue = pv.max(dataValues.map(function(d) pv.max
  (d)));
var minValue = pv.min(dataValues.map(function(d) pv.min
  (d)));
var numberOfObservations = dataValues[0].length;

var upperQuartile = function(d){ d.sort(pv.naturalOrder);
  return d[Math.round(numberOfObservations*3/4)]; }
var lowerQuartile = function(d){ d.sort(pv.naturalOrder);
  return d[Math.round(numberOfObservations/4)]; }

var xScaleMapping = pv.Scale.ordinal(experiments).split
  (0,width),
  yScaleMapping = pv.Scale.linear(minValue,maxValue)
.range(0,height);

var chartPanel = new pv.Panel().width(width).height
  (height).margin(20);

chartPanel.add(pv.Rule)
   .data(yScaleMapping.ticks())
   .bottom(function(d) yScaleMapping(d))
   .strokeStyle("lightgray")
   .anchor("left").add(pv.Label)
   .text(function(d) d);

chartPanel.add(pv.Rule)
   .data(experiments)
   .bottom(0)
   .height(4)
   .left(function(d) xScaleMapping(d))
   .strokeStyle("lightgray")
   .anchor("bottom")
   .add(pv.Label).text(function(d) d);
```

```
chartPanel.add(pv.Dot)
   .data(dataValues)
   .left(function() xScaleMapping(experiments
     [this.index]) - (tailWidth/2))
   .shape("tick")
   .size(tailWidth)
   .angle(Math.PI/2)
   .strokeStyle(lineColor)
   .bottom(function(d) yScaleMapping(pv.max(d)));

chartPanel.add(pv.Dot)
   .data(dataValues)
   .left(function() xScaleMapping(experiments
     [this.index]) - (tailWidth/2))
   .shape("tick")
   .size(tailWidth)
   .angle(Math.PI/2)
   .strokeStyle(lineColor)
   .bottom(function(d) yScaleMapping(pv.min(d)));

chartPanel.add(pv.Dot)
   .data(dataValues)
   .left(function() xScaleMapping(experiments
     [this.index]))
   .shape("tick")
   .strokeStyle(lineColor)
    .size(function(d) yScaleMapping(pv.max(d)) -
     yScaleMapping(pv.min(d)))
   .bottom(function(d) yScaleMapping(pv.min(d)));

chartPanel.add(pv.Bar)
   .data(dataValues)
   .left(function() xScaleMapping(experiments
     [this.index]) - (barWidth/2))
   .bottom(function(d) yScaleMapping(lowerQuartile(d)))
   .height(function(d) yScaleMapping(upperQuartile(d)) -
     yScaleMapping(lowerQuartile(d)))
   .fillStyle(fillColor)
   .strokeStyle(lineColor)
   .width(barWidth)

   .title(function(d) "Min: " + numberFormat (pv.min(d)) +
     " Max: " + numberFormat (pv.max(d)) + " LQ: " +
     numberFormat (lowerQuartile(d)) + " Median: " +
     numberFormat (pv.median(d)) + " UQ: " + numberFormat
```

```
    (upperQuartile(d)) + " Mean: " + numberFormat
    (pv.mean(d)));

chartPanel.add(pv.Dot)
   .data(dataValues)
   .left(function() xScaleMapping(experiments
     [this.index]) - 10)
   .shape("tick")
   .size(20)
   .strokeStyle(lineColor)
   .angle(Math.PI/2)
   .bottom(function(d) yScaleMapping(pv.median(d)));

chartPanel.add(pv.Dot)
   .data(dataValues)
   .left(function() xScaleMapping(experiments
     [this.index]))
   .size(20)
   .strokeStyle(lineColor)
   .fillStyle(lineColor)
   .angle(Math.PI/2)
   .bottom(function(d) yScaleMapping(pv.mean(d)));

chartPanel.render();

   </script>
  </body>
</html>
```

5.8.6.2 Modify the scatterplot visualization described in Section 5.8.5 where, by clicking on the scatterplot point, the user is redirected to a Web page with the details on the specific tranquilizing agent. The CID property is a unique identification for each of the agents. The PubChem database contains the underlying information on each CID record. For example, the URL http://pubchem.ncbi.nlm.nih.gov/summary/summary.cgi?cid=441233 will display the record for the agent whose CID is 441233 (the first observation in the list).

The following is an example implementation with Fig. A. 53 showing the resulting screenshot and Fig. A.54 showing the results of clicking on one of the scatterplot dots.

```
<html>
 <head>
  <script type="text/javascript" src=
    "../protovis.js"></script>
```

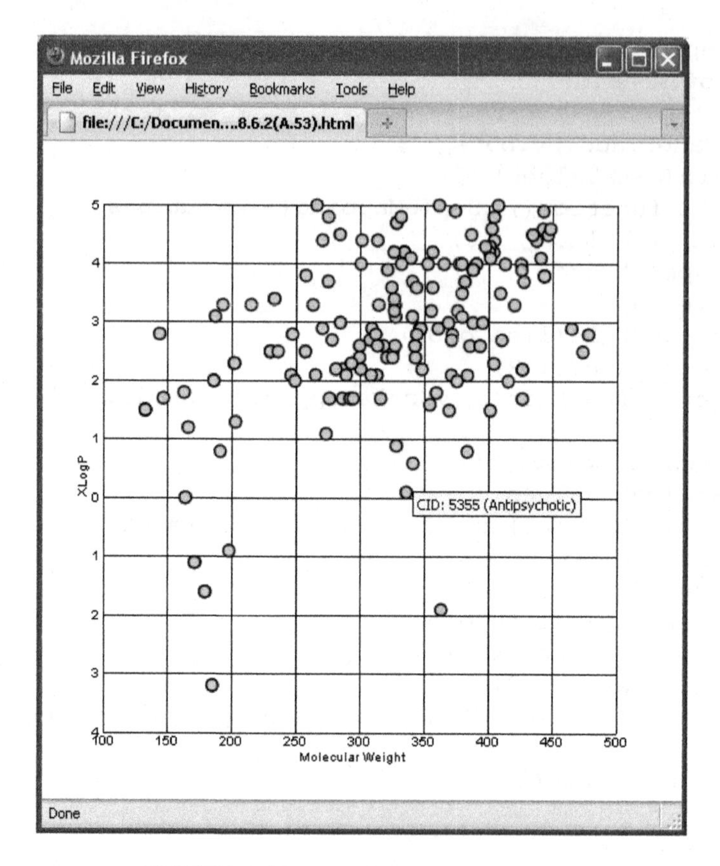


**FIGURE A.53** Exercise 5.8.6.2 screenshot

```
    <script type="text/javascript" src=
      "Tranquilizers.js"></script>
  </head>
  <body>

    <script type="text/javascript+protovis">

var panelWidth    = 400,
    panelHeight   = 400,
    margin        = 30
    maxMolWtValue = pv.max(tranquilizingAgents,
      function(d) d.MW)
    xScale = pv.Scale.linear(0, maxMolWtValue )
      .range(0, panelWidth).nice(),
    minALogPValue = pv.min(tranquilizingAgents,
      function(d) d.XLogP)
   maxALogPValue = pv.max(tranquilizingAgents, function(d)
     d.XLogP)
```

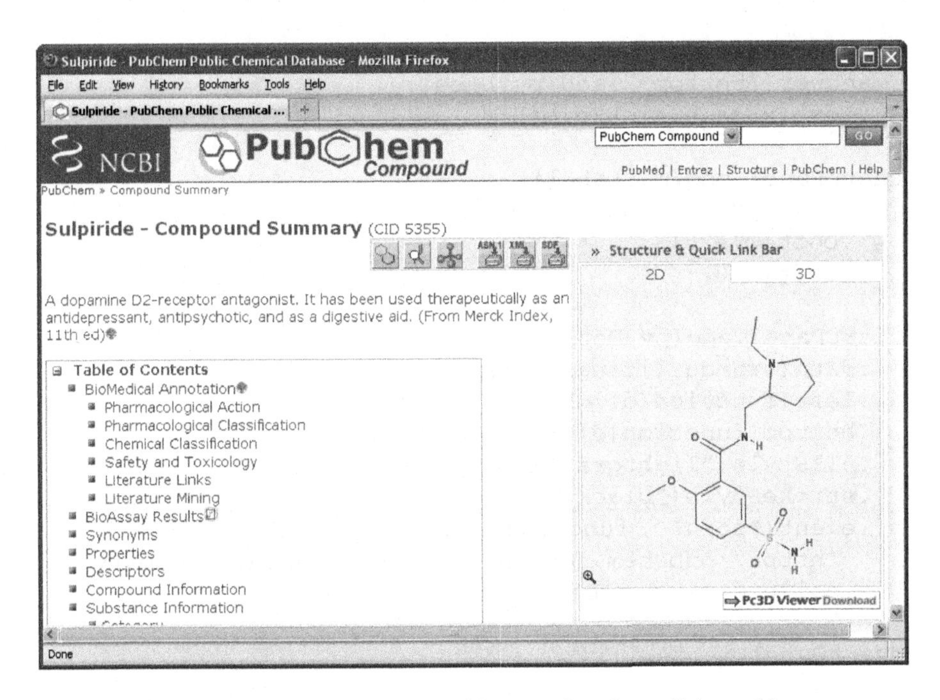


**FIGURE A.54** Exercise 5.8.6.2 screenshot (hyperlink result)

```
    yScale = pv.Scale.linear(minALogPValue, maxALogPValue)
      .range(0, panelHeight).nice();

var chartPanel = new pv.Panel()
   .width(panelWidth)
   .height(panelHeight)
   .margin(margin)
   .event("all")
   .event("mousedown", pv.Behavior.point());

chartPanel.add(pv.Rule)
   .data(yScale.ticks())
   .bottom(yScale)
  .anchor("left").add(pv.Label)
   .text(yScale.tickFormat);

chartPanel.add(pv.Label)
  .text("XLogP")
  .left(-10)
  .bottom(175)
  .textAngle(-Math.PI/2);
chartPanel.add(pv.Rule)
   .data(xScale.ticks())
```

```
    .left(xScale)
  .anchor("bottom").add(pv.Label)
    .text(xScale.tickFormat);

chartPanel.add(pv.Label)
     .text("Molecular Weight")
     .bottom(-25)
     .left(150);

chartPanel.add(pv.Dot)
    .data(tranquilizingAgents)
    .left(function(d) xScale(d.MW))
    .bottom(function(d) yScale(d.XLogP))
    .fillStyle("lightgray")
    .strokeStyle("black")
    .event("point", function(d) self.location =
      "http://pubchem.ncbi.nlm.nih.gov/summary/summary.
      cgi?cid=" + d.CID);

chartPanel.render();

    </script>
  </body>
</html>
```

5.8.6.3 Using the ROC plot created in Exercise 5.6.4.4, generate an interactive ROC plot so that moving a cursor close to a point in the plot provides more information on the cutoff and the quality of the underlying logistic regression model at that point, as shown in Fig. 5.84.

The following is an example implementation with Fig. A.55 showing the resulting screenshot.

```
<HTML>
  <head>
    <script type="text/javascript" src=
      "../protovis.js"></script>
     <script type="text/javascript" src=
      "actualAndProbability.js"></script>
  </head>
  <body>
    <script type="text/javascript+protovis">

var width = 300, height = 200, margin = 40;

var cutoffs = pv.range(0,1.01,0.01);
var selectedCutoff = -1;
```

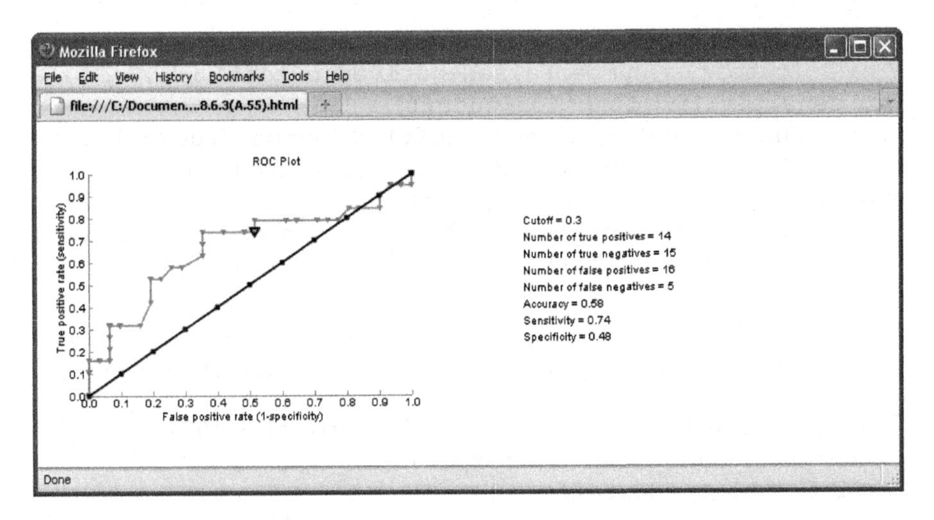


**FIGURE A.55** Exercise 5.8.6.3 screenshot

```
var twoDPFormat = pv.Format.number().fractionDigits(0,2);

var truePositive = function(actual, predicted, cutoff)
   ((predicted >= cutoff) && (actual == 1)) ? 1 : 0;
var trueNegative = function(actual, predicted, cutoff)
   ((predicted < cutoff) && (actual == 0)) ? 1 : 0;
var falsePositive = function(actual, predicted, cutoff)
   ((predicted >= cutoff) && (actual == 0)) ? 1 : 0;
var falseNegative = function(actual, predicted, cutoff)
   ((predicted < cutoff) && (actual == 1)) ? 1 : 0;

var numberTruePositives = function(cutoff)
  pv.sum(actualVSPredicted.map(function(d) truePositive
    (d.actual, d.predicted, cutoff)));
var numberTrueNegatives = function(cutoff)
  pv.sum(actualVSPredicted.map(function(d) trueNegative
    (d.actual, d.predicted, cutoff)));
var numberFalsePositives = function(cutoff)
  pv.sum(actualVSPredicted.map(function(d)
     falsePositive(d.actual, d.predicted, cutoff)));
var numberFalseNegatives = function(cutoff)
  pv.sum(actualVSPredicted.map(function(d)
     falseNegative(d.actual, d.predicted, cutoff)));

var accuracy = function(cutoff)
    (((numberTruePositives(cutoff) + numberTrueNegatives
      (cutoff)) / (numberTruePositives(cutoff) +
```

```
      numberTrueNegatives(cutoff) + numberFalsePositives
       (cutoff) + numberFalseNegatives(cutoff) ) ) );
var sensitivity = function(cutoff)
     (numberTruePositives(cutoff) / (numberTruePositives
      (cutoff) + numberFalseNegatives(cutoff)));
var specificity = function(cutoff)
     (numberTrueNegatives(cutoff) / (numberTrueNegatives
      (cutoff) + numberFalsePositives(cutoff)));

var xScaleMapping = pv.Scale.linear(0,1).range(0,width),
  yScaleMapping = pv.Scale.linear(0,1).range(0,height);

var chartPanel = new pv.Panel().width(width + 400)
  .height(height).margin(margin)
  .def("selectedCutoff", -1)
  .event("mousemove", pv.Behavior.point());

chartPanel.add(pv.Label)
  .left(150)
  .top(-5)
  .text("ROC Plot");

chartPanel.add(pv.Rule)
 .strokeStyle("lightgray")
 .left(0);
chartPanel.add(pv.Rule)
 .strokeStyle("lightgray")
 .width(xScaleMapping(1))
 .bottom(0);

chartPanel.add(pv.Rule)
     .data(yScaleMapping.ticks(10))
     .bottom(yScaleMapping)
     .left(0)
     .width(4)
     .strokeStyle("lightgray")
    .anchor("left").add(pv.Label)
     .text(yScaleMapping.tickFormat);

chartPanel.add(pv.Label)
 .text("True positive rate (sensitivity)")
 .left(-20)
 .bottom(35)
 .textAngle(-Math.PI/2);
```

```
chartPanel.add(pv.Rule)
     .data(xScaleMapping.ticks(10))
     .left(xScaleMapping)
     .bottom(0)
     .height(4)
     .strokeStyle("lightgray")
    .anchor("bottom").add(pv.Label)
     .text(yScaleMapping.tickFormat);

chartPanel.add(pv.Label)
   .text("False positive rate (1-specificity)")
   .bottom(-25)
   .left(65);

chartPanel.add(pv.Line)
   .data(cutoffs)
   .bottom(function(d) yScaleMapping(sensitivity(d)))
   .left(function(d) xScaleMapping(1-specificity(d)))
   .strokeStyle("lightblue")

   .add(pv.Dot)
    .event("point", function(d) { selectedCutoff =
       this.index; return chartPanel; } )
     .event("unpoint", function(d) { selectedCutoff = -1;
       return chartPanel; } )
    .shape("triangle")
    .size(function() selectedCutoff == this.index ? 12 : 2)
    .strokeStyle(function() selectedCutoff == this.index ?
      "darkblue" : "lightblue");

chartPanel.add(pv.Line)
   .data(pv.range(0,1.1,0.1))
   .bottom(function(d) yScaleMapping((d)))
   .left(function(d) xScaleMapping((d)))
   .strokeStyle("green")
   .add(pv.Dot)
    .shape("square")
    .size(2);

chartPanel.add(pv.Label)
 .left(400)
 .bottom(height-50)
 .text(function() selectedCutoff != -1 ? "Cutoff = " +
    twoDPFormat(cutoffs[selectedCutoff]) : "");
```

```
chartPanel.add(pv.Label)
 .left(400)
 .bottom(height-65)
  .text(function() selectedCutoff != -1 ? "Number of true
    positives = " + numberTruePositives
    (cutoffs[selectedCutoff]) : "");

chartPanel.add(pv.Label)
 .left(400)
 .bottom(height-80)
  .text(function() selectedCutoff != -1 ? "Number of true
    negatives = " + numberTrueNegatives
    (cutoffs[selectedCutoff]) : "");

chartPanel.add(pv.Label)
 .left(400)
 .bottom(height-95)
  .text(function() selectedCutoff != -1 ? "Number of false
    positives = " + numberFalsePositives
    (cutoffs[selectedCutoff]) : "");

chartPanel.add(pv.Label)
 .left(400)
 .bottom(height-110)
  .text(function() selectedCutoff != -1 ? "Number of false
    negatives = " + numberFalseNegatives
    (cutoffs[selectedCutoff]) : "");


chartPanel.add(pv.Label)
 .left(400)
 .bottom(height-125)
  .text(function() selectedCutoff != -1 ? "Accuracy = " +
    twoDPFormat(accuracy(cutoffs[selectedCutoff])) : "");

chartPanel.add(pv.Label)
 .left(400)
 .bottom(height-140)
 .text(function() selectedCutoff != -1 ? "Sensitivity = " +
   twoDPFormat(sensitivity(cutoffs[selectedCutoff])) : "");

chartPanel.add(pv.Label)
 .left(400)
 .bottom(height-155)
```

```
  .text(function() selectedCutoff != -1 ? "Specificity = " +
    twoDPFormat(specificity(cutoffs[selectedCutoff])) : "");


chartPanel.render();


   </script>
  </body>
</html>
```

# BIBLIOGRAPHY


Adelson, E. (n.d.). Illusions and demos. Retrieved March 2011 from http://web.mit.edu/persci/people/adelson/illusions_demos.html

Ahlberg, C., & Shneiderman, B. (1999). Visual information seeking: Tight coupling of dynamic query filters with starfield displays. In *Readings in information visualization: Using vision to think (interactive technologies)*. Boston: Morgan Kaufmann.

Baddeley, A. (2010). Working memory. *Scholarpedia*, *5*(2), 3015, revision #76180.

Becker, R. A., Eick, S. G., & Wilks, A. R. (1999). Visualizing network data. *In Readings in information visualization: Using vision to think (interactive technologies)*. Boston: Morgan Kaufmann.

Bederson, B. B., Grosjean, J., & Meyer, J. (2004). Toolkit design for interactive structured graphics. *IEEE Transactions on Software Engineering*, *30*(8), 535–546.

Bederson, B. B., Stead, L., & Hollan, J. D. (1994). Pad++: Advances in multiscale interfaces. In Proceedings of CHI 94: Conference on Human Factors in Computing Systems,(pp. 315–316). New York: ACM Press.

Berrar, B. P., Dubitzky, W., & Granzow, M. (2003). A practical approach to microarray data analysis. Heidelberg: Springer.

Bertin, J. (1983). *Semiology of graphics: Diagrams, networks, maps.* (William J. Bergtrans.). Madison, WI: The University of Wisconsin Press.

Bostock, M., & Heer, J. (2009). Protovis: A graphical toolkit for visualization. *IEEE Transactions on Visualization and Computer Graphics*, *15*(6), 1121–1128.

Brooks, F. P. (1996). The computer scientist as toolsmith II. *Communications of the ACM*, *39*(3), 61–68.

Card, S. K., Mackinlay, J., & Shneiderman, B.(Eds.) (1999). *Readings in information visualization: Using vision to think (interactive technologies)*. Boston: Morgan Kaufmann.

Chandler, D. (2007). *Semiotics: The basics*. London: Routledge.

Cleveland, W. S. (1993). *Visualizing data*. Summit, NJ: Hobart Press.

Cleveland, W. S. (1994). *The elements of graphing data* (2nd ed.). Summit, NJ: Hobart Press.

Cockburn, A., Karlson, A., & Bederson, B. B. (2008). A review of overview+detail, zooming, and focus+context interfaces. *ACM Computing Surveys*, *41*(1:2), 1–31.



*Making Sense of Data III: A Practical Guide to Designing Interactive Data Visualizations*, First Edition. Glenn J. Myatt and Wayne P. Johnson.


© 2011 John Wiley & Sons, Inc. Published 2011 by John Wiley & Sons, Inc.

Codd, E. F. (1990). *The relational model for database management* (2nd ed.). Menlo Park, CA: Addison-Wesley.

Constantine, L. L., & Lockwood, L. A. D. (1999). *Software for use: Aa practical guide to the models and methods of usage-centered design*. New York: ACM Press.

Eichenbaum, H. (2008). Memory. *Scholarpedia, 3*(3), 1747, revision #72202.

EPA. (2011). 2011 fuel economy guide data. Available online at http://www.fueleco nomy.gov/feg/download.shtml

Fekete, J. D., & Plaisant, C. (2002). Interactive information visualization of a million items. In Proceedings of IEEE Symposium on Information Visualization 2002, 117–124; (DOI:10.1109/INFVIS.2002.1173156). Available online at http://www.cs.umd.edu/local-cgi-bin/hcil/rr.pl?number=2002-01

FGED Society. (2011). Minimum information about a microarray experiment - MIAME. Retrieved March 2011 from http://www.mged.org/Workgroups/MIAME/miame.html

Green, M. (1998). Toward a perceptual science of multidimensional data visualization: Bertin and beyond. Available online at http://graphics.stanford.edu/courses/cs448b-06-winter/papers/Green_Towards.pdf

Heer, J., Stuart, C. K., & Landay, J. A. (2005). Prefuse: A toolkit for interactive information visualization. In *Proceedings of the SIGCHI conference on Human factors in computing systems* (CHI '05), 421–430. New York: ACM Press. (DOI:10.1145/1054972.1055031)

Heer, J., & Agrawala, M. (2006). Software design patterns for information visualization. *IEEE Transactions on Visualization and Computer Graphic, 12*(5), 853–860. (DOI:10.1109/TVCG.2006.178)

Heer, J., & Bostock, M. (2010). Declarative language design for interactive visualization. *IEEE Transactions on Visualization and Computer Graphic, 16*(6), 1149–1156. (DOI:10.1109/TVCG.2010.144)

Hey, T., Tansley, S., & Tollè, K. (2009). The fourth paradigm: Data-intensive scientific discovery. Redmond, WA: Microsoft Research.

Hoaglin, D. C., Mosteller, F., & Tukey, J. W. (2000). *Understanding robust and exploratory data analysis*, library classics ed. Hoboken, NJ: John Wiley and Sons, Inc.

Holtzblatt, K., & Beyer, H. (1998). *Contextual design: Defining customer-centered systems*. Boston: Morgan Kaufmann.

Hotzblatt, K., Wendell, J. B., & Wood, S. (2005). *Rapid contextual design: A how-to guide to key techniques for user-centered design*. Boston: Morgan Kaufmann.

Johnson, J. (2008). *GUI bloopers: Common user interface design don'ts and dos* (2nd ed.). Boston: Morgan Kaufmann.

Johnson, J. (2010). *Designing with the mind in mind: Simple guide to understanding user interface rules*. Boston: Morgan Kaufmann.

Johnson J., & Henderson, A. (2002). Conceptual models: Begin by designing what to design. *Interactions, 9*(1), 25–32. (DOI:10.1145/503355.503366)

Kelley, D. (1996). The designer's stance: An interview with David Kelley by Bradley Hartfield. In Bringing design to software. New York: ACM Press.

Kossyln, S. M. (2006). *Graph design for the eye and mind*. New York: Oxford University Press.

Kuniavsky, M. (2003). *Observing the user experience: A practitioner's guide to user research*. Boston: Morgan Kaufmann.

Luck, S. J. (2007). Visual short-term memory. *Scholarpedia, 2*(6), 3328, revision #47721.

Mack, A., & Rock, I. (2000). *Inattentional blindness.* Cambridge, MA: The MIT Press.

Malacara, D. (2002). *Color vision and colorimetry: Theory and applications.* Bellingham, WA: SPIE Publications.

Marr, D. (2010). *Vision.* Cambridge, MA: The MIT Press. (Original work published 1982)

Mullet, K., & Sano, D. (1995). Designing visual interfaces: Communication oriented techniques. Englewood Cliffs, NJ: Prentice Hall.

Myatt, G. J. (2006). *Making sense of data: A practical guide to exploratory data analysis and data mining.* Hoboken, NJ: John Wiley and Sons, Inc.

Myatt, G. J., & Johnson, W. P. (2009). *Making sense of data II: A practical guide to data visualization, advanced data mining methods, and applications.* Hoboken, NJ: John Wiley and Sons, Inc.

Nielsen J. (1993). *Usability engineering.* Boston: Morgan Kauffman.

Norman, D. (1994). *Things that make us smart.* New York: Basic Books.

Norman, D. (2002). *The design of everyday things.* New York: Basic Book.

Norman, D. (2004). *Emotional design.* New York: Basic Book.

Norman, D. (2010). *Living with complexity.* Cambridge, MA: The MIT Press.

Palmer, S. (1999). *Vision science: Photons to phenomenology.* Cambridge, MA: The MIT Press.

Pinker, S. (1997). *How the mind works.* New York: W.H. Norton and Company.

Protovis. (n.d.). Retrieved March 2011 from http://vis.stanford.edu/protovis/

Pylyshyn, Z. W. (2003). *Seeing and visualizing: It's not what you think.* Cambridge, MA: The MIT Press.

Reingold, E. M., Charness, N., Pomplun, M., Stampe, D. M. (2001). The perceptual aspect of skilled performance in chess: evidence from eye movements. *American Psychological Society, 12*(1), 48–55.

Reisberg, D. (1987). External representations and the advantages of externalizing one's thoughts. In *Proceedings of the Eighth Annual Conference of the Cognitive Science Society.* Hillsdale, NJ. Erlbaum Associates, Inc.

Roth, S. F., Chuah, M. C., Kerpedjiev, S., & Kolojejchick, J. A. (1997). Toward an information visualization workspace: Combining multiple means of expression. *Human-Computer Interaction, 12,* 131–185. Mahwah, NJ: Lawrence Erlbaum Associates, Inc.

Sharp, H., Rogers, Y., & Preece, J. (2007). *Interaction design: Beyond human-computer interaction* (2nd ed.). Hoboken, NJ: John Wiley and Sons, Inc.

Shneiderman, B., Plaisant, C. (2010). *Designing the user interface: Strategies for effective human-computer interaction* (5th ed.). Boston, MA: Addison-Wesley.

Spence, R. (2001). *Information visualization.* New York: ACM Press.

Stenning, K., & Oberlander, J. (1995). A cognitive theory of graphical and linguistic reasoning: Logic and implementation. *Cognitive Science, 19,* 97–140.

Theus, M., & Urbanek, S. (2008). Interactive graphics for data analysis: Principles and examples. Boca Raton, FL: Chapman and Hall/CRC.

Thomas, J. J., & Cook, K. (Eds.). (2005). *Illuminating the path: The research and development agenda for visual analytics.* Richland, WA: National Visualization and Analytics Center.

Tidwell, J. (2005). Designing interfaces. Sebastopol, CA: O'Reilly.

Tufte, E. R. (1983). *The visual display of quantitative information.* Cheshire, CT: Graphics Press.

Tufte, E. R. (1990). *Envisioning information.* Cheshire, CT: Graphics Press.

Tversky, B. (1997). Cognitive principles of graphic displays. AAAI Technical Report FS-97-03.AAAI. Available online at http://www.aaai.org/Papers/Symposia/Fall/1997/FS-97-03/FS97-03-015.pdf

Tversky, B. (2003). Spatial schemas in depictions. In *Spatial schemas and abstract thought.* Cambridge, MA: The MIT Press.

Unwin, A., Theus, M., & Hofmann, H. (2006). *Graphics of large datasets.* Singapore: Springer.

Unwin, A. (2008). Good graphics? In Handbook of data visualization. Berlin: Springer.

Velmans, M. (1999). When perception becomes conscious. *British Journal of Psychology, 90*(4), 543–566. Available online at http://cogprints.org/838/

Wainer, H. (1997). *Visual revelations: Graphical tales of fate and deception from Napolean Bonaparte to Ross Perot.* New York: Copernicus.

Wainer, H. (2005). *Graphic discovery: A trout in the milk and other visual adventures.* Princeton, NJ: Princeton University Press.

Ward, W., Grinstein, G., & Keim, D. (2010). *Interactive data visualization: Foundations, techniques, and applications.* Natick, MA: A. K.Peters, Ltd.

Ware, C. (2000). *Information visualization: Perception for design.* Boston: Morgan Kaufmann.

Ware, C. (2008). *Visual thinking for design.* Boston: Morgan Kaufmann.

Wertheirmer, M. (1923).Laws of organization in perceptual form. In The Internet Resource Classics in the History of Psychology. Available online at http://psychclassics.yorku.ca/Wertheimer/Forms/forms.htm

Wickham, H. (2009). Ggplot2: Elegant graphics for data analysis. Heidelberg: Springer.

Wilkinson, L., Rubin, M., Rope, D., & Norton, A. (2001).nViZn: An algebra-based visualization system. In International Symposium on Smart Graphics 2001. Available online at http://www.cs.uic.edu/~wilkinson/Publications/ibm.pdf

Wilkinson, L. (2005). *The grammar of graphics.* Canada: Springer.

Yaffa, J. (2007 August 12). The road to clarity. The New York Times Magazine. Available online at http://www.nytimes.com/2007/08/12/magazine/12fonts-t.html?pagewanted=1&_r=1

Yarbus, A. L. (1967). *Eye movements and vision.* (Basil Haigh, Trans.). New York: Plenum Press. Available online at http://wexler.free.fr/library/files/yarbus%20%281967%29%20eye%20movements%20and%20vision.pdf

Zhang, J. (1997). The nature of external representations in problem solving. *Cognitive Science, 21*(2),179–217. Hillsdale, NJ: Lawrence Erlbaum Associates, Inc. (DOI:10.1016/S0364-0213(99)80022-6)

Zhang, J. (2001). External representations in complex information processing tasks. In *Encyclopedia of Microcomputers.* New York: Marcel.

# INDEX




*Making Sense of Data III: A Practical Guide to Designing Interactive Data Visualizations*, First Edition. Glenn J. Myatt and Wayne P. Johnson.


© 2011 John Wiley & Sons, Inc. Published 2011 by John Wiley & Sons, Inc.

Printed in the United States
By Bookmasters